VOLUME FOURTEEN

Radioactivity in the Environment

REMEDIATION OF CONTAMINATED ENVIRONMENTS

RADIOACTIVITY IN THE ENVIRONMENT

A companion series to the Journal of Environmental Radioactivity

Series Editor
M.S. Baxter
Ampfield House
Clachan Seil
Argyll, Scotland, UK

Volume 1: Plutonium in the Environment
(A. Kudo, Editor)

Volume 2: Interactions of Microorganisms with Radionuclides
(F.R. Livens and M. Keith-Roach, Editors)

Volume 3: Radioactive Fallout after Nuclear Explosions and Accidents
(Yu.A. Izrael, Author)

Volume 4: Modelling Radioactivity in the Environment
(E.M. Scott, Editor)

Volume 5: Sedimentary Processes: Quantification Using Radionuclides
(J. Carroll and I. Lerche, Authors)

Volume 6: Marine Radioactivity
(H.D. Livingston, Editor)

Volume 7: The Natural Radiation Environment VII
(J.P. McLaughlin, S.E. Simopoulos and F. Steinhäusler, Editors)

Volume 8: Radionuclides in the Environment
(P.P. Povinec and J.A. Sanchez-Cabeza, Editors)

Volume 9: Deep Geological Disposal of Radioactive Waste
(R. Alexander and L.E. McKinley, Editors)

Volume 10: Radioactivity in the Terrestrial Environment
(G. Shaw, Editor)

Volume 11: Analysis of Environmental Radionuclides
(P.P. Povinec, Editor)

Volume 12: Radioactive Aerosols
(C. Papastefanou, Author)

Volume 13: U–Th Series Nuclides in Aquatic Systems
(S. Krishnaswami and J. Kirk Cochran, Editors)

Volume 14: Remediation of Contaminated Environments
(G. Voigt and S. Fesenko, Editors)

VOLUME FOURTEEN

RADIOACTIVITY IN THE ENVIRONMENT

REMEDIATION OF CONTAMINATED ENVIRONMENTS

Editors

G. VOIGT
International Atomic Energy Agency,
Agency's Laboratories Seibersdorf, Austria

S. FESENKO
International Atomic Energy Agency,
Agency's Laboratories Seibersdorf, Austria

ELSEVIER
Amsterdam • Boston • Heidelberg • London • New York • Oxford
Paris • San Diego • San Francisco • Singapore • Sydney • Tokyo

Elsevier
Linacre House, Jordan Hill, Oxford OX2 8DP, UK
Radarweg 29, PO Box 211, 1000 AE Amsterdam, The Netherlands

First edition 2009

British Library Cataloguing in Publication Data
A catalogue record for this book is available from the British Library

Library of Congress Cataloging-in-Publication Data
A catalog record for this book is available from the Library of Congress

ISBN: 978-0-08-044862-6
ISSN: 1569-4860

For information on all Elsevier publications visit our website at elsevierdirect.com

Printed and bound in Hungary

09 10 11 12 13 10 9 8 7 6 5 4 3 2 1

Contents

Contributors

Rudolph M. Alexakhin
Russian Institute of Agricultural Radiology and Agroecology

Kasper G. Andersson
Risø – DTU National Laboratory for Sustainable Energy

Mark Audet
Atomic Energy of Canada Ltd. (AECL)

Ingrid Bay-Larsen
Department of Plant and Environmental Sciences, Norwegian University of Life Sciences

Bao-Dong Chen
Research Center for Eco-Environmental Sciences, Chinese Academy of Sciences

Piero R. Danesi
Independent Consultant, Arsenal, Vienna

W. Eberhard Falck
Consultant, Cornelis Pronklaan 102, NL-1816NR Alkmaar

Horst M. Fernandes
International Atomic Energy Agency, NEFW, Waste Technology Section

Sergey Fesenko
International Atomic Energy Agency, Agency's Laboratories

Abel J. González
Autoridad Regulatoria Nuclear (Argentine Nuclear Regulatory Authority)

Ivan Kryshev
Research and Production Association (RPA) "Typhoon"

Tatyana Lavrova
Radiation Monitoring Department of the Ukrainian Research Institute for Hydrometeorology

Deborah Oughton
Department of Plant and Environmental Sciences, Norwegian University of Life Sciences

Mike Pearl
UKAEA, The Manor Court

Per Roos
Risø – DTU National Laboratory for Sustainable Energy

Peter Schmidt
Wismut GmbH

Roger Seitz
Savannah River National Laboratory

Natalya Semioshkina
Helmholtz Zentrum München-Institute of Radiation Protection

Vyacheslav Shershakov
Research and Production Association (RPA) "Typhoon"

Gabriele Voigt
International Atomic Energy Agency, Agency's Laboratories

Oleg Voitsekhovych
Radiation Monitoring Department of the Ukrainian Research Institute for Hydrometeorology

Lisa Zeiller
International Atomic Energy Agency, Agency's Laboratories

Yong-Guan Zhu
Research Center for Eco-Environmental Sciences, Chinese Academy of Sciences

Introduction

Because of operation of numerous nuclear facilities, nuclear weapons testing, radiation accidents and so on, vast areas across the globe have been recognised as contaminated sites which require remediation.

In some cases, especially in relation to inadequate practices for disposal and waste management which were specific for military nuclear activities in the period 1940–1960, these were planned activities, and releases of radionuclides into the environment were authorised, keeping in mind that such sites would be recovered after completion of the operation.

It should be recognised that such practices are also in current use, and remediation is an obligatory component of decommissioning nuclear facilities, and the cost of these actions are normally included in the total operational costs of any nuclear facility or nuclear installation. The main difference with the past bad waste management practice is the well-developed regulation and well-justified extent of the planned contamination of the environment.

In some cases, former contaminated sites, released for unrestricted use, can be reclassified because of an underestimation of radiation risk or because of a lack of the relevant radiation safety regulations at the time when the decision was taken. So, in the past, some nuclear facilities were processed based on former criteria for radiation protection that were not as strict as they are at present. Therefore, nowadays such contaminated sites are to be considered as sites which may present a hazard to human populations and the environment and are considered to require remediation. Often a need for the remediation of such contaminated sites may be recognised only after long-term use (and disposal) of radioactive materials for a variety of medical, industrial or research purposes. There have also been several severe radiation accidents which have resulted in a need for recovery of vast areas.

Thus, radioactively contaminated sites have been generated for many reasons, and they were contaminated to varying extents and by different radionuclides. This has led to a necessity for the development of a variety of approaches, tools for remediation planning, remedial measures and technologies.

The purpose of this book is to present a state-of-the-art summary of modern technologies for remediation of contaminated sites. As mentioned earlier, a classification of the site in terms of its potential hazard to human beings and the environment is always based on the general requirements and criteria for radiation protection of the public. Furthermore, recommendations of the relevant international bodies (ICRP, 1991,

1999, 2007, 2003, 2009) and radiation safety standards, requirements and guides (IAEA, 1996, 2000, 2002, 2003, 2004, 2006, 2007) provide clear guidelines on the remediation process, including the objectives, legal and regulatory frameworks to take into account in remediation situations, the requirements for and recommendations on remediation programmes and so on. Therefore, Chapter 1 of this book describes current international approaches to the remediation of contaminated sites, detailing the distinctive features of terminology used in defining regulatory aspects of remediation; the statements and recommendations of the International Commission on Radiological Protection; and the safety standards, recommendations and guidelines issued on the subject by the International Atomic Energy Agency.

Site characterisation represents the first necessary stage in the remediation of any contaminated environment. Indeed, all decisions for evaluation of need for site remediation, further remediation planning and implementation of remedial actions and, post-remediation, for the assessments (to ensure that there is compliance of the residual concentrations of radionuclides in the environment with the expected levels) are based on the site characterisation. Such aspects are described in Chapter 2, which also covers the necessary technical aspects of site characterisation, including classification of sites for remediation purposes, sampling and measurement strategies and quality assurance issues.

The book also describes modern decision-aiding technologies and Environmental Decision Support Systems (EDSSs) intended for remediation planning and optimising remediation strategies. Following the Chernobyl accident, the application of EDSSs became an important decision-support tool for both remediation and emergency planning. Their fast development and implementation into the practice of remediation of the areas affected by the Chernobyl accident (as well as the implementation for decision support of specially developed methods and technologies) were directly connected to the corresponding recommendations of the ICRP (ICRP, 1983, 1989) and to the rapid introduction of personal computers into research, research development and industrial applications. As a result, a number of the EDSSs were developed at both international (mainly in the framework of EC-funded projects) and national levels; an overview of these EDSSs is given in Chapter 3. The chapter also describes current scientific approaches to the decision-aiding technologies, with special attention to the applications of Multi-Attribute Utility Analysis (MAUA) for remediation purposes, as this follows from the recent ICRP Publication 103 (ICRP, 2007).

Remediations of vast areas affected by the major radiation accidents (unrestricted releases of radionuclides to the Techa River, and the Kyshtym, Chernobyl, Goiânia and Palomares accidents) are major sources of data on the effectiveness of large-scale remedial actions, and Chapter 4 provides a historical overview both of the development of the radiological situations after these accidents and of the effectiveness of the remedial actions implemented.

Remediation of former nuclear weapon testing sites at the Bikini and Enewetak Atolls and Maralinga (Chapter 5) represents another important practical example of the successful implementation of remedial actions. Although the remedial measures implemented in these three cases were rather similar, the diversity of the radiological conditions encountered gave important information on the potential effectiveness of these actions.

The term 'decommissioning' refers to administrative and technical actions taken to allow the removal of some or all of the regulatory controls from a nuclear facility (except for a repository which is 'closed' rather than 'decommissioned') (IAEA, 2001). These actions involve different options: decontamination, dismantling and removal of radioactive materials, waste, components and structure; that is, they involve a variety of remedial actions. Six case studies, demonstrating successful experience in remediation as a part of the decommissioning of different nuclear facilities, are considered in Chapter 6, while Chapter 7 provides an example of remediation planning for the largest uranium-mining facility in Ukraine.

Chapters 8 and 9 review a wide spectrum of potentially effective remedial measures, applicable to different categories of contaminated locations and contaminants, comprising areas contaminated by radiation accidents and those contaminated by natural radionuclides associated with nuclear fuel cycle and fossil material mining.

Finally, Chapter 10 considers human-based remedial measures, comprising the perception of this activity by the population and the associated side-effects. Experience gained after the major radiation accidents, following the remediation of areas contaminated as a result of past activities and nuclear testing sites, has shown that, in addition to the radiological factors, environmental, sociological, psychological and economic factors produce a strong impact on the decision-making process with regard to the choice of remediation strategies (Voigt et al., 2000; WHO, 1996; Oughton, 2003; ICRP, 2007). It has also been recognised that the involvement of all interested parties is a key element in the successful justification of remediation strategies. In spite of the fact that Chapter 10 of the book considers all these issues based largely on the Chernobyl-related experience, the conclusions and recommendations made by this chapter are applicable for any remediation activities and are of paramount importance to successful remediation planning.

Every chapter is accompanied by numerous references to the original papers and technical documents.

Overall the book provides a comprehensive consideration of remediation as one of the main tools that provide sustainable development of various modern industries, covering a wide spectrum of related issues such as the site characterisation of contaminated sites before and after remediation, the safety requirements, remediation planning, the effectiveness of individual measures in different environments and so on. The aim of this book is also to serve as a supportive material for education purposes, training and fellowships in radioecology and radiation protection, and it provides reference information for both nuclear industry staff and environmentalists involved in remediation of the environment.

REFERENCES

IAEA. (1996). *International Basic Safety Standards for Protection Against Ionizing Radiation and for the Safety of Radiation Sources*. IAEA Safety Series No. 115. IAEA, Vienna.

IAEA. (2000). *Restoration of Environments Affected by Residues from Radiological Accidents: Approaches to Decision Making*. IAEA-TECDOC-1131. IAEA, Vienna.

IAEA. (2001). *Decommissioning of Nuclear Fuel Cycle Facilities Safety Guide*. IAEA Safety Standard No. WS-G-2.4. IAEA, Vienna.

IAEA. (2002). *Non-Technical Factors Impacting on the Decision Making Processes in Environmental Remediation*. IAEA-TECDOC-1279. IAEA, Vienna.

IAEA. (2003). *Remediation of Areas Contaminated by Past Activities and Accidents Safety Requirements*. IAEA Safety Standards Series No. WS-R-3. IAEA, Vienna.

IAEA. (2004). *Planning, Managing and Organizing the Decommissioning of Nuclear Facilities: Lessons Learned*. IAEA-TECDOC-1394. IAEA, Vienna.

IAEA. (2006). *IAEA Safety Glossary: Terminology Used in Nuclear, Radiation, Radioactive Waste and Transport Safety (Version 2.0)*. IAEA Booklet. IAEA, Vienna.

IAEA. (2007). *Remediation Process for Areas Affected by Past Activities and Accidents*. IAEA Safety Standards Series No. WS-G-3.1. IAEA, Vienna.

ICRP. (1983). *Cost-Benefit Analysis in the Optimisation of Radiation Protection*. ICRP Publication 37. *Annals of the ICRP*, 10 (2/3). Pergamon Press, Oxford.

ICRP. (1989). *Optimisation and Decision-Making in Radiological Protection*. ICRP Publication 55. *Annals of the ICRP*, 20 (1). Pergamon Press, Oxford.

ICRP. (1991). *1990 Recommendations of the International Commission on Radiological Protection*. ICRP Publication 60. *Annals of the ICRP*, 21 (1–3). Elsevier, Amsterdam.

ICRP. (1999). *Protection of the Public in Situations of Prolonged Radiation Exposure*. ICRP Publication 82. *Annals of the ICRP*, 29 (1/2). Elsevier, Amsterdam.

ICRP. (2003). *A Framework for Assessing the Impact of Ionising Radiation on Non-Human Species*. ICRP Publication 91. *Annals of the ICRP*, 33 (3). Elsevier, Amsterdam.

ICRP. (2007). *The 2007 Recommendations of the International Commission on Radiological Protection*. ICRP Publication 103. *Annals of the ICRP*, 37 (2–4). Elsevier, Amsterdam.

ICRP. (2009). *Application of the Commission's Recommendations to the Protection of Individuals Living in Long Term Contaminated Territories after a Nuclear Accident or a Radiation Emergency*. Elsevier, Amsterdam (in press).

Oughton, D. H. (2003). Protection of the environment from ionising radiation: ethical issues. *Journal of Environmental Radioactivity*, 66, 3–18.

Voigt, G., K. Eged, B. J. Howard, Z. Kis, A. F. Nisbet, D. H. Oughton, B. Rafferty, C. A. Salt, J. T. Smith, and H. Vandenhove. (2000). A wider perspective on the selection of countermeasures. *Radiation Protection Dosimetry*, 92, 45–48.

WHO. (1996). *Health Consequences of the Chernobyl Accident. Results of the IPHECA Pilot Projects and Related National Programmes*. Scientific Report. WHO, Geneva.

G. Voigt
S. Fesenko
Editors

CHAPTER 1

International Approaches to Remediation of Territorial Radioactive Contamination

Abel J. González*, ☆

Contents

*Corresponding author. Tel.: +541 1632 31758
E-mail address: agonzale@sede.arn.gov.ar

☆ Mr. A. J. González, senior adviser to the Argentine Nuclear Regulatory Authority, is a representative to the United Nations Scientific Committee on the Effects of Atomic Radiation (UNSCEAR), vice president of the International Commission on Radiological Protection (ICRP) and member of the Commission of Safety Standards of the International Atomic Energy Agency (IAEA).

Autoridad Regulatoria Nuclear (Argentine Nuclear Regulatory Authority), Avenida del Libertador 8250, 1429 Buenos Aires, Argentina

Radioactivity in the Environment, Volume 14
ISSN 1569-4860, DOI 10.1016/S1569-4860(08)00201-5

1. Introduction

The so-called *remediation* of territories[1] experiencing *contamination* with radioactive substances has been one of the more elusive issues for the radiation protection community to tackle and regulate. Following the presence of radioactive *residues*[2] over a territory, radiation protection experts have generally been unable to respond to a simple and straightforward question from anxious members of the general public: *Is it safe for me and my family to live here*? Providing non-conclusive and consistent answers to such a simple enquiry was most unhelpful. Experts tried to explain that, while the territory was in fact *contaminated*, *remediation* had to be 'optimised', and depending on many factors (generally incomprehensible for the common public), they might or might not remain there. Moreover, some experts, dishonouring their professional responsibilities, implicitly advised members of the public that it was ultimately their decision to leave or to remain in a 'contaminated' territory (this was often done in reaction during so-called 'stakeholders involvement' meetings).

The terms *remediation* and *contamination* are purposely italicised in this introductory chapter because their meaning is vague and ambiguity in understanding has been part of the problem in solving this controversial issue. Practical solutions for the conundrum of whether a *contaminated* territory needs *remediation* have been unconvincing for a growingly sceptical public, *inter alia* because the arguments were unimpressive and puzzling. There have been common misunderstandings on the basic concepts, not only by the public but also among the 'experts' themselves. While this book will mainly address the technical aspects of the problem, this initial chapter is intended to present some conceptual misapprehensions and to describe the radiation protection paradigm that is internationally recommended for tackling the issues and the regulatory approaches.

2. Misunderstandings

2.1. Contamination

The term (radioactive) 'contamination' is widely misunderstood and its misinterpretation has had enormous effects in radiation protection strategies.

[1]The term *territory*, from Latin *territorium*, from *terra* 'land', and its derivatives, is used to mean just an area of ground or land rather than its usual connotation of an area under the jurisdiction of a ruler or state.

[2]The term (radioactive) *residues* is used for radioactive materials that have remained in the environment from early operations and accidents involving the use of radioactive substances. The term could also, in principle, be used to describe the presence on land of primordial radioactive materials that have accumulated over time as a result of natural processes, such as the territorial deposits of radium caused by the drying of underground water released from natural springs.

Surprisingly, the term derives from a historical religious background for describing *impurity*. Contamination originates from the Latin *contaminat-*, *contaminare*, or 'make impure'; from *contamen*, or 'contact, pollution'; from *con-*, or 'together with', plus the base of *tangere*, or 'to touch'. This grammatical acceptance becomes particularly important in some languages where the term is translated as *impurity*, the obvious connotation being that something that is 'contaminated' is automatically unacceptable regardless of the quantification of such 'contamination'. A typical example of this use is the religious understanding of *contaminated* (e.g. non-kosher) food, namely food not satisfying the requirements of religious law with regard to its origin and preparation.

Perhaps, this religious nuance is one of the reasons why the term 'contamination' may have reached a connotation that was not intended when introduced by radiation protection specialists. The experts' original intention was to refer only to the presence of any (radioactive)materials expressed by the quantity [radio]*activity*, namely describing an amount or concentration of radionuclides in a given energy state at a given time; they did not intend to give any indication of impurity or dirtiness, nor even of the magnitude of the hazard involved. However, in the public mind, 'contamination' became a quasi-synonym for dangerously undesirable [radio]*activity*. In sum, while the term is commonly used by experts to quantify the presence and distribution of radioactive material in a given environment, it became widely misinterpreted as a measure of radiation-related danger.

Moreover, the term strictly refers to radioactive substances on surfaces and within solids, liquids or gases (including those in the human body), where their presence is unintended or undesirable, or to the process giving rise to their presence in such places. But unfortunately, the term is used more informally (even by experts) to refer to the quantity [radio]*activity* on a surface, and is misinterpreted and misunderstood as a *dangerous* level of [radio]*activity*.

The misunderstanding of the term 'contamination' may also originate in misunderstandings of the terms describing the originators of contamination, such as 'pollution', 'release', 'discharges' and 'source term', which are widely used and also misperceived. 'Pollution' is sometimes used to denote the 'release' (controlled or uncontrolled) of radioactive materials into the environment. This contrasts with the widely used term radioactive 'discharge', which means planned and controlled release (usually in gaseous or liquid form) of radioactive material into the environment. Strictly, discharge is the act or process of releasing material, but it is nowadays also misused to characterise the material released. International conventions consider discharges a legitimate practice within limits authorised by a competent regulatory body (IAEA, 1994, 1997b), which can be used for managing liquid or gaseous radioactive materials that originate from regulated facilities during their normal operations. Obviously, the intention of the conventioneers was to regulate and control properly and in a safe manner those unavoidable discharges rather than to contaminate territories.

Another puzzling terminology is *source term*, which strictly refers to the amount and isotopic composition of radioactive materials that assumably might be released from a facility should a postulated accident actually occur. The concept is used in modelling hypothesised releases of radionuclides into the environment, particularly in the context of accidents at nuclear installations and of imagined protracted releases from radioactive waste in repositories over long periods of time. Again, the intention of safety experts is to model hypothetical releases in order to check safety features rather than to sanctify the contamination of territories.

2.2. Remediation

The term (radioactive) 'remediation' became closely associated with the misinterpretations of 'contamination', as the former is a consequence of the latter. The term may be used in a variety of contexts and, as a result, it can be badly misunderstood. In common parlance, it means providing a *remedy*, namely a pharmaceutical product, cure or treatment, for a medical condition. Not surprisingly, members of the public became extremely anxious when informed that the place where they are living will be subject to 'remediation' because of a radiation-related 'contamination'!

Environmental radiation protection specialists, however, use 'remediation' to mean the removal or reduction of radioactive substances from environmental media such as soil, groundwater, sediment or surface water. The ultimate purpose of 'remediation' is protecting human health and the environment against potential detrimental effects from radiation exposure, rather than eliminating 'contamination' *per se*.

In international standards (IAEA, 2006a), the term 'remediation' has been formally defined as any measures that may be carried out to reduce the radiation exposure from existing contamination of land areas through actions applied to the contamination itself (the source) or to the exposure pathways to humans. Notably, the formal definition underlines that the term does not imply 'complete removal of the "contamination" [*sic*]', a key concept that is usually forgotten.

The untranslatable and more informal English term *cleanup* has been used as a synonym of 'remediation', and this usage has added to the misunderstanding. The term implies making a place *clean* and is taken to mean making a place absolutely free from dirt or harmful substances. The confusion arises because a decision to shrink a given level of radioactive 'contamination' may be taken simply because radioactivity is measurable and not because it is *dirty* or *harmful*. Moreover, the term *clean* can also be tacitly equated to 'morally pure', which again has religious implications. This acceptance combined with the misinterpretations of the term 'contamination' described heretofore may have played an important role in the misunderstanding.

The terms *rehabilitation* and *restoration* have also been used within the context of 'remediation'. And again, their usage has been confusing. These terms may be taken to imply restoring the conditions that prevailed before the 'contamination', presuming that such restoration is feasible, which is not normally the case (e.g. owing to the effects of the remedial actions themselves). For this and other reasons, the use of these terms as alternatives to remediation has been discouraged.

It therefore seems that there is a strong connection between the misunderstandings of 'contamination' and 'remediation'. In simple terms, remediation should be expected if there is contamination, and there will be contamination if and only if the levels of [radio]*activity* per unit area are above the given values considered unsafe.

In spite of the confusion created by the disparaged use of the term 'remediation' and its precursor 'contamination', this book will continue to use them to describe the basic subjects of the book. It is considered that the usage of this terminology is so entrenched in radiation protection practice that changing it at this stage into a more precise language may produce more harm than good.

3. Scenarios

There have been many scenarios where remediation of territorial contamination has been considered and an array of growing radiation protection has been built. Confusingly, for members of the public, however, regulations for remediation are not described "based" on levels of activity per unit area. The main regulatory quantity has been the radiation doses to be expected from the contamination rather than the activity level of the contamination itself.[3] Increasing the confusion is the fact that these doses can be expressed as integrated doses (e.g. doses to be incurred over lifetime) or as dose rates (e.g. annual doses). Moreover, the term 'dose' can refer to the *total* dose (usually referred to as *existing* or *extant* dose, be it lifetime dose or dose rate) being incurred by people as a result of living in a contaminated territory, or to the *additional* doses (namely doses added over the background doses) that are attributable to the contamination and which in turn can be projected doses or avertable doses.

Therefore, while the presence of radioactive residues in human habitats is usually described using the term 'contamination', the outcome is expressed as

[3]Internationally, a solitary exception to this rule has been the regulation for safe transport of radioactive materials, which clearly defines a 'contamination' level as the presence of a radioactive substance on a surface in quantities in excess of 0.4 Bq cm^{-2} for beta and gamma emitters and *low toxicity alpha emitters*, or 0.04 Bq cm^{-2} for all other alpha emitters. (The numeral 4 results from the definition of contamination in the old unit of activity, the Curie, to Becquerel by rounding the conversion numeral 3.7 to 4.)

the dose resulting from exposure to the situation. The nature of such exposure could be either 'certain', that is most likely to occur, or 'potential', that is possible but not definite to happen.

Within this conundrum of descriptors, this part will attempt to portray more common scenarios of territorial contamination.

Radioactive residues can originate from several causes. Occasionally, they may have been generated by the accumulation of radionuclides from normal discharges of radioactive effluents into the environment from planned and properly authorised human activities, so-called *practices*. They may also be radioactive remnants from the termination and decommissioning of a practice. Most commonly, radioactive residues are the result of human activities that have been carried out in the past without being regulated, where the termination of the activity and the handling of the remaining residues would most probably not have been adequately considered when the activity was initiated. Examples of such ancient activities are the old industry of luminising with radium compounds and ancient mining and milling operations of ores containing natural radioactive substances. Radioactive residues may also remain from past events that may have been unforeseeable at the time of occurrence, such as accidents releasing long-lived radioactive materials to the environment. Finally, the largest amount of radioactive residues in the human habitat are a legacy from past military operations that were both foreseeable and avoidable: nuclear weapons testing, for example, resulted in the release of huge quantities of radioactive materials that were dispersed over vast areas.

It should be noted that the complexity of the situations created by territorial contamination was not recognised early enough. The many assessments of the aftermath of the Chernobyl accident (IAEA, 1988, 1991, 1996b, 1996c, 1997a, 2006b, 2006c) have shown the difficulties in dealing with this type of situations. However, it was not before 2002 that the IAEA issued a report where governments and international organisations documented the severity of the problem. The 2002 Proceedings of the 2000 RADLEG International Conference[4] addressed comprehensively the issue of environmental remediation within the context of the radiation legacy from the 20th century (IAEA, 2002a). This was perhaps one of the first international intergovernmental gatherings making out the intricacies of the problem. The conference recognised that:

- 'As a result of events in the last century, mainly related to the development of nuclear energy, mankind has been forced to deal with the restoration of the environments that contain radioactive residues.

[4]RADLEG 2000 was held in Moscow, Russian Federation, from 30 October to 2 November 2000 and was organised by the Ministry of the Russian Federation for Atomic Energy in co-operation with the IAEA, the European Commission and the Russian Academy of Sciences.

- Historically, the first areas requiring environmental restoration were those where the mining and milling of uranium and thorium ores were conducted and those affected by the processing and application of concentrated natural radionuclides, such as radium-226.
- In the second half of the century, when technologies were being developed and radiation hazards were not clearly understood, a number of substantial discharges of fission products, some accidental, others deliberate, occurred, resulting in the contamination of both production sites and local inhabited areas, for example, the Mayak facility in Urals, Russian Federation, and the Sellafield facility in the United Kingdom.
- Nuclear weapons tests conducted in the 1950s and 1960s led to radioactive contamination of some large continental areas (Semipalatinsk, Nevada, Maralinga) and of islands in the Pacific Ocean. The largest nuclear reactor accident, which occurred at the Chernobyl Nuclear Power Plant in 1986, caused the radioactive contamination of extensive territories in Europe.
- The operation of nuclear facilities has led to the accumulation of large amounts of spent nuclear fuel used for both civil and military purposes as well as to the production of high-level radioactive waste.
- In some facilities, a significant fraction of the spent fuel and radioactive waste, mainly originating from the early period of nuclear power, is stored in conditions that do not meet present safety requirements, for example, surface water bodies and underground cavities.
- Of the nuclear facilities now undergoing decommissioning, some are in conditions which threaten to create environmental contamination, for example, floating disused nuclear submarines.
- Accidents at some of these facilities could lead to contamination of both local and distant areas due to river, marine and atmospheric transport [*sic*]'.

The items if this impressive enumeration would be the concern of the radiation protection community in years to come.

4. The International Radiation Protection Paradigm

4.1. The recommendations of the ICRP

Radiation protection is not a science but a paradigm, namely a model for keeping people safe from the potential detriment that radiation exposure may cause. Notwithstanding, it is of course based on solid scientific knowledge that characterises radiation exposure and its health effects.

Internationally, the United Nations Committee on the Effects of Atomic Radiation (UNSCEAR), which reports to the United Nations General Assembly, provides the epistemological basis of this knowledge. In order to benefit from a worldwide paradigm, the universal consensus on the science underlining radiation protection should be complemented by a global agreement on commonly accepted protection ethics.

The currently accepted international radiation protection paradigm is surprisingly homogenous. It is recommended by the International Commission on Radiological Protection (ICRP). The ICRP is a charity operating as an advisory body that offers its recommendations to regulatory and advisory agencies, mainly by providing guidance on the fundamental principles on which appropriate radiological protection can be based. Since its inception in 1928, the ICRP has regularly issued enumerated reports containing recommendations regarding protection against the hazards of ionising radiation. The first report in the current series, ICRP Publication 1, contains the recommendations adopted in 1958 (ICRP, 1959). ICRP Publication 26 (ICRP, 1977) contains the recommendations adopted in 1977. The recommendations that are still used in current standards appeared as Publication 60 (ICRP, 1991) in 1990. Recently, in 2007, the ICRP adopted new recommendations that were issued as ICRP Publication 103 (ICRP, 2007).

International organisations and national authorities responsible for radiological protection, as well as users of radiation and radioactive substances, have taken the recommendations and principles issued by ICRP as a key basis for their protective actions. As such, virtually all international standards and national regulations addressing radiological protection are based on the commission's recommendations, currently on those contained in ICRP Publication 60. The relevant international standards, namely the *International Basic Safety Standards for Protection Against Ionizing Radiation and for the Safety of Radiation Sources* (IAEA, 1996a), are also based on these recommendations. At the time of the edition of this book, a process is underway to revisit the current national and international standards vis-à-vis the new recommendations contained in ICRP Publication 103 (ICRP, 2007).

Following ICRP Publication 103, the ICRP has recently issued specific recommendations for defining what situations require radiation protection control measures. Issued as ICRP Publication 104 (ICRP, 2008) and entitled *Scope of Radiological Protection Control Measures*, the new report recommends approaches to national authorities for their definition of the scope of radiological protection control measures through regulations. This is particularly relevant within the context of remediation.

The ICRP has approved additional recommendations on the *Application of the Commission's Recommendations to the Protection of Individuals Living in Long Term Contaminated Territories after a Nuclear Accident or a Radiation Emergency*, which are expected to clarify further the ICRP position on the issue of remediation. The recommendations in this report complement those in ICRP Publication 82. They develop further the role of stakeholders, introduced for the first time in this publication by the ICRP, recognising that those concerned with this type of situation should be involved and be given the opportunity to participate directly in the implementation of protective actions to control their exposures. They also take into account the evolution introduced by the 2007 recommendations in ICRP Publication 103 from the previous process-based approach of practices and intervention to an approach based on the characteristic of radiation exposure situations (see next section). They particularly emphasise the new approach of the ICRP, which re-enforces its principle of optimisation of protection (see Section 4.3) to be applied in a similar way to all exposure situations with restrictions on individual doses. The new report is still in the editing process.

4.2. Characterisation of exposure situations

Within the context of remediation, there is an important presentational difference between the ICRP Publications 60 and 103. As indicated before, the former uses a process-based approach distinct through the concepts termed *practices* and *intervention* – a *practice* being defined as a human endeavour that can increase the overall exposure to radiation and an *intervention* being defined as human actions that decrease the overall exposure to radiation. Thus, 'remediation', in the language of ICRP Publication 60, is an archetypical *intervention*.

Conversely, ICRP Publication 103 uses a situation-based approach to characterise the possible situations where radiation exposure may occur. It considers that the term *planned exposure situations* better characterises its previous intentions for defining *practices*, and *emergency exposure situations* and *existing exposure situations* for *interventions*. The new characterisation is defined as follows:

- *Planned exposure situations* are situations involving the deliberate introduction and operation of sources.
- *Emergency exposure situations* are situations that may occur during the operation of a planned situation or from a malicious act or from any other unexpected situation and require urgent action in order to avoid or reduce undesirable consequences.
- *Existing exposure situations* are exposure situations that already exist when a decision on control has to be taken, including prolonged exposure situations after emergencies.

Thus, contaminated territories requiring 'remediation' would be a case of *existing exposure situations* (see hereinafter).

Notwithstanding these differences, it should be noted that both the concepts of 'practice' and 'intervention' and of planned, emergency and existing situations are widely used in radiation protection and will be used in this book.

4.3. The basic principles of the protection paradigm

Three fundamental principles provide the basis for the ICRP radiological protection paradigm, namely *justification*, *optimisation* and *individual dose limitation*. These principles form the basis of the ethics of the paradigm, which is particularly relevant to situations of remediation. It should be recognised that exposure situations requiring remediation usually give rise to societal problems and to discussions about the ethical principles on which the radiation protection approach should be based. Although no specific philosophical doctrine has been explicitly referenced by the ICRP in the formulation of its basic recommendations, the principles on which its paradigm is based are examples of two commonly accepted ethical principles. On the one hand, the system requires that adequate radiological protection of identified individuals be ensured; for instance, the principle of individual dose limitation ensures that deterministic radiation effects on individuals are prevented and that individual risk of stochastic effects is restricted. This could be construed to be linked to the principles of *deontological ethics*. On the other hand, the overall guiding principles of optimisation and justification ensure achieving a positive benefit for the greatest number of people in society under the prevailing social and economic circumstances of the exposure situation. This could be construed to be linked to the principles of utilitarian ethics. Consideration of both these types of ethical principles is critical for the societal acceptability of the radiation protection approach in exposure situation requiring remediation.

Within the context of remediation, these fundamental principles can be formulated as *justification of remediation*, *optimisation of remedial actions* and *restriction of residual individual doses*, and are described in the following sections.

4.3.1. Justification of remediation

Any remediation should be justified; that is the alteration that remediation generates in the radiation exposure situation of the contaminated territory should do more good than harm. This means that by reducing the existing exposure through remediation, the individual or societal benefit must offset the detriment that the remediation may cause.

4.3.2. Optimisation of remedial actions

Remediation measures in a contaminated territory should be optimised; that is the level of protection to be achieved by the remediation should be the best under the prevailing circumstances, maximising the margin of benefit over harm. Optimisation should result in the likelihood of incurring exposures, the number of people exposed and the magnitude of their individual doses all being kept as low as reasonably achievable, taking into account economic and societal factors.

While optimisation can be applied qualitatively and even intuitively, the ICRP has provided extensive guidance on how to apply this fundamental principle quantitatively. In ICRP Publication 37 (ICRP, 1983), the ICRP provided recommendations on the use of cost-benefit analysis in the optimisation of radiation protection. In ICRP Publication 55 (ICRP, 1989), the ICRP provided recommendations on optimisation and decision-making in radiological protection. More recently, in its new recommendations (ICRP, 2007), the ICRP emphasised the importance of its optimisation principle and its key role of indicating that it should be applied in the same manner in all exposure situations. Restrictions are put on doses to a nominal individual (or *reference person*). Options resulting in doses greater in magnitude than that prescribed by such restrictions should be rejected at the planning stage. Importantly, these restrictions on doses are applied prospectively, as with optimisation as a whole. If, following the implementation of an optimised protection strategy, it is subsequently shown that the value of the constraint or reference level is exceeded, the reasons should be investigated, but this fact alone should not necessarily prompt regulatory action.

4.3.3. Individual dose restrictions

In order to avoid severely inequitable outcomes of the optimisation procedure, there should be restrictions on the doses or risks to individuals remaining in the contaminated territory. The ICRP has traditionally recommended an individual-related annual dose limit of 1 mSv for planned exposures from regulated practices and has further recommended the use of source-related *dose* constraints and *reference levels*, which within the context of remediation can be described as follows:

- A dose constraint is prospective and source-related restriction on the individual dose from a specific contamination source, which provides a basic level of protection for the most highly exposed individuals from such a source and serves as an upper bound on the dose in optimisation of protection for that source.

- In contrast, if protection cannot be planned in advance but is undertaken in a *de facto* situation, reference levels should be used for deciding intervention with protective measures. Reference levels should represent the level of dose or risk, above which it is judged to be inappropriate to plan to allow exposures to occur, and below which optimisation of protection should be implemented. The chosen value for a reference level will depend on the prevailing circumstances of the exposure under consideration.

The new ICRP recommendations on the *Application of the Commission's Recommendations to the Protection of Individuals Living in Long Term Contaminated Territories After a Nuclear Accident or a Radiation Emergency* have slightly renamed the principles for these particular situations as follows: *justification of protection strategies, optimisation of protection strategies and reference levels to restrict individual exposures.* The basic concepts underlining this reformulation will be discussed later.

4.4. Prolonged exposure situations

ICRP Publication 60 was basically mute on the issue of remediation. This silence in ICRP Publication 60 caused many problems for dealing with contaminated territories in the aftermath of the Chernobyl accident. Not surprisingly, national authorities and international organisations used different and often inconsistent remediation approaches, causing much public confusion.

ICRP responded to these concerns by issuing ICRP Publication 82 (ICRP, 1999), which contains recommendations for dealing with *prolonged exposure* situations. Prolonged exposures were defined as exposures adventitiously and persistently incurred by the public over long periods of time and incidental to situations in which members of the public may find themselves, the average annual dose associated with prolonged exposures being more or less constant or decreasing slowly over the years. This definition fits very well the type of situations expected from a contaminated territory.

4.4.1. Dealing with radioactive residues

The ICRP thus recommended national authorities to consider options for dealing with radioactive residues remaining from uncontrolled early operations and events. In principle, decisions on the need for intervention and on the scale and extent of any required protective action should be made on a case-by-case basis, as no general solutions were recommended. It was recognised that the necessary actions may vary greatly in complexity and scale, involving site rehabilitation through in situ treatment of residues (covering of residues, deep ploughing, soil treatment to prevent uptake by plants, etc.) or scrapping and removing residues for storage and ultimate disposal. The methods recommended for justifying intervention and for optimising protective actions in prolonged exposure situations were to be applied in each individual situation. The ICRP recommended the use of generic reference levels in terms of total dose rate being incurred (see hereinafter), which were expected to provide guidance for the solution of difficult problems.

An interesting issue addressed in ICRP Publication 82 is whether the additional annual doses attributable to radioactive residues from earlier unregulated human activities and events should be subject to any restriction criterion. In principle, there are no impediments in these situations to restricting the attributable individual doses to arbitrary levels. But, in many situations, the origins (and originators) of these activities and events are not even traceable, and it might not be reasonable or even feasible to impose on society the current criteria of radiological protection for practices, which were not available at that time.

ICRP Publication 82 clearly states that if the radioactive residues are the result of a *practice*, the residual prolonged exposures attributed to the practice should be restricted by, among other things, application of individual dose constraints and limits. If the activity has not been controlled according to these requirements, intervention should be considered and, if necessary, implemented.

The ICRP always recognised that there are differences of perception between the residual doses remaining after the application of the system of radiological protection to practices and those for which intervention may be considered. Moreover, as the system of radiological protection is applied on a case-by-case basis, with the prevailing conditions being taken into account, the final residual prolonged annual dose can be different in different cases. In addition, the exposures can be heterogeneous and even uncertain to occur. All these situations create practical problems, concluded the ICRP, including those of public acceptance of different levels, types and even degrees of certainty of residual annual doses.

In cases of radioactive residues that are attributable to current practices, the recommended dose limits and constraints are applicable to the residues remaining after the discontinuation of operation of the sources within the practice. Therefore, in ICRP Publication 82, the ICRP considers that the recommended dose limits and constraints should be applied prospectively to the prolonged exposure from the radioactive residues expected to remain in human habitats after the discontinuation of a practice – for instance, at the site of a decommissioned installation. In principle, the applicable dose constraint may be expected to be no higher than that applied to the operational phase of the practice. In fact, it might appear unreasonable to allow the practice to pose a greater individual risk than before after it has ceased operation. However, the two phases do not necessarily share a common set of circumstances on the basis of which to prescribe equality between the dose constraint applied before the discontinuation of a practice and that applied afterwards. If the operational dose constraint was very low, maintaining it in the post-decommissioning phase could introduce an unreasonable restriction. Should the site of a former practice be shown to satisfy the dose constraint for all its future plausible uses, the site may be released for unrestricted use and the decommissioning phase of the practice terminated. However, if this is not feasible, the site may still be released, but only for restricted use. The restriction can be considered a type of intervention because some form of institutional control will be required.

4.4.2. Generic reference levels of extant dose

A major contribution of ICRP Publication 82 to remediation is the suggestion for using *generic* reference levels of total dose (i.e. existing or extant), the dose rate being incurred in a situation of prolonged exposure. These levels can conveniently be expressed in terms of the existing annual dose, and they are particularly useful when intervention is being considered in some situations, such as exposures to radioactive residues from natural origin or to those that are a legacy from the distant past. Thus, ICRP Publication 82 recommends that an existing annual dose approaching about 10 mSv may be used as a generic reference level below which intervention is not likely to be justifiable for some existing exposure situations. Below this level, protective actions to reduce a dominant component of the existing annual dose are still optional and might be justifiable. In such cases, action levels specific to particular components can be established on the basis of appropriate fractions of the recommended generic reference level. Above the level below which intervention is not likely to be justifiable, intervention may possibly be necessary and should be justified on a case-by-case basis. Situations in which the annual (equivalent) dose thresholds for deterministic effects in relevant organs could be exceeded should require intervention. An existing annual dose rising to 100 mSv will almost always justify intervention, and this may be used as a generic reference level for establishing protective actions under nearly any conceivable circumstance.

However, the ICRP stressed that the generic reference levels recommended in ICRP Publication 82 should be used with great caution. If some controllable components of the existing annual dose are clearly dominant, the use of generic reference levels should not prevent protective actions from being taken to reduce these dominant components. These actions can be triggered by either specific reference levels or case-by-case decisions. The use of the generic reference levels should also not encourage a 'trade-off' of protective actions among the various components of the existing annual dose. A low level of existing annual dose does not necessarily imply that protective actions should not be taken for any of its components; conversely, a high level of existing annual dose does not necessarily require intervention.

In ICRP Publication 82, the ICRP insisted that challenging situations of prolonged exposure include those where high levels of natural background radiation are present and where the exposure is controllable (one such situation is the presence of natural gamma-emitting radionuclides on the ground). It therefore recommended that concerned national authorities and, as appropriate, relevant international organisations should use appropriate fractions of the recommended generic reference levels of existing annual dose as guidance for solving practical problems. The ICRP further considered that for radioactive residues from other past human activities and events that were not regulated as practices, the need, form, scale and duration of protective actions should be determined on a case-by-case basis. This should be done following the recommended principles of justification of intervention and optimisation of the protective actions, rather than through pre-selected individual dose restrictions. ICRP Publication 82 makes clear that, if necessary, the recommended generic reference levels of existing annual dose may be used as guidance. However, in cases where the origins of the situation are traceable and where those who produced the residues can still be made retrospectively liable for the protective actions, national authorities may consider applying a specific *ad hoc* restriction to the individual doses attributable to these residues, constraining the resulting doses to levels below those resulting from the optimisation process. For this purpose, additional protective actions may be required from those who created the situation. Such specific dose restrictions, however, may still be higher than the dose constraints and dose limits applied to practices. Residues that are deemed not to require protective actions should not be subject to further restrictions.

4.4.3. Potential exposure situations

Another important issue addressed by ICRP Publication 82 is that of remediation in case of *potential* exposure situations. In some circumstances, radioactive residues can be very sparsely distributed in the environment, usually as 'hot particles', giving rise to situations of prolonged potential exposure. These are situations where there is the potential but not the certainty that the exposure will actually occur. For these situations, action levels should be derived on the basis of the unconditional probability that members of the public would develop fatal stochastic health effects attributable to the exposure situation. That probability should be assessed by combining the following probabilities: the probability of being exposed to the hot particles, the probability of incorporating a hot particle into the body as a result of such exposure, the probability of incurring a dose as a result of such incorporation and the probability of developing a fatal stochastic effect from that dose. (These probabilities should be integrated over the full range of situations and possible doses.) In establishing such action levels, consideration should be given to the possibility that localised deterministic effects may also occur as a result of the incorporation of hot particles.

4.4.4. Disruptive remediation

ICRP Publication 82 notes that *disruptive remediation*, that is restrictions in the 'normal' living conditions of people, may be required after accidents that release radioactive substances into the environment. Eventually, in order to return to 'normality', such actions may need to be discontinued at some stage in spite of the continuous presence of a residual prolonged exposure. The simplest basis for justifying the discontinuation of intervention after an accident is to confirm that the exposures have decreased to the action levels that would have prompted the intervention. If such a reduction in exposure is not feasible, the generic reference level of existing annual dose below which intervention is not likely to be justifiable could provide a basis for discontinuing intervention. However, it may be difficult to discontinue protective actions that have been in force for many years: the decision may not be acceptable to the exposed population, and the social pressures may override the benefit of discontinuing the intervention. In these cases, the participation of the stakeholders in the decision-making process becomes essential. After intervention has been discontinued, the remaining existing annual dose should not influence the normal living conditions in the affected area (including decisions about the introduction of new practices), even if this dose is higher than that prevailing in the area before the accident.

4.4.5. The issue of commodities

ICRP Publication 82 also addresses the major by-product of environmental contamination, namely the presence of long-lived radionuclides in commodities for public use. When the radionuclides are attributable to a practice, their levels in the commodities should be controlled through the principles of the commission's system of radiological protection for practices. In other cases, they should conceptually be subject to intervention. Mainly due to the globalisation of markets, intervention exemption levels of radionuclides in commodities cannot be established on a case-by-case basis; rather, they need to be standardised. It is not likely that several types of commodities would be simultaneous sources of high, prolonged exposure to any given individual. On the basis of this presumption, a generic intervention exemption level of around 1 mSv is recommended for the individual annual dose expected from a dominant type of commodity, such as some building materials that may in some circumstances be a significant cause of exposure. On the basis of this recommendation, concerned national authorities and, as appropriate, relevant international organisations should derive radionuclide-specific intervention exemption levels for individual commodities, in particular for specific building materials. It should be noted that intervention exemption levels should not be used, either explicitly or implicitly, for relaxing the limits imposed on the activity of radionuclides that may be released from practices. In particular, they should not be used for clearing the recycling of materials resulting from the decommissioning of practices (these situations are better handled with the criterion of exemption for practices).

4.4.6. The quantitative recommendations in ICRP Publication 82

The numerical recommendations in ICRP Publication 82 are summarised in Figure 1.

4.5. The newer ICRP general recommendations

At this stage of development, the ICRP decided that the time was ripe for issuing a revised comprehensive set of recommendations on radiation protection. Thus, the ICRP issued ICRP Publication 103 (ICRP, 2007), which clarifies and simplifies the commission's previous recommendations, particularly those that form the bases for remediation. These new

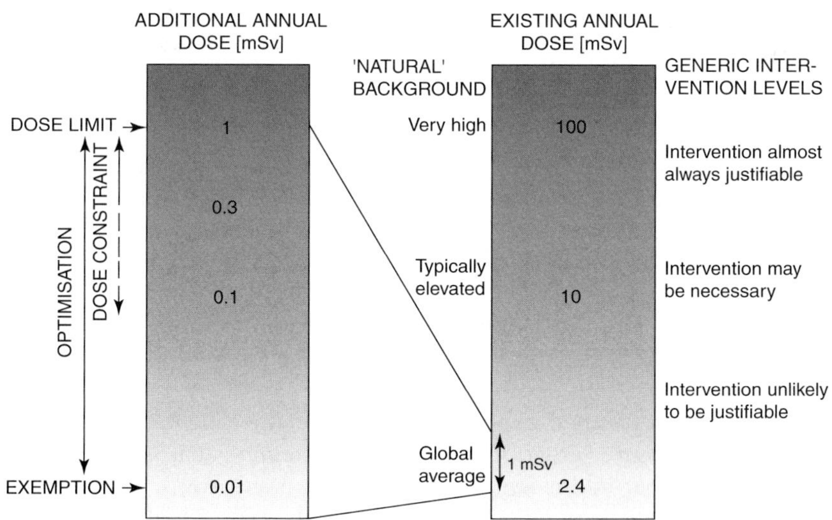

Figure 1 Quantitative recommendations of ICRP Publication 82.

recommendations, which will probably be used in future national and international standards, present some major features applicable to remediation:

- maintaining the commission's three fundamental principles of radiological protection, namely justification, optimisation and the application of dose limits, and clarifying how they apply to radiation sources delivering exposure and to individuals receiving exposure;
- evolving from the previous process-based protection approach using practices and interventions and moving to a situation-based approach applying the fundamental principles of justification and optimisation of protection to all controllable exposure situations, which are now characterised as planned, emergency and existing exposure situations;
- maintaining the commission's individual dose limits for effective dose and equivalent dose from all regulated sources in planned exposure situations – these limits represent the maximum dose that would be accepted in any planned exposure situations by regulatory authorities;
- re-enforcing the principle of optimisation of protection, which should be applicable in a similar way to all exposure situations, with restrictions on individual doses and risks, namely dose and risk constraints for planned exposure situations and reference levels for emergency and existing exposure situations; and
- significantly, including an approach for developing a framework to demonstrate radiological protection of the environment.

Thus, as indicated before, the new recommendations recognise three types of exposure situations that replace the previous categorisation of exposure situations into practices and interventions. As indicated before, these three exposure situations are: *planned exposure situations*, which are situations involving the planned introduction and operation of sources (this type of exposure situation includes situations that were previously categorised as practices); *emergency exposure situations*, which are unexpected situations such as those that may occur during the operation of a planned situation, or from a malicious act, requiring urgent attention; and *existing exposure situations*, which are exposure situations that already exist when a decision on control has to be taken, such as those caused by natural background radiation.

While a situation of contaminated territories can, in principle, be framed under any of the three situations described above, it can usually be considered an existing exposure situation even if it is an undesired evolution from a planned exposure situation or the long-term result of an emergency exposure situation. Thus, the category that is relevant for remediation is that addressing existing exposure situations. ICRP defines existing exposure situations as those that already exist when a decision on control has to be taken. There are many types of existing exposure situations that may cause exposures high enough to warrant radiological protective actions, or at least their consideration. It may also be necessary to take radiological protection decisions concerning existing manmade exposure situations such as residues in the environment resulting from radiological emissions from operations that were not conducted within the commission's system of protection, or contaminated land resulting from an accident or a radiological event. There are also existing exposure situations for which it will be obvious that action to reduce exposures is not warranted (see next section).

The ICRP has repeatedly warned that existing exposure situations can be complex in that they may involve several exposure pathways, and they generally give rise to wide distributions of annual individual doses ranging from very low to, in rare cases, several tens of milliSieverts. In many cases the behaviour of the exposed individuals determines the level of exposure – for instance, the distribution of individual exposures in a long-term contaminated territory, which directly reflects differences in the dietary habits of the affected inhabitants. The multiplicity of exposure pathways and the importance of individual behaviour may result in exposure situations that are difficult to control.

The ICRP now recommends that reference levels, set in terms of individual dose, should be used in conjunction with the implementation of the optimisation process for exposures in existing exposure situations. The objective is to implement optimised protection strategies, or a progressive range of such strategies, which will reduce individual doses to below the reference level. However, exposures below the reference level should not be ignored; these exposure circumstances should also be assessed to ascertain whether protection is optimised or whether further protective measures are needed. An endpoint for the optimisation process must not be fixed *a priori*, and the optimised level of protection will depend on the situation. It is the responsibility of regulatory authorities to decide on the legal status of the reference level, which is implemented to control a given situation. Retrospectively, when protective actions have been implemented, reference levels may also be used as benchmarks for assessing the effectiveness of the protection strategies. The use of reference levels in existing situation is illustrated by Figure 2, which shows the evolution of the distribution of individual doses with time as a result of the optimisation process.

According to the new ICRP recommendations, reference levels for existing exposure situations (such as those situations that may be candidates for remediation) should be set typically in the 1–20 mSv band of projected dose. The individuals concerned should receive general information on the exposure situation and the means to reduce their doses. In situations where individual life-styles are key drivers of the exposures, individual monitoring or assessment as well as education and training may be important

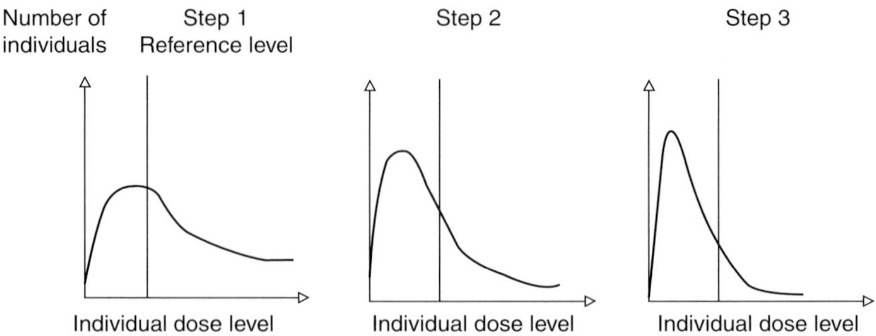

Figure 2 ICRP presentation on the use of a reference level in existing exposure situation and the evolution of the distribution of individual doses with time as a result of the optimisation (ICRP, 2009) process.

requirements. Living in contaminated land after a nuclear accident or a radiological event is a typical situation of that sort.

The ICRP recommends that the main factors to be considered for setting the reference levels for existing exposure situations are the feasibility of controlling the situation and the past experience with the management of similar situations. In most existing exposure situations, the exposed individual as well as the authorities desires to reduce exposures to levels that are close or similar to situations considered as 'normal'. This applies particularly in situations of exposures from material resulting from human actions.

The current recommended values for protection criteria are compared in the following table with those provided by the previous recommendations in ICRP Publication 60 (ICRP, 1991) and the derivative ICRP Publication 82. The comparison shows that the current recommendations are essentially the same as the previous recommendations encompassing the previous values but are wider in their scope of application.

Intervention	Previous reference levels	Current reference level
Unlikely to be justifiable	$< \approx 10$ mSv year^{-1}	Between 1 and 20 mSv year^{-1} according to the situation
May be justifiable	$> \approx 10$ mSv year^{-1}	
Almost always justifiable	Towards 100 mSv year^{-1}	

4.6. Remediation and environmental protection

The traditional position of the ICRP on the remediation of contaminated territories for purposes of protection of the environment has evolved over time. Usually, radiation protection practice has previously been concerned with mankind's environment only with regard to the transfer of radionuclides through it, primarily in relation to planned exposure situations, because this directly affects the radiological protection of human beings. Consequently, in ICRP Publication 60, the ICRP considered that the standards of environmental control needed to protect the general public would ensure that other species are not put at risk. Therefore, if remediation is not needed for humans, it should not be needed for other species.

While the ICRP continues to believe that this is likely to be the case, it also recognises that interest in the protection of the environment has greatly increased in recent years, in relation to all aspects of human activity, which has been accompanied by the development and application of various means of assessing and managing the many forms of human impact upon it. The growing need for advice and guidance on such matters in relation to radiological protection has, however, not arisen from any new or specific concerns about the effects of radiation on the environment. There seemed to be a lack of consistency at international level with respect to addressing such issues in relation to radioactivity. The ICRP is also aware of the needs of some national authorities to demonstrate, directly and explicitly, that not only humans but also the overall environment is being protected.

The ICRP therefore decided to develop a clearer framework to assess the relationships between exposure and dose, and between dose and effect, and the consequences of such effects, for non-human species, on a common scientific basis. This issue was first discussed in ICRP Publication 91 (ICRP, 2003), and it was concluded that for the protection of human beings, it was necessary to draw upon the lessons learned from the development of the systematic framework. This framework is based on an enormous range of knowledge that the ICRP attempts to convert into pragmatic advice that will be of value in managing different exposure situations, bearing in mind the wide range of errors, uncertainties and knowledge gaps of the various databases. ICRP is currently issuing a report on reference animals and plants (see www.icrp.org/draft_animals.asp).

It should be noted that in contrast to human radiological protection, the objectives of environmental protection are both complex and difficult to articulate. There are global needs and efforts to maintain biological diversity, to ensure the conservation of species and to protect the health and status of natural habitats, communities and ecosystems. These objectives may be met in different ways, ionising radiation – depending on the environmental exposure situation – being only a minor consideration. A sense of proportion is necessary in trying to achieve these objectives.

4.7. Excluding and exempting

The ICRP has long recognised that there may be exposure situations for which it will be obvious that remediation to reduce exposures is either unfeasible or not warranted. While many prolonged exposures to contaminated territories are controllable, a number of situations can be either uncontrollable or essentially unamenable to control (e.g. exposure to undisturbed levels of natural radioactivity). Exposure situations that are uncontrollable or unamenable to control are generally subject to *exclusion* from the scope of radiological protection measures. Other situations may be controllable but considered trivial by the authorities and unwarranted to be controlled. Exposure situations that are unwarranted to control are subject to *exemption*. According to ICRP, the decision as to what components of existing exposure are either not amenable to control or unwarranted to be controlled requires a judgement by the regulatory authority that will depend on the controllability of the source or exposure and also on the prevailing economic, societal and cultural circumstances.

ICRP Publication 104 provides advice for deciding the radiation exposure situations that should be covered by the relevant regulations because their regulatory control can be justified and, conversely, those that may be considered for *exclusion* from the regulations because their regulatory control is deemed to be unamenable and unjustified. It also provides advice on the situations resulting from regulated circumstances, but which may be considered by regulators for *exemption* from complying with specific requirements because the application of these requirements is unwarranted and exemption is the optimum option. Thus, the report describes exclusion criteria for defining the scope of radiological protection regulations, exemption criteria *inter alia* in existing exposure situations. The report also addresses specific exposure situations including situations of contamination to naturally occurring radioactive materials and low-level radioactive waste. The quantitative criteria in the report are intended only as generic suggestions to regulators for defining the regulatory scope, in the understanding that the definitive boundaries for establishing the situations that can be or need to be regulated will depend on national approaches. The principles for exclusion and exemption of radiation sources recommended by the ICRP are relevant for decisions on remediation.

It is interesting to note that ICRP Publication 104 suggests that, for existing exposure situations, the definition of scope should address whether the extant exposure is high enough for regulatory intervention to be justified and whether the justified control measures are warranted or the protection is already optimised. The issue therefore is not whether regulations are justified or the expected increase in exposure is large enough to warrant the application of regulatory requirements (as in the case of planned exposure situations). Many existing exposure situations involving natural radiation and radioactive materials may be either excluded from the regulatory scope on the basis that regulation is not justified or exempted from the application of regulatory requirements that are not deemed to be warranted. In these situations, regulations or their application would not be expected to lead to an improvement in protection sufficient to offset the societal efforts and possible detriment arising from regulatory enforcement and implementation. However, in some situations, regulations may specify levels defining a type of non-action ceiling above which some regulatory requirements would apply. For the existing exposure situations that may remain in the long-term aftermath following an emergency, consideration should be given to specifying optimum levels of activity of residual radioactive material above which regulatory requirements would apply to the legal person responsible for remediation. Importantly, the ICRP Publication 104 recommends that, for levels corresponding to a residual annual dose of the order of 1 mSv and higher, control measures are likely to be justified, but higher or lower values may be appropriate in particular circumstances.

4.8. Non-technical factors

It should be emphasised that non-technical factors have an enormous influence on remediation policies. For this reason, the ICRP has always cautioned that its recommendations are based on objective assessments of the health risks associated with exposure levels and on radiological protection attributes of various exposure situations. However, members of the public (and sometimes their political representatives) may have personal and distinct views on radiation risks, for instance those attributable to artificial sources of exposure in relation to those due to natural sources.

This usually results in differently perceived needs for response and a different scale of protection, depending on the origin of the exposure. The public claim for protection is generally stronger when the source of exposure is a technological by-product rather than when it is considered to be of natural origin. Typically elevated prolonged exposures due to natural radiation sources are usually ignored by society, while relatively minor prolonged exposures to artificial long-lived radioactive residues are a cause of concern and sometimes prompt remediation actions that are unnecessary in a radiological protection sense.

This reality of social and political attributes, generally unrelated to radiological protection, usually influences the final decision on remediation. Therefore, while the ICRP reports should be seen as a provider of decision-aiding recommendations mainly based on scientific considerations on radiological protection, the outcome of the commission's advice is expected to serve just as an input to a final (usually wider) decision-making process, which may include other societal concerns and considerations and the participation of relevant *stakeholders* rather than radiological protection specialists alone.

4.9. Living in long-term contaminated territories after a nuclear accident or a radiation emergency

As indicated before, the ICRP is just issuing new recommendations on the application of its recommendations to the protection of individuals living in long-term contaminated territories after a nuclear accident or a radiation emergency (ICRP, 2009). The new recommendations recognise that nuclear accidents and radiation emergencies are managed according to guidance covering short-, medium- and long-term actions. The most recent guidance related to the management of the short- and medium-term actions is provided by recently approved ICRP recommendations on the *Application of the Commission's Recommendations for the Protection of People in Emergency Exposure Situations*. The post-accident rehabilitation situation covered by this report corresponds to the long-term actions that may be necessary to implement in case of nuclear accident or radiological events resulting in long-lasting contamination of large inhabited territories. The transition from an emergency exposure situation to a following existing exposure situation is characterised by a change in management, from strategies mainly driven by urgency – with potentially high levels of exposures and predominantly central decisions – to more decentralised strategies aiming to improve living conditions and reduce exposures as low as reasonably achievable, given the circumstances. These strategies must take into account the long-term dimension of the situation with the direct involvement of exposed individuals in their own protection. The ICRP recommends in its new report that this transition should be undertaken in a co-ordinated and fully transparent manner and be agreed and understood by all the affected parties. The decision to allow people to live in contaminated territories that mark the transition between the emergency and existing exposure situations will be taken by the authorities. This will mark the beginning of the post-accident rehabilitation phase, on which the new recommendations focus.

The types of exposure situations considered in these new recommendations are the result of dispersive events that lead to radioactive contamination over relatively extended areas. The pattern of deposition is dependent on the magnitude of the dispersive event, in terms of both activity and energy release, and on prevailing meteorological conditions at the time of the release, in particular the wind direction and any rainfall occurring during the passage of the plume. For an extended release, wind direction can be expected to vary over time. In the longer term, rainfall and weathering will allow penetration of deposited radionuclides into soil and some migration via water pathways or through resuspension. In plants, uptake of radionuclides from soils may vary seasonally. The levels of deposition may vary greatly from one area to another. For instance, after the Chernobyl accident, surface contamination (activity per unit surface area) varied by factors of up to 10–100 within the same village. Generally, in the longer term, one or a few radionuclides will dominate as the principal contributors to human exposure.

The new recommendations explore the various exposure pathways that can be distinguished following the contamination of the environment in these situations: external exposure due to deposited radionuclides or intake via consumption or inhalation of contaminated material. Intakes of radionuclides by humans may result from consumption of vegetables, meat or milk from animals from affected areas, and fish. Its transfer to animals will depend on the intake of the animal and the metabolism of the various radionuclides by the animal. Radionuclides deposited directly on plants or in soil may be bound to insoluble particles and be less available for intestinal absorption than radionuclides incorporated in feedstuffs. There may be considerable variation in intakes by the population with time, depending on season of the year and resulting agricultural practices, and the types of soil and vegetation. Certain areas such as alpine pastures, forests and uplands may show longer retention in soils than agricultural areas, and high levels of transfer to particular foods, for example berries and mushrooms in forests, may give rise to elevated intakes.

The new recommendations also re-characterise the exposure in these situations. In most existing exposure situations affecting the place of living of the population, the level of exposure is mainly driven by individual behaviour and is difficult to be controlled at the source. This generally results in a very heterogeneous distribution of exposures. The main consequence of living in a contaminated territory is that it is difficult to escape from the contamination. The day-to-day life or work in such a territory inevitably leads to some exposures. The exposure situation prevailing after the termination of large-scale protective actions implemented during the early or intermediate phases of a nuclear accident or a radiological event will generally show a very broad range of individual exposures, both for the doses already received and for the projected residual doses. The range of individual exposure may be affected by many individually related factors. These include:

- location (of home and work) with respect to the contaminated areas (after clean-up);
- profession or occupation and therefore time spent at work, undertaken in particular areas affected by the contamination; and
- individual habits, particularly the diet of each individual, which could be dependent on her/his socio-economic situation.

Experience has shown that the use of 'average individual' is not adapted for the management of exposure in a contaminated territory. Large differences may exist between neighbouring villages, within families inside the same village or even within the same family according to the diet, the living habits and the occupation. These differences generally result in a highly skewed dose distribution among the affected population.

It can be noted that individual doses attributable to the contamination are generally extremely small but very variable. This may create a perception problem of inequity among the population. Another example presented by the ICRP relate to the exposure from ingestion of contaminated foodstuffs. This may result from both chronic and episodic intakes according to the relative importance of locally produced foodstuffs in the diet.

In fine, for the same total intake, the resulting whole body activity at the end of the period is significantly different. This illustrates the intrinsic different burden between daily ingestion of contaminated foodstuff and periodic ingestion. In practice, for people living in contaminated territories, the whole body activity results from a combination of daily and episodic intakes depending on the origin of foodstuffs and the dietary habits.

The new recommendations also precisely explore the principles of the ICRP paradigm as they apply to this particular situation.

4.9.1. Justification of protective strategies

In its main recommendations, the ICRP had considered that for existing exposure situations, protection strategies – carried out to reduce individual exposures – should achieve sufficient individual or societal benefit to offset the detriment that is caused [ICRP, 2007, §203]. Justification of protective strategies, however, goes far beyond the scope of radiological protection as they may also have various economic, political, environmental, social and psychological consequences. The social and political value of reducing exposure and limiting inequity in the exposure received by those living in the contaminated areas needs to be included when justification of protection strategies is carried out. The proper consideration of many of these non-radiological factors may require expertise other than radiological protection and could dominate decisions on protection strategies. The ICRP recognises that justification is concerned with the cumulative benefits and impacts of individual protective actions composing the protection strategy. A range of individually justified actions may be available, but may not provide a net benefit when considered as an overall strategy because, for example, collectively they bring too much social disruption for the considered exposed population as a whole, or they are too complex to manage. Conversely, a single protective action may not be justified alone, but may contribute to an overall net benefit when included as part of a protection strategy.

The ICRP restates that the responsibility for ensuring an overall benefit to society as well as to individuals when populations are allowed to stay in contaminated territories lies with governments or national authorities. However, the new recommendations provide an important role to the individuals concerned also. In existing exposure situations, the ICRP says, justification should be considered for all protective actions that may be included in a protection strategy: those implemented centrally and locally by authorities, experts and professionals and those directly implemented by the exposed individuals as self-help protective actions with the support of the authorities. The protection strategy defined by the authorities should take into account both categories of protective actions and should, in fact, enable affected individuals to take self-help initiatives. However, as far as inhabitants implement – and thus largely decide – self-help protective actions themselves, they must be properly informed and, if relevant, trained (to use the means and equipment provided by the authorities) in order to take informed decisions concerning their own protection, with a net benefit. The balance to be considered by the individuals includes, on the one hand, their desire to improve the situation and, on the other hand, the 'burden' induced by the implementation of protective actions.

4.9.2. Optimisation of protection strategies

The ICRP recalls that the process of optimisation of protection is intended for application to those situations for which the implementation of protection strategies has been justified. The principle of optimisation of protection with a restriction on individual dose is central to the system of protection as it applies to existing exposure situations. Due to its judgemental nature, there is a strong need for transparency of the optimisation process. All the data, parameters, assumptions and the values that enter into the process should be presented and defined very clearly. This transparency assumes that all relevant information is provided to the involved parties and that the traceability of the decision-making process is documented properly, aiming for an informed decision (ICRP, 2007, §34).

The ICRP recognises that the case of an existing exposure situation following an emergency exposure situation comprises some specificity. The fact that the population will stay in a contaminated territory is *per se* a compromise for them and their relatives (family, friends). The optimisation process in such a case faces many specific challenges. The challenges recognised by the ICRP are the following:

- Consumer versus producer interest: to live in a contaminated territory supposes that an economic activity is maintained on the spot with local production and trade of goods including foodstuffs. Optimisation

strategies should balance the need to protect individuals against distribution of radioactivity and the need for the local economy to exist and to be integrated in the global market;
- Local population versus national and international population: the conditions to restore a kind of 'normal' life in the contaminated territories suppose solidarity in sharing some disadvantages of the situation between local and non-local populations (mainly related to the movement of goods and people). Optimisation strategies should favour equity, taking into account national regulation and plans as well as international recommendations (e.g. on trade of foodstuffs);
- The multiple decisions taken by the inhabitants in their day-to-day life: in most cases, the level of exposure is driven by individual behaviour. The authorities should facilitate processes to allow inhabitants to define, optimise and apply their own protection strategies if required. A positive aspect is that individuals regain control on their own situation. However, self-help protective actions may be disturbing (e.g. paying constant attention to the food one eats, the places one goes, the material one uses and the things one touches in order to avoid as much as possible internal and external exposures). This supposes that affected individuals are fully aware of the situation and well informed. To support this, various local individuals may also need to be properly equipped and possibly trained (for the use of equipment provided by the authorities). Authorities should also be prepared to assist segments of the population with particular needs (elderly, mentally handicapped and so on).

4.9.3. Reference levels to restrict individual exposures

In case of an existing exposure situation following an emergency exposure situation, the radiation source is under control, but the controllability of the situation may remain difficult and request a constant vigilance from the inhabitants in their day-to-day life. This constitutes a burden for the individuals living in contaminated territories and for the society as a whole. However, both may find a benefit in continuing to live in the affected areas. Countries generally cannot afford to lose a part of their territory, and in general, most inhabitants would prefer staying in their home rather than being relocated (voluntarily or not) to non-contaminated territories. As a consequence, when the level of contamination is not too high to prevent sustainable human activities, authorities will preferably implement all the necessary protective measures to allow people to continue to live in contaminated territories instead of abandoning them. These considerations suggest that appropriate reference levels should be preferably chosen in the '1 to 20 mSv' band proposed by the ICRP in its Publication 103. As the long-term objective for existing exposure situations is 'to reduce exposures to levels that are close or similar to situations considered as normal', the ICRP recommends to select the reference level for the optimisation of protection of individuals living in contaminated territories in the lower range of the 1–20 mSv year^{-1} band recommended in Publication 103 for the management of such category of exposure situations. Past experience has demonstrated that typical values used for constraining the optimisation process in long-term post-accident situations fall in the range of the dose limit for planned exposure situations, namely an additional dose of around 1 mSv year^{-1}. National authority may take into account the prevailing circumstances and also usefully take advantage of the timing of the overall rehabilitation programme to adopt intermediate reference levels to progressively improve the situation.

It should be noted that these reference levels are not in contradiction with the generic levels recommended in ICRP Publication 82. It should be recalled that the latter are expressed in terms of (total) extant dose, while the former are expressed in terms of the additional dose attributable to the contamination.

5. International Standards on Remediation

5.1. The BSS

The standards that govern general international requirements on radiation protection are the *International Basic Safety Standards for Protection Against Ionizing Radiation and for the Safety of Radiation Sources*, or *BSS* (IAEA, 1996a). However, these standards are basically mute about remediation of contaminated territories. They include only generic requirements for intervention in what at the time was termed *chronic* exposure situations.

The BSS presumed that the state would have determined the allocation of responsibilities for the management of interventions in chronic exposure situations between regulatory authorities, national and local intervening organisations and even registrants or licensees. Under this proviso, the BSS require that generic or site-specific remedial action plans for chronic exposure situations shall be prepared by intervening organisations, as appropriate. The plans shall specify remedial actions and action levels that are justified and optimised, taking into account (a) the individual and collective exposures, (b) the radiological and non-radiological risks and (c) the financial and social costs, the benefits and the financial liability for the remedial actions. They also require that action levels for intervention through remedial action shall be specified in terms of appropriate quantities, such as the annual average ambient dose-equivalent rate or the suitable average activity concentration of radionuclides that exist at the time remedial action is being considered. However, the BSS fail to prescribe numerical action levels for remediation.

The many assessments of the Chernobyl accident (IAEA, 1988, 1991, 1996b, 1996c, 1997a, 2006b, 2006c) clearly demonstrated that the BSS had to be complemented with specific guidance. In the year 2000, the IAEA issued guidance on restoration of environments affected by residues from radiological accidents, with approaches to decision making (IAEA, 2000). At the same time, RADLEG 2000 (IAEA, 2002a) concluded that 'the task of the natural environment preservation in areas of functioning of the nuclear industry's enterprises and nuclear power plants, remediation of lands contaminated with radionuclides, remains to be one of the top priorities. It is evident that efforts on restoration and prevention of eventual activity releases into the biosphere will be linked with huge expenditures and will take many decades. Moreover, the economic situation in states with radiation inheritance dictates that the expenditures for the environmental risk reduction are possible only if it is of vital importance. And this means that a very difficult and controversial choice should be made in setting priorities on remediation and prevention policy [*sic*]'. Areas that need remediation from technologically enhanced natural radiation were also discussed at various fora (see, e.g., IAEA, 2002c, 2003b, 2004a, 2004b, 2005a, 2005b, 2006b, 2006d). After the accident in Goiania, Brazil, the IAEA started to publish a review of major radiological abnormal situations around the world, many of them requiring remediation (IAEA, 1998, 2006a, 2006b). Last but not least, the IAEA started to tackle the controversial issue of decommissioning of nuclear installations and its consequent remediation of sites (IAEA, 2003b, 2004c). It seems that the time was ripe for achieving an international consensus on standards on remediation!

5.2. First safety requirements for remediation

It was not until November 2003 (IAEA, 2003a) that the IAEA finally established safety requirements for the remediation of areas contaminated by past activities and accidents. The new requirements did not introduce any fundamental change in the remediation philosophy. The objectives of remediation were now formulated as:

- to reduce the doses to individuals or groups of individuals being exposed;
- to avert doses to individuals or groups of individuals who are likely to get exposed in the future; and
- to prevent or reduce environmental impacts from the radionuclides present in the contaminated area.

Reductions in the doses to individuals and environmental impacts were to be achieved by means of interventions aimed at:

- removing the existing sources of contamination;
- modifying the pathways of exposure; and/or

- reducing the numbers of individuals or other receptors exposed to radiation from the source.

Remediation was expected to be established on a site-specific basis, and the remedial measures and protective actions shall be justified and the levels shall be optimised. The justification requirement should be implemented by means of a decision-aiding process requiring a positive balance of all relevant attributes relating to the contamination, such as the avertable annual doses, both individual and collective, the health detriments attributable to the intervention, the expected reduction in the anxiety caused by the situation and the social costs, disruption and environmental effects that may result from the implementation of remedial measures. The optimisation requirement is aimed at determining the best-under-the-circumstance intervention level, namely the optimum nature, scale and duration of the remedial measures, which should be selected from a set of justified options for remediation. In some cases, the restricted use of human habitats may be the outcome of the optimisation process for remediation (IAEA, 2007).

The new requirements also noted that the results of such a decision-aiding process for justification and optimisation shall be used as an input for the final decision-making process which may encompass other considerations (such as the remaining residual doses) and may involve relevant concerned parties, the so-called stakeholders.

The requirements established a generic reference level for aiding decisions on remediation as an existing annual effective dose of 10 mSv from all environmental sources, including the natural background radiation.

If remediation is justified for dose levels below the generic reference level to reduce a dominant component of an existing annual dose, a reference level specific to particular components can be established on the basis of appropriate fractions of the generic reference level. Such specific reference levels (such as intervention levels and action levels) shall be subject to the approval of national authorities for particular situations of prolonged contamination that are amenable to intervention on the basis of the optimisation process. Specific reference levels can be expressed in terms of the avertable annual dose or a subsidiary quantity such as activity concentration ($Bq\,g^{-1}$) or surface contamination density ($Bq\,cm^{-2}$).

5.3. Regulating non-technical factors

As indicated before, the IAEA also analysed the non-technical factors impacting on the decision-making processes in environmental remediation (IAEA, 2002b). It concluded that a range of non-technical factors will influence the choice of technologies to be employed in remediation and the strategy for their implementation. These factors include:

- economy, employment and infrastructure;
- costs, funding and financing;
- regulatory and institutional aspects;
- stakeholder perception and participation;
- project implementation–related risks;
- co-contamination issues;
- future land use; and
- stewardship issues.

These factors may have both positive and negative impacts on the decision-making process for choosing appropriate remediation technologies and strategies and on the timeliness with which the chosen technologies and strategies can be implemented. The relevance of any given factor depends on the specific problem and context, which are likely to vary from site to site and across member states. Some factors may be more subject to 'control' than others. The weighting of the different factors in the choice of a solution will therefore be affected by a variety of considerations, including technological, economic and sociological concerns.

The IAEA provides an illustration of the way in which these factors can be addressed, where significant, for remediation decisions. It outlines the range of formal decision-aiding methods that can be useful for organising information and making comparisons between different options. It is emphasised that a formal decision-aiding tool is not a substitute for the judgement and deliberation that build towards a decision (IAEA, 2002b). However, formal decision-aiding methods can, in themselves, constitute an important element of quality control and quality assurance. The formalised process helps to make transparent whether all relevant aspects of the process have been addressed and gives a framework for the documentation of inputs to and outputs from the process.

The way that members of the public perceive the contamination situation and the approach to remediation will influence the decision-making process in a variety of ways. Good awareness of the perceptions of stakeholders and the public at large is important for identification of issues and for evaluation of risks and acceptability of possible solutions. It is also the starting point for building participation processes. Through communication between experts, decision makers and members of the stakeholder communities, participatory processes and negotiation between different interest groups can sometimes be used effectively as mechanisms for exploring solutions. Table 1 lists a range of objectives and considerations that need to be taken into account for remediation decision making.

Table 1 Objectives and considerations forming the basis of an integrated assessment for remediation decision making.

Objectives	Considerations
Consider a full range of possible effects, across health, environmental, socio-cultural and economic disciplines.	Combining all effects into a single metric is probably not possible.
Apply a standard approach that reconciles different methodologies, assumptions and data used previously and anticipated to be used in the future.	Tailor the assessment process to site conditions; no single approach is appropriate for all applications.
Reflect existing environmental, sociocultural and economic conditions.	Focus on potential changes in levels rather than on attempting to establish absolute risk/impact levels of existing conditions.
Employ a consistent approach for evaluating the same types of risks/impacts for different population groups.	Do not assume common values for all affected groups; rather, solicit their input.
Consider cumulative effects of multiple sources and interactive effects of multiple contaminants.	Conduct screening analyses and establish cut-off points to exclude minor sources from the full assessment and incorporate emerging toxicity data.
Evaluate risks/impacts at several geographic scales: local through regional.	Develop different conceptual models to capture local and regional effects.
Evaluate risks/impacts in the near-, intermediate- and long-term time frames.	Address the near term quantitatively, while addressing the longer term for some risks/impacts qualitatively (at least for now).
Consider the individual and cumulative effects of uncertainties.	Focus on major uncertainties, as determined by sensitivity analyses.

The intention is to ensure a technically sound and also socially acceptable decision that meets norms of adequacy or satisfactory performance in relation to the whole range of different concerns. Stakeholder participation in itself does not always guarantee success. But lack of participation may contribute to difficulties in implementing technically sound remediation solutions.

Although the present publication discusses the non-technical factors influencing remediation decision making, it should be emphasised that there are always critical engineering and scientific considerations. If a technology is not viable or is not reasonably expected to perform for the problem in hand, this limits the solutions. However, failure to include relevant non-technical factors may derail an otherwise technically effective solution. Bringing together technical and non-technical factors is thus a critical element in successful implementation of a remediation solution.

5.4. Revising the BSS

The BSS are in process of revision. They will incorporate the new ICRP recommendations and will surely provide definitive international requirements to regulate remediation of contaminated territories.

A draft of the revised BSS has been posted in the web for comments. It addresses the issue of remediation under regulations for existing exposure situations. Within the context of remediation, the new requirements will apply to:

- exposure due to contamination of areas by residual radioactive material from past activities that were never subject to regulatory control or were not regulated according to the BSS and a nuclear or radiological emergency, after an emergency exposure situation has been declared ended;
- exposure to commodities, including food, feed and drinking water, incorporating radionuclides coming from these contaminated areas; and
- exposure to natural sources, including radionuclides of natural origin in commodities including food, feed, drinking water, agricultural fertiliser and soil amendments, and so on.

The draft establishes *generic* requirements such as that the government shall include in the legal framework protection and safety provision for the management of existing exposure situations and that the framework shall:

- specify the types of situations that are included in its scope;
- specify the general principles underlying the strategies developed to reduce existing exposure, avert potential exposure or reduce the likelihood of occurrence of such exposure when such actions have been determined to be justified (i.e. for remediation);
- assign responsibilities for the establishment and implementation of strategies for the management of existing exposures to the regulatory body and other relevant authorities and, as appropriate, to registrants, licensees and other parties involved in the implementation of remedial and protective actions; and
- provide for the involvement of stakeholders in decisions regarding the development and implementation of strategies for managing exposures, as appropriate.

Moreover, the government shall ensure that a programme is established to identify and evaluate existing exposure situations and to determine which public exposures are of concern for radiation protection. In addition, the regulatory body or other relevant authority assigned to establish a strategy for managing an existing exposure situation shall ensure that it defines the objectives pursued by the strategy and appropriate reference levels. The regulatory body or other relevant authority shall implement such strategy, including arranging for the evaluation of the available remedial and protective actions for the achievement of the objectives and of the efficiency of planned and implemented actions and ensuring that information is available to exposed individuals about the potential health risks and about the available means for reducing their own exposure.

Generally, for members of the public, the revised draft establishes that the government and the regulatory body or other relevant authority shall ensure that the strategy for the control of existing exposure situations is commensurate with the risks associated with the existing exposure situation and that remedial or protective actions yield sufficient benefit to outweigh the detriments associated with taking them, including detriments in the form of radiation risks. Moreover, the regulatory body or other relevant authority and other parties responsible for remedial or protective actions shall ensure that the form, scale and duration of such actions are optimised. While this optimisation process is aimed at providing optimised protection to all exposed individuals, priority shall be given to those groups of individuals whose residual exposure exceeds the reference level, and all reasonable steps shall be taken to avoid intake of doses above the reference levels. Reference levels shall typically be expressed as an annual effective dose to the representative person in the range 1–20 mSv or other equivalent quantity, the actual value depending on the feasibility of controlling the situation and past experience in managing similar situations. In addition, the regulatory body or other relevant authority shall periodically review the reference levels to ensure that they remain appropriate in the light of prevailing circumstances.

Specifically, in the case of remediation of areas contaminated by residual radioactive material from past activities or from nuclear or radiological emergencies, the government shall ensure that provision is made in the legal framework for:

- the identification of all legal persons responsible for the contamination and for financing the remediation programme and of appropriate arrangements for alternative sources of funding should such legal persons be unable to meet their liabilities;
- the identification of the legal persons responsible for planning, implementing and verifying the remedial actions;

- the establishment of any restrictions on the use of or access to the area before, during and, if necessary, after remediation;
- an appropriate system for archiving, retrieving and amending records that cover the nature and extent of contamination, the decisions made before, during and after remediation and information on verification, including the results of all monitoring and surveillance programmes after completion of the remedial work.

Moreover, the government shall ensure that an appropriate waste management strategy is established to deal with any waste caused by the remedial work and that provision for such a strategy is made in the legal framework.

The legal persons responsible for the planning, implementation and verification of remedial actions shall ensure that a remedial action plan is prepared and submitted to the regulatory body for approval. The remedial actions are aimed at the timely and progressive reduction of the hazard and eventually, if possible, the removal of restrictions on the use of or access to the area; any additional exposure received temporarily by members of the public as a result of the remedial work is justified on the basis of the resulting net benefit, including the final reduction of the annual dose. In the choice of the optimised remediation option, the radiological and non-radiological impacts on health, safety and the environment are considered, together with technical, social and financial factors, and the costs of transport and disposal of the waste; the radiation exposure of, and other risks to, the workers handling it; and subsequently, the exposure of the public associated with its disposal are all taken into account. A mechanism for public information should be in place, and the stakeholders affected by the existing exposure situation are involved in the planning, implementation and verification of the remedial actions, including any post-remediation monitoring and surveillance.

In addition, the regulatory body shall take responsibility for:

- approval of the remedial action plan and granting of any necessary authorisation;
- establishment of criteria and methods for assessing safety;
- review of work procedures, monitoring programmes and records;
- review and approval of significant changes in procedures or equipment that may have an environmental impact or may alter the exposure conditions of remediation workers or of members of the public;
- receipt and assessment of reports of abnormal occurrences;
- performance of regular inspections and, where necessary, any enforcement actions;

- verification of compliance with the legal and regulatory requirements, including criteria for waste management and discharges established for the remediation programme; and
- where necessary, establishment of regulatory requirements for post-remediation control measures (IAEA, 2007).

For its part, the legal person responsible for carrying out the remedial work shall:

- ensure that the work, including the management of the resulting radioactive waste, is conducted in accordance with the approved remedial action plan;
- take responsibility for all aspects of safety, including the performance of a safety assessment;
- monitor and survey the area regularly during remediation so as to verify the levels of contamination, to ensure compliance with the requirements for waste management and to enable any unexpected levels of radiation to be detected and the remedial action plan to be modified accordingly, subject to approval by the regulatory body;
- perform a survey after completion of the remedial work to demonstrate that the endpoint conditions, as established in the remedial action plan, have been met; and
- prepare and retain a final remediation report for submission to the regulatory body (IAEA, 2007).

After the remedial work has been completed, the regulatory body or other relevant authority shall:

- review, amend as necessary and formalise the nature, extent and duration of any post-remediation control measures already identified in the remedial action plan with due consideration of the residual risk;
- identify the legal person responsible for any post-remediation control measures;
- where necessary, impose specific restrictions on the remediated area, to control the access by unauthorised individuals, the removal of radioactive material or the use of such material – including its use in commodities and in future, for instance the use of water resources and use of material for the production of food or feed – and the consumption of food from the area; and
- periodically review the conditions in the remediated area and, if appropriate, amend or remove any restrictions.

The legal person responsible for the post-remediation control measures shall establish and maintain for as long as necessary an appropriate programme, including any necessary provisions for monitoring and surveillance, to verify the long-term effectiveness of the completed remedial actions in areas in which controls are required after remediation. The conditions prevailing after the completion of the remedial actions, if the regulatory body has imposed no restrictions or controls, shall be considered to constitute the background conditions for new practices or for habitation of the land.

For areas with long-lasting residual contamination in which the government had decided to allow habitation and the resumption of social and economic activities, the government shall ensure, in consultation with stakeholders, that arrangements are in place, as necessary, for the ongoing control of exposure with the aim of establishing living conditions that can be considered as normal, including establishment of reference levels consistent with day-to-day life and of an infrastructure to support continuing self-help protective actions in the affected areas, such as information provision, advice and monitoring.

REFERENCES

IAEA. (1988). *Proceedings of the All-Union Conference on the Medical Aspects of the Chernobyl Accident.* IAEA-TECDOC 516. International Atomic Energy Agency, Vienna.

IAEA. (1991). *The International Chernobyl Project: Assessment of Radiological Consequences and Evaluation of Protective Measures.* International Atomic Energy Agency, Vienna.

IAEA. (1994). *Convention on Nuclear Safety.* IAEA Legal Series No. 16. International Atomic Energy Agency, Vienna.

IAEA. (1996a). *International Basic Safety Standards for Protection Against Ionizing Radiation and for the Safety of Radiation Sources.* IAEA Safety Series No. 115. International Atomic Energy Agency, Vienna.

IAEA. (1996b). *Declaration of Participants of the First International Conference of the European Commission, Belarus, Russian Federation and Ukraine on the Radiological Consequences of the Chernobyl Accident.* IAEA Document INFCIRC/511. International Atomic Energy Agency, Vienna, 11 June 1996, http://www.iaea.org/Publications/Documents/Infcircs/1996/inf511.shtml

IAEA. (1996c). *Proceedings of the International Conference: One Decade After Chernobyl, Summing Up the Consequences of the Accident*, Vienna, Austria, 8–12 April 1996, International Atomic Energy Agency, Vienna.

IAEA. (1997a). *Dosimetric and Biomedical Studies Conducted in Cuba of Children from Areas of the Former USSR Affected by the Radiological Consequences of the Chernobyl Accident.* IAEA-TECDOC-958. International Atomic Energy Agency, Vienna.

IAEA. (1997b). *Joint Convention on the Safety of Spent Fuel Management and on the Safety of Radioactive Waste Management.* Reproduced in document IAEA INFCIRC/546. International Atomic Energy Agency, Vienna.

IAEA. (1998). *Dosimetric and Medical Aspects of the Radiological Accident in Goiânia in 1987.* IAEA-TECDOC–1009. International Atomic Energy Agency, Vienna.

IAEA. (2000). *Restoration of Environments Affected by Residues from Radiological Accidents: Approaches to Decision Making.* IAEA-TECDOC-1131. International Atomic Energy Agency, Vienna.

IAEA. (2002a). *Radiation Legacy of the 20th Century: Environmental Restoration.* IAEA-TECDOC-1280. International Atomic Energy Agency, Vienna.

IAEA. (2002b). *Non-Technical Factors Impacting on the Decision Making Processes in Environmental Remediation.* IAEA-TECDOC-1279. International Atomic Energy Agency, Vienna.

IAEA. (2002c). Technologically Enhanced Natural Radiation (TENR II), *Proceedings of an International Symposium Held in Rio de Janeiro*, Brazil, 12–17 September 1999. IAEA-TECDOC-1271. International Atomic Energy Agency, Vienna.

IAEA. (2003a). *Remediation of Areas Contaminated by Past Activities and Accidents Safety Requirements.* IAEA Safety Standards Series No. WS-R-3. International Atomic Energy Agency, Vienna.

IAEA. (2003b). Safe Decommissioning for Nuclear Activities Proceedings of an International Conference in Berlin, Germany, 14–18 October 2002, *IAEA Proceedings Series*, International Atomic Energy Agency, Vienna.

IAEA. (2004a). *Testing of Environmental Transfer Models Using Data from the Remediation of a Radium Extraction Site.* IAEA BIOMASS-7. International Atomic Energy Agency, Vienna.

IAEA. (2004b). *The Long Term Stabilization of Uranium Mill Tailings.* IAEA-TECDOC-1403. International Atomic Energy Agency, Vienna.

IAEA. (2004c). *Planning, Managing and Organizing the Decommissioning of Nuclear Facilities: Lessons Learned.* IAEA TECDOC-1394. International Atomic Energy Agency, Vienna.

IAEA. (2005a). Environmental Contamination from Uranium Production Facilities and Their Remediation, *Proceedings of an International Workshop*, Lisbon, February 2004, IAEA Proceedings Series, International Atomic Energy Agency, Vienna.

IAEA. (2005b). *Remediation of Sites with Dispersed Radioactive Contamination.* IAEA Technical Reports Series No. 424. International Atomic Energy Agency, Vienna.

IAEA. (2006a). IAEA Safety Glossary: Terminology Used in Nuclear, Radiation, Radioactive Waste and Transport Safety (Version 2.0), *IAEA booklet*, International Atomic Energy Agency, Vienna

IAEA. (2006b). International Conference Chernobyl: Looking Back to Go Forward, Vienna, 6-7 September 2005, Organized by IAEA on behalf of the Chernobyl Forum, http://www.iaea.org/NewsCenter/Focus/Chernobyl/pdfs/05-28601_Chernobyl.pdf

IAEA. (2006c). *Environmental Consequences of the Chernobyl Accident and Their Remediation: Twenty Years of Experience Report of the UN Chernobyl Forum Expert Group "Environment".* IAEA Radiological Assessment Reports. International Atomic Energy Agency, Vienna.

IAEA. (2006d). *Remediation of Sites with Mixed Contamination of Radioactive and Other Hazardous Substances.* IAEA Technical Reports Series No. 442. International Atomic Energy Agency, Vienna.

IAEA. (2007). *Remediation Process for Areas Affected by Past Activities and Accidents.* IAEA Safety Standards Series No. WS-G-3.1. International Atomic Energy Agency, Vienna.

ICRP. (1959). Recommendations of the International Commission on Radiological Protection. ICRP Publication 1. Pergamon Press, Oxford.

ICRP. (1977). Recommendations of the International Commission on Radiological Protection. ICRP Publication 26, *Annals ICRP* 1(3), Pergamon Press, Oxford.

ICRP. (1983). Cost-Benefit Analysis in the Optimisation of Radiation Protection. ICRP Publication 37, *Annals ICRP* 10(2/3), Pergamon Press, Oxford.

ICRP. (1989). Optimisation and Decision-Making in Radiological Protection. ICRP Publication 55, *Annals ICRP* 20(1), Pergamon Press, Oxford.

ICRP. (1991). 1990 Recommendations of the International Commission on Radiological Protection. ICRP Publication 60, *Annals ICRP* 21(1–3), Elsevier, Amsterdam.

ICRP. (1999). Protection of the Public in Situations of Prolonged Radiation Exposure. ICRP Publication 82, *Annals ICRP* 29(1/2), Elsevier, Amsterdam.

ICRP. (2003). A Framework for Assessing the Impact of Ionising Radiation on Non-Human Species. ICRP Publication 91, *Annals ICRP* 33(3), Elsevier, Amsterdam.

ICRP. (2007). The 2007 Recommendations of the International Commission on Radiological Protection. ICRP Publication 103, *Annals ICRP* 37(2–4), Elsevier, Amsterdam.

ICRP. (2008). Scope of Radiological Protection Control Measures. ICRP Publication 104, *Annals ICRP* 37(5), Elsevier, Amsterdam.

ICRP. (2009) Application of the Commission's Recommendations to the Protection of Individuals Living in Long Term Contaminated Territories after a Nuclear Accident or a Radiation Emergency (ICRP has approved these new recommendations at meeting of the ICRP Main Commission held in Buenos Aires, Argentina, 24–26 October 2008), Elsevier, Amsterdam (in press).

CHAPTER 2

SITE CHARACTERISATION AND MEASUREMENT STRATEGIES FOR REMEDIATION PURPOSES

Sergey Fesenko*, Lisa Zeiller *and* Gabriele Voigt

Contents

*Corresponding author. Tel.: +431 2600 28248; Fax: +431 2600 28222
E-mail address: s.fesenko@iaea.org

International Atomic Energy Agency, Agency's Laboratories Seibersdorf, Wagramer Strasse 5, PO Box 200, A-1400 Vienna, Austria

Radioactivity in the Environment, Volume 14
ISSN 1569-4860, DOI 10.1016/S1569-4860(08)00202-7

1. The Role of Site Characterisation in Remediation of Contaminated Environments

The first stage of remediation of any contaminated environment starts with an extensive evaluation of its need, followed by a comprehensive remediation planning, if considered necessary. These tasks are achievable only when information on the site under evaluation is available, reliable and sufficient for making decisions. The final stage of remediation is a final evaluation with the overall aim of demonstrating that the objectives of remediation were successfully achieved and to ensure that the site can be released for the planned use. Simple examples clearly demonstrate an important role of the site characterisation in the general process of remediation of contaminated sites (Table 1), and Figure 1 shows its role in the whole process of the remediation and release for use of contaminated sites (according to IAEA, 1990).

The first stage of site characterisation is to obtain data for the assessment of the current or potential impacts of radiation on the population and biota inhibiting the site or surrounding areas. The long-term presence of adverse effects requiring remediation, potential spread of contamination outside of the contaminated site and other transboundary effects are also within the focus of the initial analysis (IAEA, 1999a).

The second stage is devoted to remediation planning based on the optimisation of necessary resource utilisation and effectiveness, and with this receiving maximum benefit from the use of available remedial technologies keeping in mind stakeholder involvement and their response to the problem. This stage is also based on the information from contaminated pathway analysis, which is being used as input to the further assessments. Matters which should be considered here include the following items:

- significance of different exposure pathways;
- availability of technological solutions and resources and
- public perception.

To achieve these objectives, an optimised sampling (measurement) programme is required. The development of such programme commences with the determination of baselines and comprises the study on the history of all nuclear activities at the site. In case the source of the radioactive contamination is known and there are available records about the radionuclides involved and their contamination densities and chemical form, a more rapid reaction and post-release management is possible. Alternatively, the presence of contamination has to be identified by measurements and expert judgement if no other information is available. The recommended structure of the successive actions with special attention to the pathway analysis is given in Table 1. This action normally allows the justification of the list of radionuclides of concern, their potential ambient concentrations and physico-chemical status of key radionuclides in the environment. One of the tasks at this stage is to study site-specific parameters of radionuclide transfer and site-specific factors governing radionuclide behaviour in the environment.

To ensure that the site can be reused following remediation, compliance of residual concentrations of radionuclides in the environment with the expected levels must be verified (IAEA, 1999a). Thus, post-remediation monitoring is to be performed mainly for the purpose of proving

Table 1 Illustrative example of successive steps in site characterisation.

Step	Actions taken
Discovery of contamination	A routine radiological check of a disused site shows high levels of radioactivity in some areas. At this stage, results usually include only some count rates from a radiation detector at several locations.
Confirmation and initial determination of scope	A grid survey using properly calibrated equipment is carried out. The property boundaries and exits of the natural drainage system are given special attention. This survey shows that dose limits for members of the public might be exceeded under some circumstances within the boundaries of the site. No evidence is found that contamination has moved off-site. The level of protection necessary for workers undertaking further characterisations is determined.
Primary characterisation	If a significant problem exists, the primary site characterisation survey is to be undertaken. The aim is to determine the exact spatial extent of the problem, including depth profiles, and to gather sufficient information to allow a full dose assessment for potential site occupiers and off-site populations. Any radiological threats to the environment from the existing situation or from possible remediation actions would be identified in such an assessment.
Secondary characterisations	If consideration of the results of the primary characterisation led to a decision to undertake remediation, then a detailed characterisation would be necessary to allow decisions to be made about the exact remediation method, and then on details of that action. At this stage, some characterisation is needed to allow the full engineering design of the remediation. In addition, characterisation of any waste stream may be necessary before the transport and disposal options for the waste are approved.
Remediation monitoring and quality control	As removal, treatment or fixation of the contamination proceeds, further monitoring and characterisation continues to update remediation plans and to provide quality control information.
Verification and end-point dose assessment	Following the completion of remediation, a survey should be carried out to document the radiological status of the site and to allow a site for any future use.

Source: Adapted with permission from IAEA (1998b).

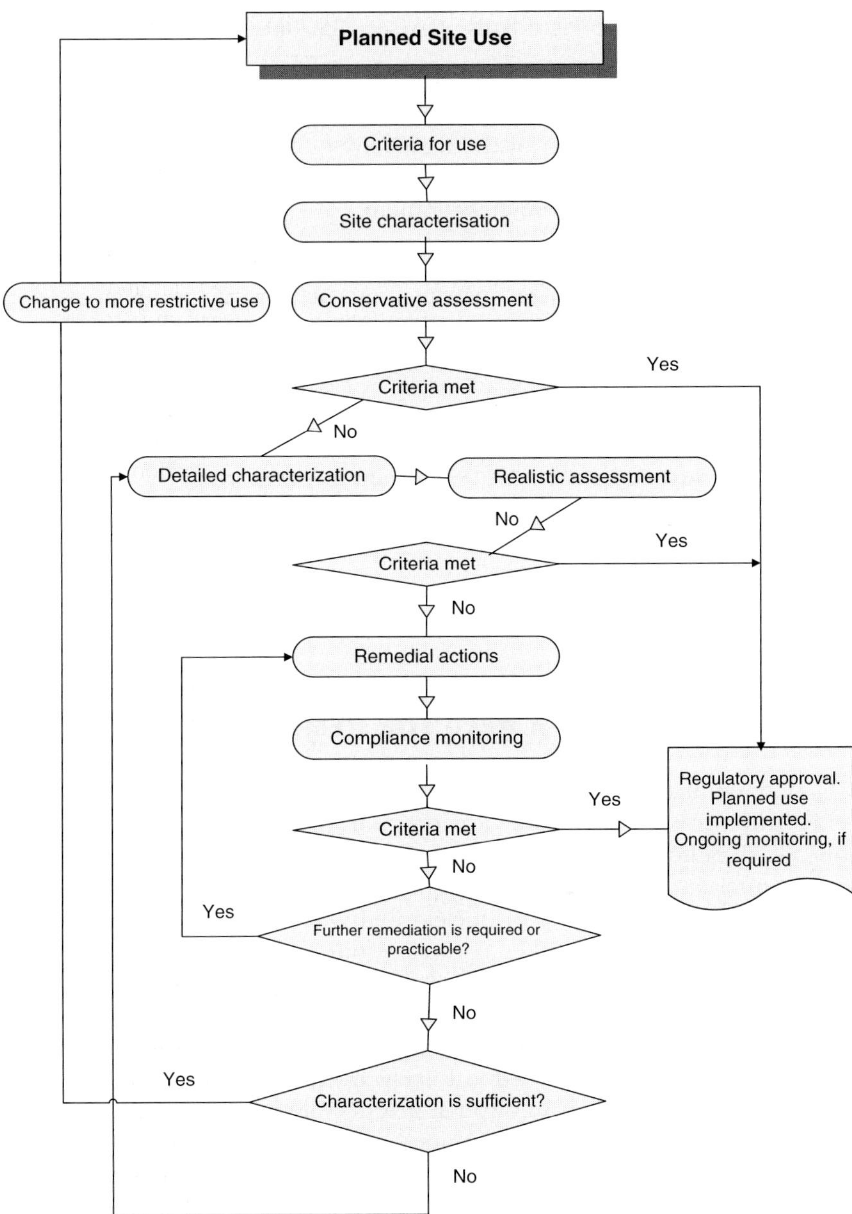

Figure 1 Key stages in remediation and release of contaminated sites (adapted with permission from IAEA, 1999a).

compliance with the standards (IAEA, 1999a) and also for public reassurance:

- Any area with residual contamination must be identified, and the nature, quantity and distribution of the radioactivity determined.
- Radionuclide dispersion and migration will not have deleterious effects on the population and the environment.

The site characterisation during the post-remediation stage should be able to suggest appropriate management actions if compliance with radiation safety criteria was not confirmed. Post-remediation characterisation of a site to be released for future use should consist of several key activities:

- definition of an appropriate strategy;
- planning and management of post-cleanup monitoring;
- site monitoring;
- impact assessment and
- reporting results.

2. Sources of Radionuclides in the Environment Determining a Need for Remediation

Contamination of the environment resulting in a need for remediation can occur from a variety of different sources. The main sources of radionuclides are military tests explosion; radiation accidents; various nuclear fuel cycle activities (uranium mining, processing of ore, fuel fabrication and reprocessing of nuclear fuel); electricity generation; mining and processing of some other natural resources such as gas, oil, tin and phosphate; application of radionuclides in industry, research and medicine; and loss of nuclear weapons.

Although most of the sites requiring large-scale remediation are part of the former 'radiation legacy' (local areas around sites of former radiation accidents, sites of inadequate disposal practices and sites of nuclear weapon testing), there are ongoing and planned activities (such as mining of other natural resources such as coal, oil, gas, etc., as well as processing and decommissioning of nuclear facilities) where remediation is a significant component of operation.

The radionuclides of concern are the naturally occurring uranium and thorium series and man-made radionuclides (e.g. ^{60}Co, ^{137}Cs, ^{90}Sr, ^{239}Pu, ^{241}Am and others). The primary public health threats from these radioactive elements are through inhalation, external whole body exposure to gamma radiation, and ingestion of radionuclides through consumption of food and water or skin emersions (IAEA, 1999b). Table 2 provides a list of industrial and military activities, which results in a need for remediation, and examples of such sites are also given (IAEA, 2001).

2.1. Weapons production and assembly

From the late 1940s until 1980s, facilities for nuclear reprocessing constituted the largest radiation component with impact on the environment, and were the most substantial source of environmental contamination (Aarkrog, 2001). The production of nuclear weapons includes operation with large quantities of radioactive materials. Because of weak operational and inadequate disposal and waste management practices associated with production of nuclear weapons materials during 1940s–1970s, soils, water and other media were heavily contaminated, creating substantial areas requiring remediation (Burnazyan, 1990; Fetisov et al., 1993; IAEA, 1999a, 1999b). For example, widespread disposal practice at that time was based on direct radionuclides disposals to the ground, into surface water bodies or into the sea (Techa, Hanford, Sellafield). Inappropriate storage of liquid waste often resulted in leakage of radioactive material into the soils and groundwater. Solid wastes (e.g. contaminated machinery, clothing, etc.) were commonly buried directly in trenches (IAEA, 1999b). Such practice resulted in severe environmental consequences and raised a need for application of remedial measures.

As a result of improvement of radiation safety culture and development and better nuclear technologies, the discharge of radionuclides into the environment have substantially lowered and therefore the current doses received by the population and staff, as well as impacts on the environment, are rather low.

There were some military applications of depleted uranium obtained from reprocessed fuel. These activities have also resulted in creation of contaminated sites, many of very large areas; however, the radiological impact of application of weapons containing depleted uranium was considered to be negligible.

2.2. Nuclear power generation

During normal operation of nuclear power plants, controlled release of radioactive materials into the environment occurs, resulting in a perceived radiological impact on the general public and the environment.

The nuclear fuel cycle comprises of the mining and milling of uranium ores, the fabrication of nuclear fuel, power generation, reprocessing of nuclear fuel, decommissioning of redundant plant and radioactive waste disposal. Some components of a nuclear fuel cycle, in particular the mining

Table 2 Main nuclear activities determining contamination of the environment and examples of contaminated sites.

Groups of nuclear activities	Examples of activities	Examples of sites	Key radionuclides
Nuclear weapon production and assembly	Reactor operations, irradiated fuel processing, plutonium purification, fuel fabrication, fuel enrichment, waste storage: landfills, tanks machining, warhead development, temperature testing, depleted uranium penetrators and armour	Russian Federation: Mayak, Tomsk-7 Site, Nuclear Complex at Severk, Krasnoyarsk-26 USA: Hanford Site, Rocky Flats Plant, Savannah River Site, Idaho National Engineering and Environmental Laboratory; Pantex Plant, Rocky Flats Plant, Y-12 Plant at Oak Ridge Reservation, Los Alamos National Laboratory	^{60}Co, ^{90}Sr, ^{137}Cs, ^{210}Pb, ^{226}Ra, ^{228}Ra, ^{228}Th, ^{230}Th, ^{232}Th, ^{234}U, ^{235}U, ^{238}U, ^{238}Pu, $^{239+240}$Pu, ^{241}Pu, ^{241}Am
Nuclear weapon testing	Underground test sites, above ground test sites, safety testing of weapons, radiological warfare agent testing	Australia: Maralinga site; Kazakhstan: Semipalatinsk Test Site; USA: Nevada Test Site, Tonopah Test Range	^{3}H, ^{14}C, ^{60}Co, ^{90}Sr, ^{129}I, $^{134/137}$Cs, Pu isotopes, ^{237}Np, ^{241}Am
Peaceful nuclear explosions	Oil and gas reservoir enhancement	Russian Federation: Stravropol and Orenburg regions, Komi Republic	^{60}Co, ^{90}Sr, $^{134/137}$Cs, Pu isotopes, ^{237}Np, ^{241}Am

Radiation accidents and incidents	Tank with radioactive waste explosion leakages, reactor accidents, pipe breakage, aircraft crashes, lost submarines	Russian Federation: Mayak, Tomsk, Chelyabinsk; Ukraine: Chernobyl NPP; USA: Hanford Reservation; Greenland (Tule); Spain (Palamares)	^{3}H, ^{90}Sr, ^{99}Tc, ^{129}I, ^{137}Cs, Pu and Cm isotopes, ^{241}Am
Nuclear fuel cycle activities	Uranium mining and milling, reactor operations, fuel reprocessing, waste repositories	Germany: various Wismut sites; USA: Belfield, North Dakota, Monticello Mill Tailings Site, Utah	^{60}Co, ^{90}Sr, ^{137}Cs, ^{210}Pb, ^{226}Ra, ^{228}Ra, ^{228}Th, ^{230}Th, ^{232}Th, ^{234}U, ^{235}U, ^{238}U, ^{238}Pu, $^{239+240}Pu$, ^{241}Pu, ^{241}Am
Industrial applications	Hospitals, pharmaceutical testing, nuclear research	Worldwide: USA: University of Georgia, Lawrence Berkley National Laboratory; Russia: Kurchatov Centre in Moscow	^{3}H, ^{14}C, ^{32}P, ^{35}S, ^{131}I, ^{226}Ra
Mining and processing of ore with U and Th impurities	Oil and gas mining, phosphorus industry; rare earth production; welding rods instrument manufacturing and maintenance, including luminising dials, colouring agents for glazing	Syrian oil fields; USA: Teledyne Wah Chang, zirconium production; UK: Ditton Manor Park, Stirling Army Site heavy metal	U, Th, ^{226}Ra, ^{222}Rn, ^{210}Pb, ^{210}Po

Source: Adapted from IAEA (1998a, 1999a, 1999b, 1999c, 2001).

and milling of uranium ores, have greater environmental impacts than do others and result in the need for a more extended remediation.

Mining and milling of ores provide release of heavy natural radionuclides (or naturally occurring radioactive materials (NORMs)) and other environmental stressors, that is emission of particulates, heavy metals, acids, etc. Their environmental impacts are very numerous and diverse. Its extent depends on technologies used by mining and milling facilities and distinctive features of the affected ecosystems. Radionuclides of concern mainly include uranium isotopes and its daughter products such as ^{230}Th, ^{226}Ra, ^{222}Rn, ^{210}Pb and ^{210}Pn, etc. Despite low releases of radionuclides, nuclear fuel cycle facilities can contribute to the contamination of the environments to the extent that remedial actions might be advisable even in the absence of any accidents or incidents (IAEA, 1997a). It should be noted that radioactive residues can normally be observed at the territory of nuclear facilities, but that their decommissioning always includes remediation of site and is an essential part of the decommissioning plan to assure that the area can be released for further use (IAEA, 1995a, 1995b, 1997b). The list of radionuclides of concern which can be considered for remediation planning at the different stages of nuclear fuel cycle are ^{60}Co, ^{90}Sr, ^{137}Cs, ^{210}Pb, ^{226}Ra, ^{228}Ra, ^{228}Th, ^{230}Th, ^{232}Th, ^{234}U, ^{235}U, ^{238}U, ^{238}Pu, $^{239+240}$Pu, ^{241}Pu and ^{241}Am.

2.3. Nuclear tests

Radionuclides originated from nuclear weapons testing comprise a large source of man-made radioactivity dispersed over the world. Nuclear explosives have been detonated above ground, on surface and below ground. Examples of contamination from this type of activity can be found at test locations in the South Pacific at the Atolls of Mururoa and Fangataufa (French Polynesia), Novaya Zemlya (Russia), Semipalatinsk (Republic of Kazakhstan), Maralinga (Australia) and the Nevada Test Site (USA) (Anspaugh and Church, 1986; Anspaugh et al., 1990; Yamamoto et al., 1996; Donaldson et al., 1997; KAEA, 1998; Robinson and Noshkin, 1999; Aarkrog, 2001; Carlsen et al., 2001).

Local fallout close to the sites amounts to around 12%; tropospheric fallout, which refers to deposits in a latitude band around the latitude of the test site, to around 10%; and global fallout which mainly deposited on the same hemisphere as the test site (UNSCEAR, 2000) accounts for the rest.

The testing of nuclear weapons accompanies the release of fission products and activation products; however, sometimes, radioactive materials were deliberately dispersed into the environment.

Contaminants are mainly long-lived fission products and original fissile components of test weapons. Many radionuclides are only significant in the first year or so after the nuclear explosion. Radionuclides of concern, which can potentially determine the need for remediation, are ^{90}Sr, ^{99}Tc, ^{134}Cs, ^{137}Cs, ^{237}Np, $^{238,239,240}Pu$ and ^{241}Am.

2.4. Radiation accidents

In the course of nuclear weapons fabrication and handling, exploitation at nuclear fuel cycle facilities and nuclear power generation, there have been several severe accidents resulting in substantial contamination and the need for remediation of vast areas (Alexakhin et al., 2004). These include the Kyshtym (1957), Palomares (1966), Thule (1968), Chernobyl (1986) and Goiania (1987) accidents.

The radiation accident at the military plant producing plutonium in Chelyabinsk-40, Eastern Urals, on 29 September 1957 is one the most serious accidents concerning radionuclide release and dispersion into the environment. As a result of thermal explosion of the reservoir containing radioactive waste (their radionuclide composition corresponded to one-year storage fission products with the exception that Cs radionuclides were lacking), about 100 PBq radionuclides were released beyond the reservoir site (about 10% of the total amount of radioactive substances in the reservoir). The main components of radionuclides in the mixture were: $^{144}Ce+^{144}Pr$ (66%), $^{95}Zr+^{95}Nb$ (24.9%), $^{90}Sr+^{90}Y$ (5.4%) and $^{106}Ru+^{106}Rh$ (3.7%) (Nikipelov et al., 1987).

It was for the first time that an accident had resulted in larger-scale contamination of the environment, accompanied by remedial actions for the protection of the population inhibiting the surrounding areas. Although the main radiation impact to the population was determined mainly by short-lived radionuclides, the radionuclide determining the long-term hazard specifically to the environment was ^{90}Sr; thus, remedial actions were applied mainly against this radionuclide (Burnazyan, 1990; Trapeznikov, 1993; Akleev and Kiselev, 2001).

The accident at the Chernobyl NPP (ChNPP) is considered as the most severe reactor accident in the history of nuclear energy generation. After the first few months following the accident, the high radiocaesium depositions were for the driving factor in comparison to other radionuclides, and a wide range of different remedial actions against ^{137}Cs has been used to mitigate the consequences of the Chernobyl accident (Krouglov et al., 1990; Balonov, 1993; Izrael et al., 1994; Alexakhin et al., 2004; Fesenko et al., 2007).

In the following other accident scenarios and the main radionuclides released at nuclear facilities are given (IAEA, 1989):

- reactor meltdown with or without failed containment (^{3}H, ^{90}Sr, ^{99}Tc, ^{129}I, ^{137}Cs, Pu and Cm isotopes, ^{241}Am);

- reactor meltdown with particle containment (^{3}H, ^{90}Sr, ^{99}Tc, ^{129}I, ^{137}Cs) and
- nuclear reprocessing plant release (^{90}Sr, ^{129}I, ^{137}Cs, Pu, Cm and Am isotopes).

Another group of accidents include accidents where radiation sources have been lost, stolen or discarded. Since 1962, reported accidents with sealed radiation sources have resulted in 21 fatalities among members of the general public (IAEA, 1999b).

In Goiânia (Brazil) in 1987, the environment was substantially contaminated by a stolen ^{137}Cs radiotherapy source (Rozental et al., 1989; Da Silva et al., 1991; Eisenbud and Gesell, 1997) from an abandoned hospital with total activity of 50.9 TBq. Its improper use resulted in the need for an extensive survey in an area of 67 km^{2}, and significant expenses were required for the remediation of the affected areas. In a similar case, 16.7 TBq of ^{60}Co from a broken radiotherapy source was mixed in metallic products in 1983 in Mexico that resulted in a considerable exposure of the staff and the population in touch with these products (Aarkrog, 2001).

Nowadays, only two accidents with loss of nuclear weapons from aircrafts, which resulted in contamination of the environment by transuranic elements, are known: at Palomares (Spain) in 1966 and Thule in Greenland in 1968 (Wrenn, 1974; Iranzo and Richmond, 1987). The site characterisation of the Palomares allowed the identification of 2.26 km^{2} of farmland and urban lands contaminated by $^{239/240}$Pu, while in Thule such area is around 1,000 km^{2}.

Some sites with local contamination can be associated with the loss of radioactive materials during transportation (e.g. along railroad tracks) especially during the times of disarmament and reduction of nuclear facilities. For instance, such contaminated sites have been found in some places in Ukraine (Rudy, 1993).

2.5. Other nuclear applications

It should be realised that radioactivity and radionuclides are in wide use for a variety of scientific, medical, agricultural and industrial purposes. The main consumers of unsealed radioactive sources are hospitals where nuclear practices are used mainly for both diagnostic and therapeutic purposes, but also for cure. Most of the open radionuclides used such as ^{67}Ga, ^{99m}Tc, ^{123}I and ^{201}Tl are short lived and thus hardly impact the environment requiring remediation. However, some of them (^{3}H, ^{14}C, ^{32}P, ^{35}S and ^{131}I) are found in the liquid effluents and consequently in the environment (Aarkrog, 2001). The latter radionuclides are also in use for different research applications (e.g. genetic studies, animal health or environmental studies) while ^{147}Pm is widely applied in the industry. In addition, sealed radionuclides (^{60}Co, ^{137}Cs, ^{226}Ra, ^{241}Am, etc.) are used for irradiation purposes, which in case of careless use or an accident might cause harm to the people and the environment.

In some cases, either through accidents, ignorance or a lacking radiation safety culture, sites have been left contaminated with residues from the operations. Such sites include factories where radium was used in luminescent paint and thorium was used in thorium-coated gas mantles. Recently, luminising workshops which primarily used ^{226}Ra to make luminous dials for aircraft and military vehicle instruments, as well as watches, were a source of environmental contamination. Residual contamination has also been found at the sites where luminising workshops were located, at military establishments and at scrap yards (IAEA, 1999b). At such sites, ^{226}Ra and its progeny could provide rather high impact on the environment mainly because of the ^{222}Rn emanation from the underlying ground resulting in significant inhalation hazard. Besides, radium compounds may be soluble in water and the groundwater pathway can also significantly contribute to potential exposures.

An additional source for contamination of the environment presents waste and its management during the collection, processing, storage and disposal of wastes associated with the use of radioactive isotopes for industrial, research and medical applications (IAEA, 1999b).

About 100 civilian peaceful nuclear explosions have been performed in the former Soviet Union countries and in the United States for different purposes (dam and gas reservoirs building, etc.). Despite that evaluated concentrations of some short-lived radionuclides were measured far away from those test sites, their contribution to the global dose was negligible (UNSCEAR, 2000). However, in some local areas, in particular in the Jakutsk region, contamination of the environment mainly by ^{90}Sr and ^{137}Cs requires application of remedial actions (Aarkrog, 2001; Ramzaev et al., 2008).

2.6. Mining of ore with U and Th impurities

NORMs such as ^{238}U, ^{232}Th and their progeny are most significant environmental stressors which should be evaluated at such sites including assessment of equilibrium among radionuclide of the U and Th series, their physico-chemical status in the release and critical exposure pathways for the population and the environment.

NORMs can be found in many ores containing metals such as copper, gold, niobium, coal and monazite. Residues from the production of phosphoric acid and various phosphates may also cause contamination of the environment (Cancio et al., 1993). Derived from naturally occurring phosphate-bearing ores, the residue can contain Ra and Th and their progeny. The dumping of phosphogypsum on the near surface, as well as the direct discharge of phosphogypsum into rivers, estuaries and coastal waters, are sources of increased concentration of naturally occurring radionuclides in the affected areas (Baetsle, 1993; Cancio et al., 1993).

Other mining and processing activities may also result in radioactive contamination of areas having non-regulated use or even agricultural use. So, wastes from copper mines in Germany were inadvertently used as road and building materials causing unintentional extra exposure of the population. Elevated levels of ^{226}Ra have been found in wastewater from coal mining and oil extraction activities but only until recently, these levels have been ignored. Upon its release from the mine, this wastewater is usually discharged into streams or sedimentation ponds. Once released into a stream or a river, the ^{226}Ra may quickly dilute, but precautions must be made if public access is available to areas near the release before adequate dilution occurs.

Radionuclides of the U decay series may be found, for instance in waste rock piles and slag. Sites may be contaminated especially by ^{210}Pb/^{210}Po particulates. Ores include copper, tin, silver, gold, niobium and monazite.

The combustion of coal for electricity generation, smelting processes, and residential and area heating leads to the release and dispersion of radium in the environment. Most of the ^{226}Ra found in coal (a few tens of kiloBecquerels per kilogram) will be contained in the ash (fly and bottom ash) and disposed of or used in construction; the remainder will be released as atmospheric emissions in form of gases or very fine particles. Contamination levels of soils are typically low compared to U mining areas or areas most affected by the severe nuclear accidents; however, such contaminated sites are widespread all over the world.

3. Site-Specific Factors Governing Remedial Actions

Radioactive contamination of the environment can cover surface areas of hundreds of square kilometres. These areas may include urban areas (roofs, walls, streets, yards, etc.), agriculturally used areas (crop lands, grasslands) and forested or other semi-natural regions (undeveloped, forest product areas, parks).

The importance of the environment as a source for exposure of the population depends on site-specific features of the contamination, which directly highlights the need for adequate site characterisation. Among specific characteristics of the contamination of the environment triggering the need for restoration measures, the following can be considered:

- ambient activity concentrations of radionuclides;
- physical properties of radionuclides;
- mobility and bioavailability of radionuclides in the environment and
- ecosystems and potential land use.

As noted earlier, the goal of such characterisation is to provide the information needed for an informed decision on the remediation of the contaminated environments. The parameters listed above are considered to be the most important for this type of site characterisation.

3.1. Ambient activity concentrations

The ambient activity concentrations of radionuclides are one of the most important criterions to identify a need in different restrictions on contaminated lands which are already part of remedial actions. Normally, in cases where a mixture of radionuclide is presented by one or two main dose-forming radionuclides, it is reasonable to specify these so-called 'reference' radionuclides, and further decisions can be based on the radionuclide activity concentrations of such radionuclides in different environments.

In the case of the Kyshtym accident, the reference radionuclide ^{90}Sr determined the long-term impact of contamination: the agricultural lands with a contamination level above $74\,kBq\,m^{-2}$ were excluded from economic use (Alexakhin et al., 2004).

Cs-137 was and still is the basic dose-forming radionuclide in the zone of the ChNPP accident (except for the period during and immediately after the release when short-lived and intermediate-term-lived radionuclides played an important role). Only in the 30-km zone around the ChNPP, where economic activities had to be discontinued, and in a small zone beyond, ^{90}Sr is of some importance. Therefore, the evaluation of the radiological consequences of the accidental releases from the ChNPP, as well as the planning and implementation of restoration measures, are based on information on the ^{137}Cs levels in the environment and its ecological half-lives in agricultural and semi-natural products (Fesenko et al., 2007).

Indeed, in the first period after the accident at the ChNPP, the decisions concerning the organisation of agricultural production on contaminated territories were mainly based on the level of contamination of agricultural lands by ^{137}Cs. Thus, agricultural lands with a level of contamination above $1{,}480\,kBq\,m^{-2}$ were excluded from economic use (IAEA, 2006a). Different restrictions and recommendations for agricultural production, taking into account the mobility in the soil and the land use, were introduced for agricultural lands with contamination above $185\,kBq\,m^{-2}$. In the least-contaminated zone (37–$185\,kBq/m^2$), the agricultural practice was practically unchanged; only the peaty soils distinguished by enhanced ^{137}Cs transfer to plants received remediation (application of increased rates of mineral fertilizers) (Fesenko et al., 2007).

Generally, it can be assumed that for similar ambient activity concentrations and similar mobility in the environment and physico-chemical characteristics (energy of β-particles or ψ-photons per disintegration), radionuclides with a longer physical half life will have a greater overall impact on the environment than short-lived radionuclides. Simply due to the fact that the residence time of those in the environment is longer, the total dose to the population or to biota species during their life span is greater (Shaw, 2007). However, it should be noted that even substantial releases of radionuclide to the environment requiring public protection from radiation do not necessitate remedial action. For example, short-lived radionuclides such as short-lived I, Zr, Ba, La isotopes, etc. present a major threat for both human beings and biota species in accidental or incidental situations, but are subject to short-term protective measures only (Fesenko et al., 2007). They are not of long-term concerns because of their fast elimination from the environment and hence do not call for remediation of the environment.

For instance, within the first few weeks after the Chernobyl accident, ^{131}I was the main contributor to the internal dose of people, and the main goal of implementing countermeasures was to prevent consumption of contaminated milk by the population as well as to decrease ^{131}I activity concentrations in milk. The high concentrations of ^{131}I and other short-lived radioiodine isotopes during the fallout led to a high iodine concentration in milk within a few days after the accident and, as a consequence, to high doses to the thyroid for a relatively large number of people living in affected areas. This is considered to be the main reason for an increased thyroid cancer incidence among the affected populations (Drozdovitch et al., 1997; IAEA, 2006a). At the same time, already within two months after the accident, concentrations of ^{131}I in the environment decreased by a factor of 100 and iodine no longer provided any considerable contribution to the exposure of the population. Taking into account that implementation of environmental remedial actions normally take several months to several years demonstrates the time limitations in the decrease of the overall impact of radiation.

Radionuclides should be considered also in combination with their progeny since, in some cases, the main impact of contamination on environment could be associated with the short-lived radionuclides as for example for the U and Th series. So, ^{222}Rn (progeny of ^{226}Ra) provides a substantial radiation impact on the environment, which often exceeds impact from ^{226}Ra. In turn, the contributions of ^{218}Po, ^{218}Po (highly hazardous α-emitters) and ^{214}Pb and ^{214}Bi (β and ψ emitters) cover a considerable part of the total dose from the mixture of the radionuclides of the ^{238}U series (Shaw, 2007).

The above examples clearly demonstrate the necessity of a comprehensive analysis of the potential exposure pathways of both parents' and progeny radionuclides in order to enable the justification as to which remedial actions can provide maximum effects and how to limit exposure from the most hazardous radionuclides in the case of multiple contamination.

3.2. Physico-chemical form of radionuclides in the environment

Radionuclides may be released into the environment in different physico-chemical forms depending on the release scenario and the density of radioactive falloutduring dispersion and deposition, and hence exist in the environment in different forms influencing behaviour and availability of radionuclides in their transfer along food chains. For example, radionuclides of local fallout of nuclear weapon testing are substantially present in the form of less-soluble silicate particles, while tropospheric or global fallout mainly includes finely dispersed aerosols or colloids. Similar effects were observed for radiation accidents or in the case of releases from different nuclear facilities such as nuclear reprocessing plants.

One of the peculiarities of the ChNPP accident was the overlapping of radioactive trails with different physico-chemical compositions of the fallout (Sanzharova et al., 1994). The major part of ^{137}Cs was deposited in the form of easily soluble, finely dispersed aerosols; however, coarsely dispersed particles and fuel particles were included to a different degree depending on the distance and the dispersion scenario. The fact that ^{90}Sr and ^{137}Cs deposition after the ChNPP accident occurred in several different physico-chemical forms – fuel particles from the destroyed reactor core, condensed particles and easily soluble aerosols – resulted in radionuclide fluxes into the ecosystems that strongly depended on the distance from the ChNPP. The presence of ^{137}Cs in the form of particles in soils resulted in the competition of two simultaneous but opposing processes, that is an increase in the plant 'available' amount with time due to the destruction of fuel particles, and a decrease in its 'mobility' due to the fixation of radionuclides in soil.

Results derived from studies carried out in the vicinity of the ChNPP show that the mechanism of plant uptake of fuel-particle-derived radionuclides is complicated. For instance, in 1986, the ^{137}Cs plant uptake decreased with increasing fuel components in the fallout due to a high proportion being bound in fuel particles during this period and thus not being available for plant uptake. Subsequently, as a result of destruction of fuel particles and leaching of ^{137}Cs, in the third year after the fallout, plant uptake of radionuclides is an inverse function of the fraction of the fuel component, that is an increase of ^{137}Cs availability in the zones with high concentration of fuel particles in the soil can be observed (Figure 2).

The period of 5–7 years after the deposition, the ^{137}Cs plant availability in zones with different fallout characteristics equal out, which is due to the decreasing amount of radionuclides present in the form of fuel particles and to caesium fixation in soil. It should also be noted that the degree of influence of fallout properties on the dynamics of ^{137}Cs transfer factors (TFs) to meadow vegetation depends in addition on soil characteristics. In

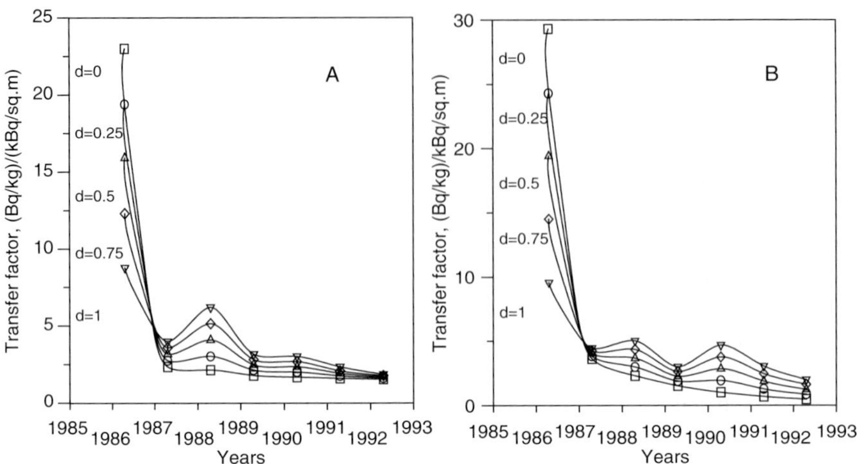

Figure 2 Variation with time in the dynamics of ^{137}Cs transfer factors (TFs) to meadow plants for different proportions (*d*) of fuel components in the fallout: (A) automorphic soils and (B) hydromorphic soils (Fesenko et al., 1996).

the case of hydromorphic soils, the influence of fuel particles is more pronounced than on automorphic ones because of the higher mobility of ^{137}Cs in hydromorphic soils. This example clearly demonstrates that in those areas where ^{90}Sr and ^{137}Cs in condensed form dominated, the contaminants were characterised by high mobility and consequently elevated availability for plant uptake. In contrast, ^{90}Sr and ^{137}Cs released into the environment as fuel particles were less available for root uptake at least during first years after the accident. This finding did not correspond to the commonly known behaviour of ^{137}Cs and ^{90}Sr, as well as to the data obtained from experimental investigations of studies conducted in the accident area of Kyshtym in South Ural in 1957.

The above-described patterns of radionuclide behaviour in the environment are important for both evaluation of the need for remediation and for further remediation planning. Overall, these facts demonstrate that evaluation of the physico-chemical form of radionuclides in the depositions plays an important role in the assessment of the environmental contamination of any contaminated site and therefore is an importance step in site characterisation.

3.3. Mobility of radionuclides in the environment

The mobility of radionuclides along the agricultural food chain, and initially within the soil, is another important factor determining consequences of radioactive contamination and remedial measures. The mobility of radionuclides in the environment depends mainly on two factors, namely the soil properties and the physical and chemical properties as discussed earlier.

Under similar contamination conditions, the exposure of the population is particularly high in regions where low-fertility soils are widespread (soils poor in nutrients and humus, with acidic pH, and of light sand or sandy loam composition). In such lands, radionuclide bioavailability is high and, in consequence, transfer rates through the soil–plant system are higher than in heavier, more fertile soils, resulting in increased levels of radionuclides in agricultural products (Fesenko et al., 2007). Examples are the biogeochemical conditions of the main zone subject to the highest contamination levels after the ChNPP accident. The zone includes the Polesyes area of Belarus, Russia and Ukraine, where light sandy and sandy loamy soddy-podzolic and hydromorphous peat soils are predominant. In this region, an increased mobility of ^{137}Cs (as well as that of ^{90}Sr) in the soil–plant system was already noted as early as in 1960s following global fallout after atmospheric nuclear weapon tests (Mare et al., 1974). According to Balonov (1994), on soddy-podzolic soils in the Polesye region, the contribution from internal irradiation to the effective equivalent dose after the ChNPP accident reached 90%, whereas on fertile, heavy chernozem soils in the same region it did not exceed 10%.

Therefore, it took considerable efforts to rank the soils according to the mobility of radionuclides. The technique, which was applied for characterisation of soils, is based on the estimation of aggregated TF values calculated as the concentration (Bq kg^{-1}) in plant per deposition density (Bq m^{-2}). There is evidence showing clear differences between the radionuclide transfers to different plants and soils, respectively. As reported in IAEA (1994) and Fesenko et al. (1997), the highest transfer rates are observed on peaty soils and mineral soils where the rate of uptake decreases with increasing clay content (Table 3).

Another general conclusion is that the extent to which environmental factors can influence the mobility of radionuclides in the environment is

Table 3 Ratio of transfer factors of ^{137}Cs to plants for different soil groups (Fesenko et al., 1997).

	Ratio	95% confidence interval	Range
Peat: Sand	3.6	3.0–4.4	2.0–8.0
Peat: Loam	10.5	7.5–14.7	3.3–37.5
Peat: Clay	27.0	20.5–35.5	8.3–48.7
Sand: Loam	2.4	1.9–2.9	0.8–6.0
Sand: Clay	5.2	4.1–6.6	1.6–14.3
Loam: Clay	2.2	1.8–2.6	0.4–4.4

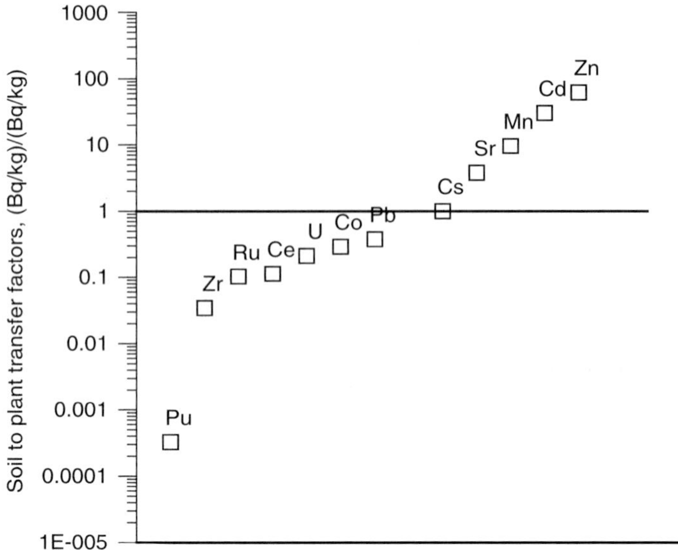

Figure 3 Transfer factors (TFs) to plant normalised to those measured for Cs.

subject to natural variability. As an example, Figure 3 shows the radionuclide TFs to grains of cereals for sandy soils normalised to calculated values based on the data given by IAEA (2009). These data show that the TF values can vary within a large range (within 6 orders of magnitude), affecting conclusions on the severity of the contamination itself, the importance of the different exposure pathways, and hence can substantially influence the technical feasibility and effectiveness of remedial actions.

Following uptake of contaminated fodder by animals, which presents another compartment of the food chain, demonstrates that the same relationships valid for uptake in soil–plant system cannot be applied in the grass–animal system. This is shown in Figure 4 where absorption coefficients in the gut measured for ruminants, available from the above IAEA-TECDOC, are normalised to the Cs plant–animal TF.

Besides the obvious high, natural variation in transfer parameter values, the data demonstrate that that the high mobility of radionuclides in some environmental compartments does not guarantee similar high mobility in other compartments. These factors have to be considered when planning and implementing environmental remediation.

3.4. Land use

It is obvious that deposition densities of soil contamination in different ecosystems result in a different impact depending on the vulnerability of the ecosystem in question and the specific features governing radionuclide

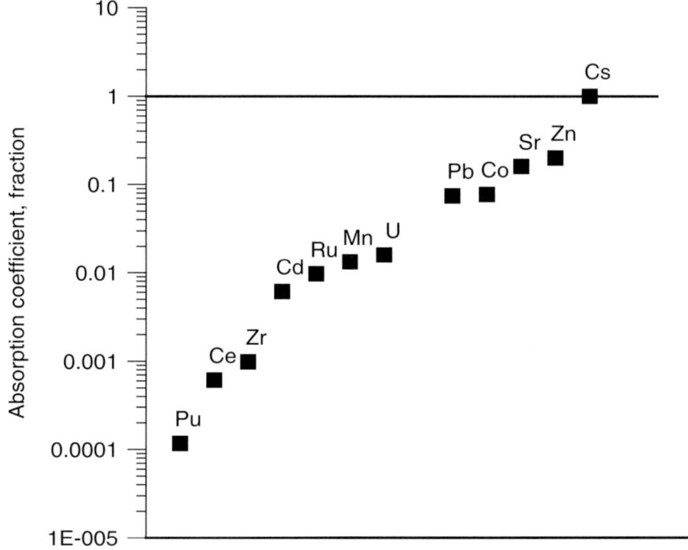

Figure 4 Absorption coefficients in the gut of ruminants normalised to that measured for Cs.

transfer in these environments. One example illustrating the influence of land use on the radioecological sensibility or vulnerability of a contaminated environment is demonstrated from the Chernobyl experience (Fesenko et al., 2000). Some plant species take up more radiocaesium than others (Alexakhin and Korneyev, 1991), and the difference can be as great as a factor of 100 dependent on soil and plant properties (Fesenko et al., 1997). Thus, the contamination of agricultural produce, and hence the need for remediation, depends on both the soil type and the plant species associated with the different types of land use. Such land-use-related variation needs to be considered when selecting a possible alternate land use for contaminated regions, together with any variation in temporary permissible levels (TPLs) for different agricultural products.

To demonstrate the effect of different land-use options, Fesenko et al. (2000) used data from the monitoring programme to estimate the ^{137}Cs reference soil contamination levels which would result in exceeding TPLs in various products from different types of land use in 1994 (Figure 5).

The reference contamination densities, calculated as a ratio of TPLs to TFs for reference plants or plant groups (Figure 5), can be considered as the levels determining the need of remedial applications and can be easily applied in practice. It is thus possible to decide whether territories with a defined land use and soil type require remediation, or whether a change in land use might be appropriate. Such options might include the conversion of arable land into pasture, or handing over agricultural land to forestry.

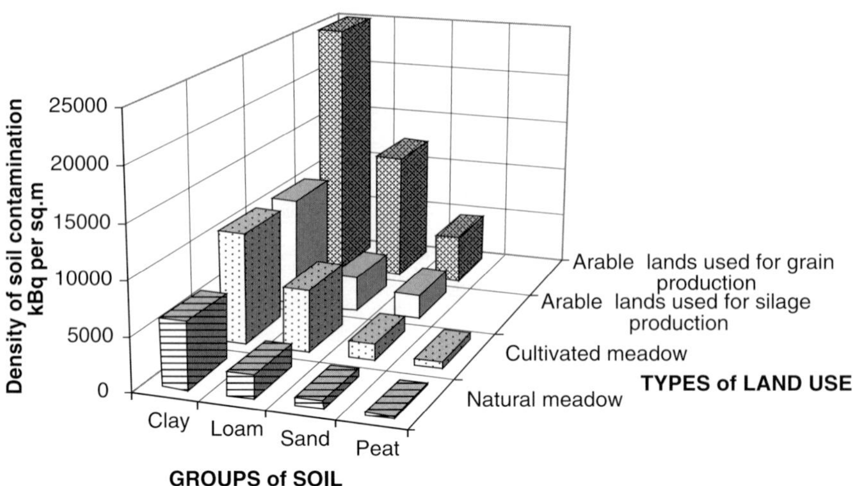

Figure 5 Densities of soil contamination by ^{137}Cs which would result in exceeding Chernobyl-related TPLs for 1994 in the various products for different types of land use and soil groups.

The final goal of any remediation strategy is to provide areas for unrestricted use; however, in the case of very large-scale contamination, there are a lot of interim solutions where areas could be realised for some restricted activities. Therefore, the data and considerations given earlier have a rather high value for time-interval remediation planning when the final goal of remediation is not achievable immediately or in a short time period.

3.5. Classification of contamination scenarios and environments

Contaminated sites can be classified according to different criteria. The origin and contamination scenarios can be taken to classify such sites assuming that for similar contamination scenarios similar environmental behaviour of radionuclides and contamination patterns can be expected. Among such categories, the following contaminated sites can be distinguished according to the radioactivity source:

(i) sites contaminated by uranium mining and milling activities, enrichment and fuel fabrication, nuclear energy production, spent fuel reprocessing;
(ii) sites of decommissioning of nuclear facilities, nuclear weapons production and handling, and radioactive waste management activities where remediation is an obligatory stage;
(iii) sites of decommissioning of other facilities where radioactive sources were used for medical, research and industrial purposes, including production of radioisotopes;

(iv) sites of nuclear weapon tests and peaceful nuclear exposures;
(v) extraction, processing of materials containing natural radionuclides, and other activities that may generate enhanced levels of radionuclides in the terrestrial environment (e.g. extraction of uranium, thorium, rare earths, gas and oil) and
(vi) sites contaminated by radiation accidents.

For each category of sites, key radionuclides and their physico-chemical forms can be preliminarily specified based on former knowledge, except for sites contaminated by radiation accidents mainly because of the diversity of possible contamination scenarios where more detailed classification is necessary to facilitate the site characterisation and further remediation planning. Indeed, in the case of Chernobyl accident, the key radionuclide was ^{137}Cs, for the Kyshtym accident it was ^{90}Sr, while for the Palomares case it was Pu. Therefore, it is advisable to identify the key radionuclides based on the study of the accident scenario. Similar considerations could be given to other nuclear activities and key radionuclides governing a need for remediation.

On the other hand, contaminated sites can be classified on the basis of preliminary assessment of the site in terms of the potential radiation risk to the population (Table 4).

While combining these two approaches with considerations of the earlier mentioned factors affecting the environment, a common classification can be presented as follows:

(a) type of nuclear activity;
(b) potential impact classification class;
(c) key radionuclides;
(d) physico-chemical properties of contaminants;
(e) environmental properties (groups according to soil type, current land use) and
(f) potential land use (for the case of restricted use).

Table 4 Site classification (IAEA, 1999a, based on US NRC, 1997).

Class 1	Areas that have or had prior to remediation, a potential for radioactive contamination or known contamination. Areas containing contamination in excess of the compliance criteria prior to remediation
Class 2	Areas that have or had prior to remediation, a potential for radioactive contamination or known contamination but are not expected to exceed the compliance criteria
Class 3	Any affected areas that are not expected to contain any residual radioactivity, or are expected to contain levels of residual radioactivity at a small fraction of the critical value

Altogether these classification elements can be of paramount importance for the first site evaluation based on historical information, as well as for further remediation planning and making decisions on the release of the site to restricted, controlled or unrestricted use.

4. Radioecological Survey of Contaminated Sites before, during and after Remediation

4.1. Planning of the survey

All decisions related to remediation, justification of the design of a remediation strategy and also the verification of its success are based on analytical data collected during site characterisation and post-remediation surveys. These data should also comprise the environmental characteristics of the contaminated site (Theocharpoulos et al., 2001). Their quality is critical since it may influence the extent, effort and method of remediation, and hence the related costs. Therefore, analytical quality and general quality assurance (QA) principles should be an integral part of all steps of characterisation and post-remediation surveys and in related survey planning processes. General QA requirements applicable for environmental measurements are given in many international standards and guides (Van der Veen and Alink, 1998; Theocharpoulos et al., 2001; ISO, 2004, 2005; IAEA, 2005a).

Planning itself is already one of the QA requirements that is mentioned in most guides and technical documents as an important step to achieve data quantity and quality needed for environmental monitoring, characterisation studies and decisions on remediation and potential land use (MARLAP, 2004; Malherbe, 2002; MARSSIM, 2000; IAEA, 1998b, 1998c, 1998d, 1998e, 1999a).

4.2. Historical information on the site

As already outlined, collection of historical site information will be the basis for the survey planning since each type of contamination scenario needs special consideration for the characterisation study and the planned remediation. The reasons for contamination vary significantly and depend on the site use, the extent and nature of accident, the type of nuclear test and many other factors. Some possible scenarios which lead to the need for remediation are described earlier. Each contamination scenario has its unique fingerprint which should be followed up during the survey. Information which needs to be available and might be of interest to support decisions on planning includes:

- maps of the site (covering the timeframe of operations, nuclear tests, etc.);
- records of authorisation to process, possess and handle radioactive materials;
- licensing records;
- operational details;
- previous classification of the site;
- accident details (in industry, NPPs, test sites);
- types of nuclear tests performed, related fission and activation product details;
- environmental measurements on individual samples (biomonitors, sediments, etc.) and data related to health and safety (dose, whole body counting, health statistics);
- pathway, migration and transfer information;
- old survey and/or background data and
- information on past remediation and/or dumping actions.

If information on past nuclear test sites is lost or badly retrievable, interviews with past employees and population of the affected area can be performed to fill gaps. Historical site information provides input for planning a survey and allows a better understanding of potential sources of contamination and related risks.

4.3. Quality of survey data

To assure good quality of analytical data, it is essential to include the data quality objectives already into the survey planning. Data quality objectives are systematic planning tools based on scientific methods. They are used to develop data collection designs and to establish specific criteria for the data to be collected. The process helps planners to identify the decision-making points for data collection activities, to determine the decisions to be made based on data collection activities and to identify the criteria used for making each decision (APHA, 1998). Such processes have been described in many documents (IAEA, 1998b; ASTM, 2006a; Keith, 1996; MARSSIM, 2000; MARLAP, 2004; EPA, 2006); it includes, besides others, development of decision rules and the specification of limits on decision errors (e.g. confidence levels of data to be envisaged, minimum detectable activities needed, maximum acceptable uncertainties of results and amount of data needed). The quantity and quality of data need to be suitable for classification after site characterisation and compliance verification and related decisions.

This required analytical and final data quality can only be achieved if all groups involved in sampling and measurement have the same, well-defined data quality criteria and agree on sampling, measuring and reporting of results and their uncertainty (see also Sections 5.7 and 6.4). Only harmonised approaches will allow summarising data for the final report that can be the basis for site evaluation and decisions on remediation.

4.4. Use of existing radiation data for planning

Also, existing data can be used to optimise the planning of the characterisation survey. Information on the heterogeneity of contamination and its spatial distribution are essential to define sample numbers, measurement points for *in situ* measurements and sampling locations including their sampling depth, since a three-dimensional distribution of radioactivity can also be expected (IAEA, 1998b) depending on the contamination scenario. Possible situations which are described using past measurements results are:

- superficial deposition;
- surface deposition which migrated into the ground;
- buried or covered activity;
- radioactive particles (hot particles) (Dale et al., 2008; Salbu, 1990; Salbu and Lind, 2005);
- localised hot spots;
- deep depositions (waste tips, leaking pipelines and containers) and
- uniform distribution over large areas and volumes (e.g. by ploughing).

If the source and quality of old data is not well defined, or their information details are not sufficient (e.g. not covering all area, no migration data from soil profiles), a verification of data accuracy and additional measurements are needed to establish a sampling plan for the survey.

4.5. Planning survey or pre-screening

In some cases, also an additional planning survey needs to be considered, especially if insufficient or no radionuclide data exists, the available data are not fulfilling the quality requirement or important parameters were not measured before, for example if no information on alpha/beta contamination is available or information on depth distribution is missing. Other authors (MARSSIM, 2000; IAEA, 1998b) also describe the need for additional surveys to support the design and optimisation of the characterisation study. The survey may confirm:

- type of scenario which led to the contamination;
- types of isotopes and isotopic ratios present;
- the initial scope for remediation;

- the location of main contamination sources and
- the distribution of contamination (e.g. local, homogeneous, depth).

The preliminary site characterisation is focusing on the discrimination of contaminated and less/not contaminated areas and the knowledge on vertical distribution of contamination. Preliminary dose information is used to adjust the health protection procedures during the survey. The scale of the planning survey is always a balance between having sufficient data for the optimisation of the characterisation planning and cost. Data of the planning surveys can only be incorporated into the characterisation data if the same quality criteria as for characterisation data are met.

4.6. Selection of characterisation techniques

Historical site data, data from previous screenings and analysis, and planning surveys are used to prepare the survey plan and to select appropriate techniques for the characterisation. Topics which need evaluation are sampling and field measurement design, number of sampling locations and field measurement points, matrix to be sampled, sampling methodology and the most appropriate equipment thereto, and laboratory analysis techniques to be applied.

The selected techniques shall produce data of the best quality to assess the contamination of the site and to optimise the remediation and future land use.

4.6.1. Sampling and field measurement designs

Sampling and field measurements can be done using different designs. Location, pattern, number of samples and sampling sequence have to be defined before conducting the characterisation survey and must be already an integral part of the sampling plan. Many designs along with their advantages and disadvantages are described (IAEA, 1999a; ICRU, 2006; MARSSIM, 2000; IAEA, 2004c; ISO, 2002a; ASTM, 2003a; EPA, 2002a). The selection of the most appropriate design for the survey depends on site-specific situations and assessment requirements. Some of the suggested designs are shown in Figure 6.

Approaches used often in remediation surveys are: complete sampling, systematic sampling, random sampling and judgemental sampling.

*4.6.1.1. **Complete sampling.*** Complete sampling or field measurements covering the whole area will give the most comprehensive information on the site, but is not practicable in most cases. For high contamination levels, for example at sites classified as Class 1 (see Table 4), and a heterogeneous distribution of the contamination, a sampling coverage of up to 100% is

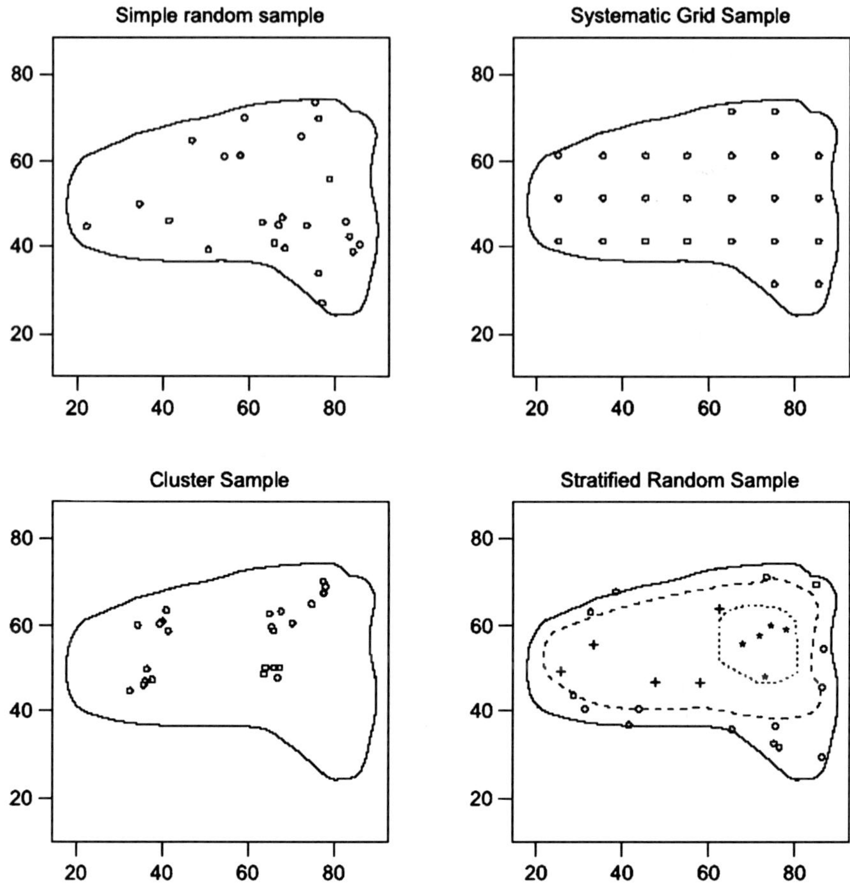

Figure 6 Examples of possible sampling designs (ICRU, 2006).

recommended (MARSSIM, 2000). The disadvantage of this sampling design is the high number of samples or measurements needed to cover larger contaminated areas.

*4.6.1.2. **Systematic sampling.*** Systematic sampling is using a sampling grid or sampling pattern (e.g. square, triangular). This type of sampling reduces the number of samples/measurement points but may give high errors if hot particles or hot spots can be expected which are smaller than the defined grid size.

*4.6.1.3. **Random sampling.*** Random sampling (selecting the sampling locations and distance between samples randomly) bears the risk of not detecting high-contamination areas and need-specific statistical considerations to assure that the required data quality objectives are met for the decisions and actions.

4.6.1.4. Judgemental sampling. Judgemental sampling needs professional experience for the selection of the sampling locations. Since an error in judgement would be very critical if the site is highly contaminated, judgemental sampling is often only proposed for sampling areas with expected low contamination (MARSSIM, 2000; IAEA, 1999a).

Factors like wind direction, erosion processes leading to natural enrichment in sediments and valley soils, enrichment in corner or at other natural barriers, soil types, transfer pathways and many others need to be considered for selecting field measurement or sampling locations. Also, areas with known pollution level and a high potential of contamination (ditches, drains) are included in the judgemental sampling design. The latter will allow a reduction of the sampling area because if no pollution will be found in these critical locations it may be concluded that the area is not contaminated.

Another advantage of judgemental sampling is that it can be used as a tool to localise the source of undeclared and unknown contamination sources. For example, sediments taken from each tributary give an indication of the direction in which contamination sources are located that are causing elevated measurement data (IAEA, 2005b). Data from judgemental sampling is often the basis for more systematic field measurements and sampling in selected parts of the site. Data that are derived from judgemental sampling shall not be included in the statistical evaluation because they have often a known and deliberately positive bias.

4.6.2. Data quality requirements related to number of field measurement and/or sampling locations

The requirements for data quality are already fixed in the early planning phase, and acceptance levels for variability and uncertainty may differ significantly during different phases of assessment. While for screening methods high uncertainties are often tolerable, measurements close to the detection limits of a method (to discriminate false negative/false positive measurements; see Section 5.7.4) and data needed for characterisation or compliance verification normally have to be of better accuracy and precision. High confidence levels for data sets and small result uncertainties require sensitive analysis techniques and a large number of field measurement points and analysed samples. The collected data should be sufficiently large to give a representative picture on site contamination and to allow the conclusion that the data quality objectives are met.

The relationships between numbers of samples and data quality objectives parameters are described in several publications (APHA, 1998;

ASTM, 2003b, 2006b; MARLAP, 2004; MARSSIM, 2000; Keith, 1996; ACS, 2008) and may cover:

- confidence levels versus numbers of false positive or false negative conclusions;
- tolerable uncertainty versus analyte concentration;
- result standard deviation and confidence levels versus sampling area grid and
- size of hot spot versus sampling grid size.

The calculation of the optimal number of measurement points and samples to be taken and analysed requires some information on method detection limits, method uncertainties which can be received from method validation data and/or heterogeneity of contaminated area (see QA in measurement) as detected in the planning surveys. The most simplistic approach for calculation of the number of sample points to be measured or to be sampled assumes a normal distribution of the contamination and includes the maximum acceptable deviation of the estimate from the true values, or the absolute limit of uncertainty of this estimation[1] (*U*) and the overall standard deviation (*s*) of a number (*m*) of samples to estimate the variability of an area and the Student's t-statistic for a given confidence level (APHA, 1998; ASTM, 2003b; RADREM, 1989):

$$N \geq \left(\frac{t_{m-1;1-\alpha} s}{U}\right)^2$$

Many statistical methods suitable for other distribution scenarios (e.g. log normal, random, stratified) and problems (e.g. detecting hot spots) are available from different sources (ICRU, 2006; ASTM, 2006b; Keith, 1996; MARSSIM, 2000) and hence it may be more realistic to describe the complex situations of most survey areas.

To avoid too many samples or measurements points for the characterisation survey, the data quality criteria set in remediation planning need to be realistic and achievable within a normal scale and cost of the project.

4.7. Selection of instrumentation and measurement technique

4.7.1. Field measurement instrumentation

The decision as to which field measurement techniques are applied in the survey depends on several factors (RADREM, 1989) and needs to be made for each remediation project:

- the type of emission or the mode of decay of the radionuclide of interest;
- the specificity of the required data (specificity can be achieved e.g. by radiochemical separation or energy discrimination);

[1]U may derive from measurement uncertainties, from heterogeneity or may be a combination of several factors.

- the sensitivity required from the measurement and
- the number of sample locations to be analysed.

Field measurement techniques are preferred, but often they are a compromise between selective radionuclide information, sensitivity and cost of a survey. Multiple types of field measurement instrumentation are available for alpha, beta and gamma radionuclides.

Detailed information for all below-mentioned methods, on their applicability to site surveys and their specificity and sensitivity, are available in many publications (MARLAP, 2004; MARSSIM, 2000; MARSAME, 2006; IAEA, 1999a; IAEA, 1998b, 1998c, 2000c; Hutter et al., 2001). Table 5 shows a compilation of hand-held field instruments and their potential applications.

Table 5 Potential applications of common hand-held instruments (MARSAME, 2006).

Instruments	Alpha	Beta	Photon	Neutron	Detectable energy range	
					Low-end boundary	Low-end boundary
Ionisation chamber detector	NA	Fair	Good	NA	40–60 keV	1.3–3 MeV
Gas-flow proportional detector	Good	Good	Poor	Poor	5–50 keV	8–9 MeV
Geiger–Muller detector	Fair	Good	Poor	Poor	30–60 keV	1–2 MeV
ZnS (Ag) scintillation detector	Good	Poor	NA	NA	30–50 keV	8–9 MeV
NaI (Tl) scintillation detector	NA	Poor	Good	NA	40–60 keV	1.3–3 MeV
NaI (Tl) scintillation detector (thin detector, thin window)	NA	Fair	Good	NA	10 keV	60–200 keV
CsI (Tl) scintillation detector	NA	Poor	Good	NA	40–60 keV	1.3–3 MeV
Plastic scintillation detector	NA	Fair	Good	NA	40–60 keV	1.3–3 MeV
BF_3 proportional detector	NA	NA	NA	Good	0.025 eV	100 MeV
^{3}He proportional detector	NA	NA	Poor	Good	0.025 eV	100 MeV

For alpha measurements in the field, alpha scintillation survey meters, alpha tract detectors, electret ion chambers, gas-flow proportional counters or long-range alpha detectors (LRAD) are used. LRAD circumvents the short range of alpha particles to be detected and uses alpha-produced ions in air (IAEA, 1999a; Hutter et al., 2001).

Beta activities are measured with electret ion chambers, gas-flow proportional counters and with Geyger Mueller survey meter with beta pancake probe.

Due to the nature of radionuclide transport through matter (soil, air) and to the attenuation of ionising radiation, *in situ* detection of alpha and beta emitter is difficult and often limited to qualitative information (IAEA, 1998b).

Especially alpha radiation is easily adsorbed by irregular, porous, moist surfaces, and this need to be considered when converting count-rate data to surface contamination levels (MARSSIM, 2000).

The best detection capability is available for gamma-emitting radionuclides. The commonly used *in situ* Germanium and NaI detectors (Reiman, 1994), combined with a multichannel analyser, have the advantage of giving also radionuclide-specific information (see Section 4.7.2). Other available techniques are electret ion chambers, pressurised ionisation chambers (PIC), Geyger Mueller survey meter with gamma probe and hand-held (pressurised) ion chamber survey meters; the latter are used also for dose-rate measurements in the 1–2,000 μS range, NaI survey meters and thermo-luminescence dosimeters.

All field measurement data need always be correlated to their measurement location. Global positioning systems (GPS) are satellite supported and are the most frequently used equipment for the purpose.

Radon measurements need to be considered for most remediation surveys due to its high impact on radiation dose and health. Radon and radon progeny emit alpha and beta particles and gamma rays. Therefore, numerous techniques have been developed for measuring these radionuclides, based on different emission scenarios (MARSSIM, 2000).

Rn-222 is the most important radionuclide (^{238}U decay chain) since it has a half life of 3.8 days and might migrate through soil and building materials. Some of the radon gas measurement methods are combined with time-consuming adsorption procedures followed by laboratory gamma measurements. Some commonly used methods are:

- activated charcoal adsorption;
- large area activated charcoal collectors methods;
- alpha track detection;
- continuous radon monitoring and
- electret ion chambers

Field X-ray fluorescence spectrometers, and low-energy gamma detectors like the Fidler probe are commonly used for field measurements. In normal usage, Fidler probes do not detect plutonium directly, but detect the 60 keV X-ray that is emitted from the decay of ^{241}Am, the progeny of ^{241}Pu. This probe type uses a thin, large area scintillation detector (IAEA, 1998b). Lee et al. (2005) developed an *in situ* method to detect fissile hot particles.

More sophisticated techniques, like chemical species laser ablation (LA), mass spectrometry or LA-ICP-OES or LA-ICP-MS are very expensive and not yet often used.

4.7.2. Mobile and aerial survey

In situ measurements in the fields are a way to obtain a large amount of measurement data, which help to reduce data uncertainty and improve the accuracy of contamination estimations. If possible, it should be done in a scanning mode using mobile laboratories mounted on tracks or other vehicles or even using low-flying aircrafts such as helicopters. Scanning data always need to be related to the specific location on the site. GPS, microwave and ultrasonic ranging are only a few techniques which can be used (IAEA, 1998b).

The type of detectors used for the monitoring depends on the remediation objectives and to what extent high method sensitivity and isotopic selectivity are needed for a decision. Normally, NaI detector with multichannel analyser and high-purity germanium gamma spectrometers are used (IAEA, 1999a; SCKCEN, 2006; Kettunen et al., 1997; Long et al., 2004; Thomé and Hurley, 1996). Most alpha and beta emitters cannot be determined directly due to self-absorption, but their gamma decay products or indicator/surrogate radioisotopes are measured instead (MARSSIM, 2000).

The minimum detectable concentration (MDC) of a scan survey depends on the intrinsic characteristic of the detectors (efficiency, physical probe area, etc.), the nature (type and energy of emission) and relative distribution of the potential contamination (point versus distributed source and depth of contamination), scan rate and other characteristics (MARSSIM, 2000).

Table 6 shows an example of typical MDCs which can be achieved using scanning surveys for two different types of NaI detector.

The total area coverage obtained using mobile and aerial monitoring helps to improve confidence in the design of the sampling plan and may also detect unknown or buried contamination sources. Changes in vegetation colour and structure of the soil surface visible in aerial screenings may also be an indication for other undeclared activities and should be considered for further investigations during the sampling campaign.

Table 6 NaI (Tl) scintillation detector scan MDCs for common radiological contaminants (MARSSIM, 2000).

Radionuclide/radioactive material	1.25 × 1.5 in. NaI detector		2 × 2 in. NaI detector	
	Scan MDC ($Bq\,kg^{-1}$)	Weighted cpm/μR/h[a]	Scan MDC ($Bq\,kg^{-1}$)	Weighted cpm/μR/h[a]
^{241}Am	1,650	5,830	1,170	13,000
^{60}Co	215	160	126	430
^{137}Cs	385	350	237	900
^{230}Th	111,000	4,300	78,400	9,580
^{226}Ra (in equilibrium with progeny)	167	300	104	760
^{232}Th decay series (sum of all radionuclides in the Th decay series)	1,050	340	677	830
^{232}Th in equilibrium with progeny in decay series	104	340	66.6	830
Depleted uranium (0.34% of ^{235}U)	2,980	1,680	2,070	3,790
Natural uranium	4,260	1,770	2,960	3,990
3% Enriched uranium	5,070	2,010	3,540	4,520
20% Enriched uranium	5,620	2,210	3,960	4,940
50% Enriched uranium	6,220	2,240	4,370	5,010
75% Enriched uranium	6,960	2,250	4,880	5,030

[a]The relationship between the detector's net count rate to net exposure rate in counts per minute per microRoentgen per hour.

4.7.3. Laboratory instrumentation and analysis methods

Laboratory measurements are always done in addition to field measurements because their capabilities to produce quantitative and isotope specific information are generally superior and give important additional information for site evaluation.

The selection of laboratory analysis techniques depends on the type (alpha, beta or gamma emitting) and emission energy of the isotopes, the concentration of the contamination, the type of matrix and the radionuclides to be measured.

Clear decisions on the analysis methods to be applied for the survey are needed before the environmental sampling. In this part of the survey planning, the analysts should be included to assure that the sample mass collected is sufficient to practically achieve the required or theoretical detection capacity of the survey methods.

The commonly used methods for this purpose are alpha spectrometry and LSC, both combined with radiochemical separation techniques (e.g. solvent extraction, extraction chromatography), gamma spectrometry and mass spectrometry, but other techniques may also be used. Tables 7–9, partly retrieved from MARSAME, 2006, show some of the detection devices which are available. Additional information, also on the principles of measurement, can be found there or in other publications (MARLAP, 2004; MARSSIM, 2000; NUREG, 2002; RADREM, 1989; Gilmore and Hemingway, 1995).

Table 7 Instruments for the detection of alpha radiation.

Detection method	Advantages/disadvantages
Alpha spectrometry with multichannel analyser	Excellent peak resolution, energy spectrum allows quantitative and qualitative radionuclide determination, needs radiochemical separation, low detection limits
Gas-flow proportional counter	Direct measurement, detector needs to be protected against contamination, radionuclide identification needs radiochemical separation prior measurement
Liquid scintillation counter	Quantification of radionuclides possible, but needs chemical separation and multichannel analyser for radionuclide identification
Alpha scintillation detector	Primarily it is used to quantify ^{226}Ra by the emanation and detection of ^{222}Rn gas using the Lucas cell
ICP-MS	Qualitative isotopic information, quantification not always easy because of lack of sensitivity or interferences (polyatomic or isobaric)

Table 8 Instruments for the detection of beta radiation.

Detection method	Advantages/disadvantages
Gas-flow proportional counter	Direct measurement, detector needs to be protected against contamination, radionuclide identification needs radiochemical separation prior measurement
Liquid scintillation counter	Quantification of radionuclides possible, but needs chemical separation and multichannel analyser for radionuclide identification
ICP-MS	Qualitative isotopic information, quantification not always easy because of lack of sensitivity or interferences (polyatomic or isobaric)
AMS	Very sensitive, isobaric interferences

Table 9 Instruments for the detection of gamma radiation.

Detection method	Advantages/disadvantages
High-purity germanium detector with multichannel analyser	Very suitable for simultaneous analysis of multiple gamma-emitting radionuclides and their quantitative determination
NaI detector with multichannel analyser	Useful when only a small number of gamma-emitting nuclides are present or for gross gamma measurements, relatively poor energy resolution, not effective for identification and quantification of individual gamma peaks in complex spectra
ICP-MS	Qualitative isotopic information, quantification not always easy because of lack of sensitivity or interferences (polyatomic or isobaric)

Many of the detection techniques require sample dissolution, a separation or enrichment of the radionuclide of interest or a separation of the interfering radionuclides or matrix elements.

4.8. Selection of sample type

During *in situ* measurements, no discrimination is made on the type of samples. Measurements cover the whole area within the reach of the detector (e.g. grass, soil, vegetation, air, water, buildings, concrete and asphalt). If additional individual samples are collected, this decision should be based on the needs for specific information and validation of the field measurement data. The main reasons for the collection of individual samples are:

- Additional information on isotopic composition and quantitative concentration on specific radionuclides for the classification of the site.
- Comparison of dose assessment data deriving from screening methods with data calculated using individual radionuclide concentration for the validation of screening methods.
- Establishment of correction factors for attenuation and absorption effects for *in situ* gamma measurement results, improvement of *in situ* calibration.

- Verification of assumptions used in the survey, for example on correlations between different radionuclides and isotopic ratios (MARSSIM, 2000).
- Evaluation of the uncertainties of the above assumptions.

Surrogate measurements and use of isotopic ratios for dose calculation are commonly used practices after the Chernobyl accident (IAEA, 1991). ^{90}Sr assessments were often based on Cs measurements using a conversion ratio from Cs/Sr. Also, Pu activity was rarely measured directly but derive from ^{241}Am data (IAEA, 1998b; MARSSIM, 2000). Before this information can be used for final conclusions on contamination, the accuracy and uncertainty of the assumption should be verified.

4.8.1. Soil samples

Soils can generally be divided into top soils, surface soils and subsurface soils. Top soils are directly exposed to the contamination and play an important role to assess re-suspension of particles. Surface soils, in addition, also cover the zone where the root systems of plants are located. Standard 40 CFR 192 (USEPA, 2001) defines the surface soils as the upper soil layer of 15 cm. But in remediation the important layer may vary depending also on the scenario which led to the contamination and the future use of the area. Surface sample depth shall be defined in the planning of the characterisation.

Factors which might be considered for decision on the most appropriate layer depth to be collected for individual samples are:

- migration behaviour of radionuclides in soil;
- main pathway of human exposure (inhalation, transfer to food, etc.);
- type of radionuclide present and
- information needed for validation and verification of screening methods.

The sample taken should be representative of the contaminated site under investigation. Pre-tests are needed to better describe the soil type and approve of the expected migration behaviour of all radionuclides of interest. Thus, if the main contamination is Pu, it can be expected that the main contamination be in the top-soil layer due to the low mobility of Pu.

It is known that migration depends on many factors such as organic content of soil, its porosity, rainfall, etc. (Walling and Quine, 1994; Bunzl et al., 2000; IAEA, 2006c). Therefore, site-specific factors need to be evaluated and included in decisions on soil sampling. If migration information is important (e.g. for making a correction for radionuclide attenuation in *in situ* gamma measurements), the sampling depth needs to be increased. Sampling depth of up to 50 cm may be useful, and often the soil is collected in individual layers to better understand the radionuclide transport.

Soil cores of deeper zones are not so frequently collected. They may be important if groundwater contamination cannot be excluded, or the contamination scenario gives indication for burying of radioactive waste or leaking of waste drums. Arial screening or Google Earth's photographs can help to identify suspicious locations. Necessary sampling depths may vary, sometimes up to 5–10 m, and need to be evaluated and adapted during sampling. Monitoring equipment, which is suitable to be inserted into bore tubes, can be used to decide on final length of cores. Normally, the detection techniques do not measure the radioactivity directly, but instead focuses on other parameters such as magnetic or electromagnetic response of buried materials. Ground penetrating radars are capable of collecting images of buried objects *in situ* if the objects are not too deep (<5 m, depending on soil matrix) (MARSSIM, 2000; IAEA, 1998b).

4.8.2. Sediments

Sediment samples are widely used as environmental indicators. They play an important role in the assessment of contaminants in natural waters as they have a high capacity to accumulate over time the low concentration of radionuclides (IAEA, 2005b). The settling of suspended particles helps remove the contaminants attached to these particles from the water (Voitsekhovitch et al., 1996; Burrough et al., 1999). The collection of sediment samples can be used for localising contamination sources, but is normally not used for site classification and characterisation since sediment samples are not representative of the site contamination and would lead to an overestimation of doses.

4.8.3. Air

Air sampling might be necessary at some sites, especially in areas with mining activities where radium, thorium and uranium are present in soil and lead to an emanation of radon.

In addition, air particulate sampling can be done to evaluate the pathway for re-suspension and the dispersion of radioactive materials (MARSSIM, 2000). Wind direction, natural boundaries like trees and buildings and other environmental conditions need to be considered for the selection of sampling location.

Air sampling in post-remediation surveys is mainly used for monitoring occupational and public health and safety.

4.8.4. Building materials

In most survey areas, existing buildings are demolished. Therefore, contamination measurements for buildings and their material are usually performed.

Due to their smooth surface, *in situ* measurements normally provide good information on the contamination. In addition, swipe samples or building surface samples may be taken. They are normally taken on a filter or fabric pad by rubbing it over a pre-determined area (normally $100\,cm^2$). Swipes can be directly countered in a proportional counter or liquid scintillation counter for alpha and beta activity determination (MARLAP, 2004), eventually followed by isotopic analyses if they show elevated radioactivity. Multi-agency radiological assessment of material and equipments (MARSAME, 2006) addresses material-related problems in detail.

4.8.5. Other samples

Water, groundwater and food samples may be collected if they have relevance for the transport pathway of the radionuclides or are significant contributors to the internal doses. In case of contamination of water and groundwater, their remediation may be of great importance and its justification can require extensive sampling (IAEA, 1999c; APHA, 1998).

Vegetation samples can give information on the bioavailability of radionuclide and radionuclide transfer and may be important if agricultural land use is planned. Biomonitors are frequently used to locate contamination areas due to their capability to accumulate radionuclides (IAEA, 1997d, 1998b, 2000a; Ellis and Smith, 1987; Sloof and Wolterbeck, 1992; Paatero et al., 1998); their sampling is valuable in the early characterisation phase. When collecting biological samples, besides the spatial, the temporal variability also needs to be considered (ICRU, 2006) for the interpretation of data.

4.9. Sampling equipment

Selection of sampling equipment and their related sample collection procedures should ensure that the sample is representative of the sample type, providing sufficient sample mass for the selected laboratory measurement method and of the required sensitivity to assure compliance with set data quality requirements. The decontamination of equipment needs to be easy to avoid cross-contamination during collection. The applications, advantages and disadvantages of some soil and sediment sampling equipments are shown in Tables 10 and 11 and can guide the selection of appropriate equipment (MARSSIM, 2000; ISO, 2002b; IAEA, 1998c, 2003, 2004c).

4.10. Final survey plan

The final survey plan should be a complete documentation on the whole decision-making process and is the basis for all practical work and analysis

Table 10 Typical soil sampling equipment.

Equipment	Application	Advantages/disadvantages
Scoop, spatula	Soft surface soil	Inexpensive, easy to use and decontaminate, difficult to use in dry and rocky soil
Top soil template	Surface up to 5 cm	Easy to use and decontaminate, defined sampling depth and area, difficult to use on rocky surface
Bulb planter	Soft surface soil	Inexpensive, easy to use and decontaminate
Soil coring device	Soils up to 60 cm, different diameter	Relatively easy to use, preserves soil core, not suitable for sandy soil, limited depth capability, can be difficult to remove and cut cores for profile information
Box coring device	Soils up to 50 cm	Preserves information of sample profile, defined collection volume and sampling depth, difficult to use
Split spoon sampler	Soil to bedrock up to 3 m	Excellent depth range, soil profile information, can be used in hard soils, in conjunction with drill rig suitable for deep cores
Spiral auger	Soil up to 1.5 m	Easy to use, suitable also for hard soil, mix layers
Bucket auger	Soft soil, up to 3 m	Easy to use, good depth range, uniform diameter and sample volume, may disrupt and mix soil horizons greater than 15 cm
Hand- or power-operated auger	Soil from 15 cm to 4.5 m	Good depth range (depends on extension length and material), generally used in conjunction with bucket auger, destroys core, difficult to use if deep cores are sampled
Subsoil probe/ linear samplers	Soft-to-medium hard soils, up to 7 m	Good depth range, undisturbed soils, profile information possible
Power-operated push corer	Flexible depths, depends on soil type	Good depth range, undisturbed soils, profile information possible, expensive

Table 11 Sediment sampling equipment.

Equipment	Application	Advantages/disadvantages
Grab sampler	Sediment surface up to 30 cm	Many types available
Gravity corer	Sediment core up to 180 cm	Many types available, limited penetration depth for low-weight corers, not suitable for soft sediments
Piston corer	Sediment core	Many types available, suitable for short (<1 m)and longer cores (up to 30 m) in fine grained unconsolidated sediments, keeps profile information, deeper coring needs hydraulic cranes
Freeze corer	Sediment with high water content up to 1 m	Keeps profile information, cooling and dry ice needed during collection and for transport

performed during the survey. It shall assure a harmonised approach used by all survey participants and data quality adequate for the assessment of the site and the conclusions to be made on remediation. Topics which need to be covered in the document are:

- background information;
- scope and objectives of the remediation and the release criterion for the future use;
- requirements on data quality during field measurements, laboratory analysis and final compliance survey;
- decision on measurement and sampling equipment;
- sample type of individual samples;
- sampling and analysis methods;
- sampling and field measurement plan (timing, sequence, locations, number and type of samples);
- list of sampling equipment and supplies for the field;
- formats for the reporting of sampling and measurement data and
- QA/QC requirements during the survey (see also Sections 5 and 6).

5. Performing Field Measurements and Sampling

Sampling and field measurements are performed in agreement with the survey planning document. The regular review of the field measurement data is important for the success of the survey, since field measurement data may give an indication that some of the assumptions during planning were not correct and the survey plan needs adaptation to the changed situation. Adaptive sampling without a pre-defined sampling plan was even proposed as a method to reduce sampling and measurement efforts in expanded sampling areas.

A harmonised approach for sampling and field measurement is mandatory especially if more than one team is performing the fieldwork and if different equipment is used for in-field measurement (see Section 5.8). Some topics which are important for sampling and field measurements are described in more detail.

5.1. Work procedures

All sampling and field measurement procedures used for the survey have to be documented. General procedures and instructions from instrument manuals or literature need to be adapted to the requirements of the survey, the site conditions and the equipment used. The procedures should be sufficiently detailed to assure that different persons are able to collect samples or perform measurements in the field without introducing a bias which influences the validity of the characterisation data and the interpretation of the results.

5.1.1. Field measurement procedures

Field measurement procedures shall include information on the placement of the equipment (height above the ground, angle, etc.), measurement time or measurement standard deviation to achieve the required detection capacity and data quality, precautions to avoid contamination (specially for alpha equipment which needs to be used close to the contamination source), cleaning and maintenance, and the instrument performance check.

5.1.2. Sampling procedures

General guidelines, standards and literature are available for the sampling of the different sample types mentioned in Section 4.8 (e.g. ISO, 1995b, 2002b, 2002c, 2002d; MARLAP, 2004; MARSSIM, 2000; APHA, 1998). This information should be adapted to the specific equipment used for sampling, the environmental and other specific conditions found during pre-tests and to the matrix sampled.

Handling and decontamination of equipment, safety rules for the sampler, preparation of the sampling area (e.g. removing of vegetation before soil sampling), selection of sample mass, preparation and preservation of sample in the field and all other factors relevant for ensuring the quality of the sample and appropriate interpretation of the results shall be addressed in the sampling procedures.

If soil cores are collected to verify the depth profile used to develop the calibration of *in situ* measurements as proposed in MARSSIM, the procedure should also define the method for the preservation of the core. Also, methods for cutting core samples, compositing samples (see Section 5.6 on composite samples), etc. need to be in place right from the beginning of the sampling campaign.

5.2. Sample containers and preservation of samples

Specifications for sample containers and necessary preservation requirements need to be set before the sampling and shall be a part of the sampling procedure. Container material quality needs to assure that the samples are not contaminated or that no sample losses occur. Also, their chemical resistance need to be considered especially if the samples were preserved with acids or if reactions with the container material are probable (MARLAP, 2004; APHA-7, 1998; RADREM, 1989). The appropriate guidance on preservation techniques is given in MARLAP (2004). Table 12

Table 12 Sample preservation and maximum storage time until analysis (APHA, 1998; DOE, 1992; USGS, 1977; EPA, 1990).

Constituent	Preservative	Container glass (G)/plastic (P)	Maximum storage time
Gross alpha	Conc. HNO_3, to $pH<2$	G or P	1 year
Gross beta	Conc. HNO_3, to $pH<2$	G or P	1 year
Radium-226	Conc. HCl or HNO_3, to $pH<2$	G or P	1 year
Radium-228	Conc. HCl or HNO_3, to $pH<2$	G or P	1 year
Radon-222	Cool 4 °C, to avoid outgasing	G with TFE lined septum	4–8 days
Uranium	Conc. HCl or HNO_3, to $pH<2$	G or P	1 year
Cs isotopes	Conc. HCl, to $pH<2$	G or P	1 year
Sr isotopes	Conc. HCl or HNO_3, to $pH<2$	G or P	1 year
I isotopes	None	G or P	14 days
Tritium	None	G	1 year

gives an example for sample handling, preservation and holding time of water samples.

Advice on the packaging and shipment of radioactive samples can be found in MARSSIM (2000). International and national regulations need to be respected for transport of samples with elevated radioactivity (IAEA, 1996, 2000b, 2002).

5.3. Sampling reports

The detailed documentation of the sampling is one important QA component of a survey. The samplers shall prepare a sampling record for each sample. It should contain all information necessary for the interpretation of the results. To assure that complete information is collected by all sampling teams, a sampling protocol form shall be prepared and filled out by each sampler. Information, which needs to be on the report, can be summarised as follows (adapted from ISO, 2002c):

- sample code;
- date of sampling;
- information on the site (name of location, GPS coordinates, photographs, land use, topographical details, weather conditions);
- description of sample (type and specification e.g. soil, layer 5–10 cm);
- sampling procedure and equipment (number of procedures if available);
- sample mass at the time of collection;
- information on sample preparation in the field (e.g. composite sample, sieving, filtering);
- information on storage and transport and
- identification and signature of sampler.

Some procedures also request information of suspected radionuclide constituents and results of field radiation measurements to optimise safety procedures in sample handling and to avoid cross-contamination.

5.4. Records of field measurements

Records of field measurements should also include similar information as for sampling, but instead of sampling equipment and methods, the field measurement equipment needs to be described. For field measurements, correlation of measurement data and field location is very important for remediation and need special attention. Techniques for acquisition of geographical information are described in IAEA (1998b) and MARLAP (2004). Back-up of field measurement data is essential to avoid loss of information.

QC measurements should be recorded as samples and be evaluated at regular time intervals (see QA in the field) to allow adjustment of equipment.

5.5. Trackability of samples

Sample trackability is an important QA requirement. It refers to the identification of the sample through the whole process of the characterisation survey, starting with field sampling, transport, through sample preparation, measurement and result reporting. It creates the link between the final results and the sampling location.

Guidelines on sample trackability are available in many publications and QA literature. A good trackability includes the sample record system, which gives unique sample codes for each sample.

The sample labels should be resistant against environmental and laboratory influences (e.g. water and acid proofed) and give information on at least sampling location and related sample code.

Also the sampling reports (Sections 5.3 and 5.4) are a part of the sample trackability. They should be transferred together with the samples, and back-up copies are retained with the project manager of the survey. In addition, a chain of custody reports should accompany the samples to track the change of location of the sample, who received the samples and when, and where they are stored. Any problems during transfer of samples (e.g. leaking of containers, breakage) are also recorded there.

Sample receipt in the laboratory and coding for sample preparation and measurement is the responsibility of the laboratories and shall be according to their internal procedures. Procedures should be in place to maintain the link from sampling location, the coding during sampling and the sampling record to the laboratory codes.

Critical for the trackability of samples are the sample preparation steps between sampling and measurement. Sieving (e.g. to <2 mm as foreseen in the IAEA (1989) procedure for soil measurement), sample drying and ashing and many other processes are modifying the physical or chemical properties of the sample and therefore may alter the interpretation of the results. Sample coding should be adjusted to reflect the preparation step at which the sample was analysed, and this information should be part of the analysis report.

Special care is required for analysis methods which require small analysis masses (e.g. for ICP-MS). Without a complete homogenisation of the whole sample, the measured sample is not representative for the sampling area since radionuclide variability within one sample may be big. This applies also if samples are fractionated to be analysed by different methods. Chain of custody documentation and trackability may then become somewhat complicated. If homogenisation methods are applied, they have to be validated to be fit for the purpose (see Section 6.4).

5.6. Replicate sampling

Replicate sampling should be performed at pre-defined intervals. Sampling locations need to be very close together. The reasons for replicate samplings are:

- evaluation of the heterogeneity of the sampling area and
- evaluation of the reproducibility of the sampling procedure and the operator (see also Section 5.8.2).

Replicate samples also may be prepared from one big sample in the field, but then it is necessary to homogenise the sample in the field prior to splitting. Practical considerations on field homogenisation and treatment of samples with high water content are explained in MARSSIM (2000).

This type of replicate sample controls the sample preparation and measurement reproducibility and may be used as a QC sample.

5.7. Composite samples

Homogeneity and trackability considerations are also important for a composite sample. The preparation of composite samples is a common practice since laboratory analysis is very laborious and costly. Composite sampling reduces sample numbers in survey studies covering large sampling areas. In this technique, a specific number of subsamples are collected from a single sampling area, homogenised and combined into a single sample before analysis (ICRU, 2006). The number of subsamples combined, the mass used, the homogenisation technique applied and other special observations need to be stated on the sampling form.

This technique reduces the variability among sample units and is normally more representative for the radiological situation (IAEA, 1999a). Several authors mention the risk of this technique to dilute high contamination samples (e.g. hot particles) (RADREM, 1989; IAEA, 2004c).

If during laboratory analysis samples become suspicious to contain hot particles, a slitting and screening of small portions might become necessary (IAEA, 1998c).

5.8. Quality assurance and quality control in sampling and field measurements

Nowadays, considerations on QA in sampling and field measurements become essential for all surveys and are already an integral part of the planning process. Guidance can be found in many publications (e.g. Ramsey, 2002; MARLAP, 2004; MARSSIM, 2000; IAEA, 1999a, 2004c; ICRU, 2006; ERAMS, 2001).

All staff involved in field measurements and sampling needs to be made aware on the QA and quality control (QC) requirements for the survey project. A good documentation of the survey planning and related sampling and working procedures as described before, previous experience, and/or training are obligatory to assure good survey data quality.

5.8.1. Training

Training needs to be foreseen in every characterisation survey plan. Only harmonised sampling and measurement approaches will guarantee comparable data, which can be used in an overall assessment and characterisation of a contaminated site. Also, experienced staff needs harmonisation efforts to improve reproducibility of sampling and measurements.

Practical field training has several advantages compared to theoretical training lessons:

- The staff can practically perform the sampling as the equipment and sampling procedures are available, the direct correction of mistakes is possible, the methodology is harmonised and the procedures are subject to improvement.
- Sampling equipment and related supply is tested and missing items will be identified during the exercise.
- Documentation and recording can be harmonised and improved.
- Measurement techniques can be practically exercised and optimised for field and environmental conditions (e.g. correction for soil density and attenuation problems in *in situ* gamma spectrometry).
- Measurement equipment used by different groups may be cross-checked, the measurement accuracy verified and improved by re-calibration (if needed).
- Method problems can be discussed and critical points of the procedures can be emphasised.

General field exercises are organised by several organisations and can be used to improve knowledge and experience in sampling and *in situ* measurements (IAEA, 2006b; Geringer et al., 2007; Paulus et al., 2003) and to define the uncertainty of the methods applied. Several publications discuss the uncertainty related to sampling (see Section 6.4.2), but a clear separation of sampling-method-related uncertainty from other factors like survey area heterogeneity, measurement and other uncertainties is very difficult.

5.8.2. Control of performance parameter

A characterisation survey needs comparable field measurement results, which are independent from the analyst and measurement instrument used. Precision and accuracy are essential parameters for ensuring the quality of measurement. The control of both parameters should be part of the QA foreseen in the survey plan. Number and type of control measurements should conform to statistical requirements and avoid the risks associated with wrong data.

The control of precision is done by replicate measurements. The precision of the operator is determined by replicate measurements of the same location with the same instrument being used by different operators, while for instrument precision the operator is the same but instruments are varied (MARSSIM, 2000). Operator-related measurement variability can be easily improved by additional training.

Accuracy can be tested, using a calibration source or a blank sample spiked with a certified radionuclide solution.

In *in situ* gamma measurements, the energy calibration data can be used for the instrument performance test (IAEA, 2003).

The results of all performance tests need to be recorded (e.g. in a control chart) as evidence for QC activities.

The sensitivity and selectivity of the scanning equipments have to be evaluated before the survey planning because these are the main criteria for the selection of the equipment. Planned tests to find buried sources of known radionuclides and known activity concentration can be important to judge the quality of the analysts and the calibration.

5.8.3. Verification of instrument calibration

Actual field conditions may differ significantly from those present during routine calibrations. Factors which may affect the calibration validity include (taken from MARSSIM, 2000):

- The energies of radioactive sources used for routine calibration may differ significantly from those of radionuclides in the field.
- The source detector geometry (e.g. point source or large area distributed source) used for routine calibration may be different than that found in the field.
- The source-to-detector distance typically used for calibration may not be always achievable in the field.
- The condition and composition of the surface being monitored and the presence of overlaying materials may result in a decreased instrument response.

If the actual measurement conditions differ significantly from the calibration assumption, the calibration needs to be adapted. Correction factors may be experimentally developed; if theoretical models are used, they need to be practically validated (IAEA, 1998b, 1998c; Boson et al., 2006). The comparison of *in situ* data with results from laboratory measurement give the best verification but often take too much time.

5.8.4. Background monitoring and blank samples

For environmental monitoring and, especially in the case of monitoring intended for justification of further remediation of contaminated site, the knowledge on background radioactivity is essential. Background measurement is a QC action to estimate a bias caused by contamination. To access the background level, the measurement or sampling is performed up-gradient of the area of potential contamination (either onside or offside) where there is little or no chance of migration of the contaminants of concern (MARSSIM, 2000).

The samples which are taken there, the so-called blank or background samples, are considered to be 'clean', although they may have an elevated level of radioactivity from naturally occurring radionuclides or from previous events (e.g. fallout) which can be regarded as a normal part of the living environment in a region. The difference between the measurement and the background gives the net residual activity. Only activities which are in excess of the natural or prevailing background are considered for remediation (IAEA, 1999a). The background levels need to be determined with the same detection sensitivity and accuracy as the main samples.

5.8.5. Contamination and losses

In order to prevent contamination arising from equipment, tools and containers, cleaning procedures should be documented and implemented. Their efficiency can be controlled by field blanks. These are samples with a known, low and homogeneously distributed contamination, which are collected and treated in the same way as the real samples. Also swipes taken after the cleaning of the sampling equipment can give indications of possible cross-contamination effects. Field blanks can be useful to document that no evaporation or absorption losses during transportation have occurred.

Contamination and recovery problems can also be present during sample preparation and measurements. They should be evaluated using field blanks, reagent blanks and during method validation (e.g. ashing losses of radionuclides).

6. Sample Preparation and Laboratory Measurements

Nearly all laboratory analysis techniques require sample preparation before measurement. This can vary from simple procedures, such as filtration (e.g. for water), filling sample to specific measurement geometries, grinding, milling and homogenisation, to complex processes as sample digestion, radionuclide enrichment and separation. General guidelines can be found in many publications (MARLAP, 2004; IAEA, 1989).

6.1. Sample screening and planning of sample preparation

All samples arriving at the laboratory shall be pre-screened if no or insufficient information on their expected radioactivity concentration and/or the type of radionuclide is available from the sampling or custody form. The established preparation plan shall consider:

- the sequence of sample preparation;
- the decontamination of preparation equipment between samples;
- the documentation of the preparation procedures (if not available already);
- review of existing procedures for their suitability for survey samples (matrix, activity concentration, etc.);
- coding of the prepared samples and link to sampling code and
- safety precautions of the laboratory technicians and analysts.

6.2. Sample processing, preparation and measurement

To avoid biases of the survey results introduced by sample processing, preparation and measurements, it is essential to harmonise the procedures used for these steps already before the measurements. Only validated or verified procedures should be used. Written procedures thereof should be available to all laboratories assisting in sample preparation and measurements. This will ensure that all data used for characterisation are comparable, of the same quality and can be combined for the final decision on remediation needs.

The following sections discuss points which need to be considered for the development or selection of methods to be used within the laboratory work.

6.2.1. Drying and ashing

During drying and ashing, care needs to be taken that the radionuclide of interest is not lost through volatilisation. This applies specifically to iodine

(Hou et al., 1998). To avoid evaporation losses during drying and ashing, samples often require special pre-treatment and preparation conditions. Table 13 gives a summary on proposed drying and ashing temperatures and related references.

The influence of the chemical form (anion, organic species, etc.) on the evaporation temperature of the radionuclide may be significant, and a wet ashing with nitric acid and/or peroxide can reduce losses. Also, adjustments of pre-ashing speed and total ashing time can hinder that samples ignite and particles containing activity get lost (IAEA, 1989; MARLAP, 2004; Bock, 1979). The selection of drying and ashing temperature may be critical for the data quality and the interpretation of the survey results and need to be harmonised for all analysing laboratories.

6.2.2. Crushing, milling, mixing and homogenisation

The reduction of sample particle size to small particles and the homogenisation of the sample after processing can be done using different equipments, and Table 14 gives some information on available equipment and its application (extracted from MARLAP, 2004; IAEA, 1998c; Retsch, 2008).

All preparation methods need to be validated, documented and agreed upon during project planning (see also Section 6.4.1).

6.2.3. Sieving

Sample sieving can be performed anytime during sample preparation. Normally, the sieving is done for the soil or sediment samples directly after sampling to separate vegetation, bigger stones and other parts which are not typical for the matrix. In case the sample is humid, sieving is often done in the laboratory after the drying process. Manual and electrical sieving equipment is available in the market. Sieving particle size depends on the procedure used and may vary with the objectives of the measurements; IAEA (1989) recommend 2 mm, while HASL-300 (DOE, 1997) and Sill et al. (1974) propose 0.25 in. (6.25 mm) and the ASTM procedure (ASTM, 2000) describes milling of the samples and using 500 μm sieves.

6.2.4. Sample digestion methods

Sample digestion methods can be divided into two different approaches: the total dissolution and the leaching method (Krey and Bogen, 1987; Smith et al., 1992; MARLAP, 2004). The advantages and disadvantages of the two approaches are summarised in Table 15.

Total dissolution should always be used if highly contaminated particles with small particle size and particles in a refractory form are expected. The

Table 13 Critical conditions for drying and ashing losses.

Isotope	Chemical form	Critical or boiling temperature	Reference
^{3}H	H_2O	100°	MARLAP, 2004
^{14}C	CO_2 (produced from CO_3^{-2})	−78.5°	Lide, 2006
^{60}Co	$CoCl_2$	~550°	Carlson, 1970 Gorsuch, 1962
	Co (no information)	450–600°	MARLAP, 2004
^{90}Sr	$Sr(NO_3)_2$	Min. 650° (not tested for higher temperature)	Li et al., 2003
	Sr (no information)	No evidence of loss at 900 °C	Sill, 1988
		Between 450–700° (depends on matrix)	MARLAP, 2004
^{129}I, ^{131}I	HOI, CH_3I	Absorption (room temp.)	IAEA, 1973
	I_2	185.2° (sublimes readily)	MARLAP, 2004
	ICl	94.4° (decompose)	Lide, 2006
	Iodine after alkali ashing	650° (90% recovery)	Hou et al., 1998
		450–500°	MARLAP, 2004
^{134}Cs, ^{137}Cs	Cs_2O	~400°	MARLAP, 2004
	CsCl, $CsNO_3$	~500°	Ritter, 1964
	Cs_2CO_3, Cs_3PO_4	~600°	Carlson, 1970
	Cs (no information)	~400°	Sill, 1988
Pu isotopes	NO_3^-	450°	MARLAP, 2004
	Pu (no information)	550° (filter), 700° (soil)	
Th isotopes	Th (no information)	750° (biological matrix)	MARLAP, 2004
		Losses during ashing in porcelain	
	ThO_2	4400°	Lide, 2006
U isotopes	U (no information)	750° (biological matrix)	MARLAP, 2004
		Losses during ashing in porcelain	
	UF4	1417°	Lide, 2006

Table 14 Typical equipment for crushing, milling, mixing and homogenisation.

Equipment	Type of preparation	Matrix
Jaw crusher	Preliminary size reduction by pressure	Hard matrices, e.g. soils, building material
Ball mill	Reduction of particle size by impact and friction	Soils, sediments, vegetation; use depends on ball and container material
Cross- and rotor- mill	Preliminary size reduction and fine grinding	All dried matrices, depends on equipment material, may be combined with sieving
Centrifugal mill	Reduction of particle size by impact and shearing	Soft and fibrous materials
Disk mill	Reduction of particle size by pressure and friction	All soft and middle hard matrices, depends on equipment material, if combined with cooling it may used for volatile matrices
Cutting mills, household and restaurant blenders	Preliminary size reduction by pressure, wet blending allows homogenisation	Vegetation or soft matrices
Mortar grinder	Fine grinding by pressure and friction	Use depends on equipment and material
Rotation mixer	Homogenisation, tumbling axis, e.g. Turbula®	All matrices with smaller particle size
V-blender	Homogenisation by splitting and remixing	All matrices with smaller particle size

role of particles within the dose assessment need to be considered and have been discussed in several articles (IAEA, 1998d; Dale et al., 2008). Dissolution, especially total dissolution, requires an understanding of the chemical processes involved: solubility, chemical exchange, decomposition and rearrangement reactions, oxidation–reduction processes, complexation and equilibrium.

The most commonly used acids are: hydrochloric acid (HCl), nitric acid (HNO_3), perchloric acid ($HClO_4$), fluoric acid (HF) and a combination

Table 15 Comparison of acid leaching and total dissolution.

Method	Advantages	Disadvantages
Acid leaching	Information on easy solvable and bioavailable radionuclides	Refractory radionuclides are not dissolved
	Less time and acid consuming	Underestimation of risks
		Bioavailability may change due to environmental influences (weathering)
Total dissolution	Gives information on the total radionuclide content of the samples	Overestimation of risks
		Time and acid consuming
		Often difficult procedures, which sometimes need special equipment

thereof. Other acids are less frequently used, especially if different radionuclides shall be analysed from one digest because the use of other acids may introduce interferences for one of them (e.g. sulphuric acid may precipitate strontium). HF dissolves silica-bound radionuclides and oxides of, for example Zr and Ti, materials often used in nuclear industry. HNO_3 oxidises metals and builds soluble nitrate forms. Perchloric acid has a very high oxidation potential which improves solubility by oxidation; for example, it oxidises U(IV) to U(VI) and UO_2^{+2} is easily soluble. If ICP-MS is used as measurement techniques, both the latter acids add chlorine interferences and need interference corrections if used (Van den Broeck and Vandecasteele, 1998). Refractory radionuclides need special treatment, especially if the radionuclide particles are formed in a high-temperature reaction.

The amount of publication on sample dissolution and fusion techniques is high and often specific for a matrix or for a radionuclide (Bock, 1979; Grindler, 1962; Sill, 1995; IAEA, 2003; MARLAP, 2004; DOE, 1997), but most of the chemical and physical principles of stable isotopes can be applied also for radionuclides.

6.2.5. Separation methods

Separation methods are complex physical and chemical processes. They aim to isolate the radionuclide of interest and to improve the detection capability of the measurement technique (e.g. by purifying the source to be measured), and to separate interferences which might disturb the measurements. The methods applied are different for different matrices and isotopes and depend strongly on the method used for the final measurement. Available separation techniques are described in MARLAP

(2004) (review and general principles), in radiochemical handbooks and in publications related to methods (see Table 16).

6.2.6. Preparation for non-destructive measurements

The preparation of samples for non-destructive measurements is normally simple. In XRF analysis, pellets are produced from homogeneous powder, with or without use of binders. More complex procedures, including embedding the sample in wax or resin, are described elsewhere (IAEA, 1997c, 2003).

For gamma measurements, the sample preparation normally comprises only grinding and homogenisation or often the sample can be placed directly on the detector. Care needs to be taken that sample geometry, density and distance to the detector are comparable with the calibration. The influence of these conditions shall be evaluated during method validation, and the attenuation effects and influence on counting efficiency need to be considered for data uncertainty (IAEA, 1997c, 2004b; Gilmore and Hemingway, 1995).

6.3. Measurements

Table 16 gives an overview on some applied laboratory analysis techniques for selected radionuclides and related references. These references give links to standard methods and references to books and publications.

Destructive analysis techniques need to be seen in conjunction with their very specific sample preparation and treatment needs. Some require separating of interferences or isolating the radionuclide of interest for the measurements.

The introduction of the traceability concept (CITAC, 2000) for the calibration and sample measurements will ensure the comparability of radionuclide results even when different methods were applied in the same or different analysing laboratories.

6.4. Quality assurance in sample processing, preparation and measurement

Harmonisation of sample processing and preparation procedures, and selection of appropriate measurement techniques are the basis for a good survey data quality. To allow a final verification of the data quality, objective evidence is required that the applied methods were appropriate for the intended use and that the measurements were under control. Therefore, QA and QC principles are essential also during sample processing, preparation, measurement and compliance verification, and are mentioned in many documents (MARLAP, 2004; MARSSIM, 2000; ISO 17025; IAEA, 1998b, 1999a, 2005a; DOE, 1997; EPA, 2002b).

Table 16 Laboratory analysis techniques for selected radionuclides.

Radionuclide	Methods of analysis	References[a]
^{3}H	Liquid scintillation counting	IAEA, 1981; MARLAP, 2004;
	Accelerator mass spectrometry (AMS)	Hou and Roos, 2008[b]; APHA, 1998; DOE, 1997
^{241}Am	Gamma spectrometry (direct measurement)	MARLAP, 2004
	Alpha spectrometry (after radiochemical separation)	Moreno et al., 1998; DOE, 1997; Hou and Roos, 2008[b]; Maxwell and Nichols, 2000
^{134}Cs and ^{137}Cs	Gamma-ray spectrometry	APHA, 1998; MARLAP, 2004[b]; Hou and Roos, 2008[b]
^{129}I	Liquid scintillation counting (co-precipitation with stable iodide)	Chao et al., 1999; APHA, 1998
	AMS	Bate and Stokely, 1982
	Neutron activation analysis	Muramatsu et al., 2008; MARLAP, 2004[b]; Hou and Roos, 2008[b]
^{238}Pu, ^{239}Pu and ^{240}Pu	Alpha spectrometry after radiochemical separation	ISO, 2007; MARLAP, 2004[b]; Hou and Roos, 2008; IAEA, 1989; Maxwell and Nichols, 2000
	ICP-MS after radiochemical separation	Kim et al., 2007[b]; DOE, 1997; Ketterer and Szechenyi, 2008; Cizdziel et al., 2008
^{241}Pu	Liquid scintillation counting after radiochemical separation	Moreno et al., 1998; MARLAP, 2004[b]; Hou and Roos, 2008[b]
^{226}Ra	Alpha spectrometry after radiochemical separation	Jia and Torri, 2007; APHA, 1998; Hou and Roos, 2008[b]; MARLAP, 2004[b]; IAEA, 1990.
	Alpha scintillation detector (^{222}Rn-emanation counting in a Lucas cell)	IAEA, 1990; DOE, 1997
	Liquid scintillation counting	Villa et al., 2005
	Gamma spectrometry (sealed container to achieve equilibrium)	Herranz et al., 2006
	ICP-MS	Makishima et al., 2008

Table 16. *(Continued)*

Radionuclide	Methods of analysis	References[a]
^{90}Sr	LSC (after precipitation or radiochemical separation)	Hou and Roos, 2008[b]; MARLAP, 2004
	LSC (Cherenkov counting)	Suomela et al., 1993
	Gas proportional counting	Heilgeist, 2000; APHA, 1998; DOE, 1997
^{230}Th, ^{232}Th	Alpha spectrometry after radiochemical separation	Oliveira and Carvalho, 2006; MARLAP, 2004
	ICP-MS (after radiochemical separation)	Ball et al., 2008; DOE, 1997; Burnett et al., 2000; Hou and Roos, 2008[b]
^{234}U; ^{235}U, ^{238}U	Fluorimetry (macro amounts) after radiochemical separation	Maxwell and Nichols, 2000; Hou and Roos, 2008; DOE, 1997
	Gamma spectrometry	Pointurier et al., 2008
	Alpha spectrometry (after radiochemical separation)	Chiappini et al., 1996; APHA, 1998; Burnett et al., 2000
	ICP-MS	MARLAP, 2004

[a]These references shall be seen as an aid to find additional information. The methods described are not a recommendation but might be of help in finding optimal methods for the intended use.
[b]Review paper.

6.4.1. Method validation and verification

In sample processing, it needs to be assured that the sample treatment did not change the radionuclide content of the measured sample compared to its concentration at the time of sampling and that the subsample is representative for the samples collected and gives unbiased and true information on the radionuclide distribution of the survey area.

Objective evidence can be provided by many approaches, for example by measurements of physical properties, the comparison of analytical results with results obtained from samples with known content (e.g. certified reference material (RM)) and through systematic investigations on performance characteristics and limitation of the methods (Emons et al., 2006). Guidance on analytical method validation is given in, for example EURACHEM/CITAC (1998) and MARLAP (2004). Method validation is also a requirement of the ISO 17025 standard on 'quality requirements for test and calibration laboratories'. Method guidebooks indicate the critical parameters of a method that need further investigation during validation (Gilmore and Hemingway, 1995). Table 17 describes some tools

Table 17 Basic tools for method validation and verification.

Problem	Tools of validation or verification	Action for optimisation or improvement
Sample processing and homogenisation sufficiently good for the sample mass needed for analysis	Spiked (before processing) sample and following the mixing during processing. Replicate analysis of several subsamples of one sample (see also Fig. 7)	Increasing length, speed and other parameters of grinding and homogenisation Change of equipment type
Contamination or cross-contamination during processing of material	Field blank-comparative measurements of original sample and sample at the end of treatment	Modification of decontamination procedure (between two samples)
	Measurement of swipe samples collected from preparation equipment	Use of different equipment or material
Contamination during sample preparation for measurement (dissolution, extraction, etc.)	Reagent blank	Change of container material or chemicals (better purity) Modification of washing procedures
Loss during processing and preparation including drying and ashing	Reference materials (RMs) or homogeneous samples with elevated radionuclide content – comparative measurements of original sample and sample after treatment (relative loss), or comparison with RM value (recovery)	Systematic experiments: modification of procedures – e.g. by reduction of heat which cause losses, modification of chemical forms
	Sample spiking, measurement of spike recovery	Testing of material which is not sensitive for absorption

Incomplete dissolution	RMs or homogeneous samples with defined radionuclide content – comparison with known radionuclide content (recovery)	Change of acids, physical conditions (e.g. temperature, pressure), use of fusion techniques
Incomplete recovery of radionuclides during enrichment and separation	RMs or homogeneous samples with defined radionuclide content	Modification of method parameters (extraction conditions e.g. time, acidity)
	Sample spiking, measurement of spike recovery	Introduction of correction factor
Influence of sample geometry, density and matrix of sample on measurement accuracy	Reproducibility measurement of same sample (e.g. different geometries, distance to detector)	Introduction of correction factor
	Fortified/spiked samples: same radionuclide content but different density and matrix	Adaptation of calibration
Interfering radionuclides or matrix elements (measurement, sample recovery, etc.)	Several different RMs or homogeneous samples with defined radionuclide content	Systematic spectra evaluation and/or experiments; modification of methods: separation of interferent, isolating radionuclide from the matrix, selection of different measurement setting (e.g. mass or energy) or implementation of correction factors
	Spiking of sample with possible interfering radionuclides or matrix elements	

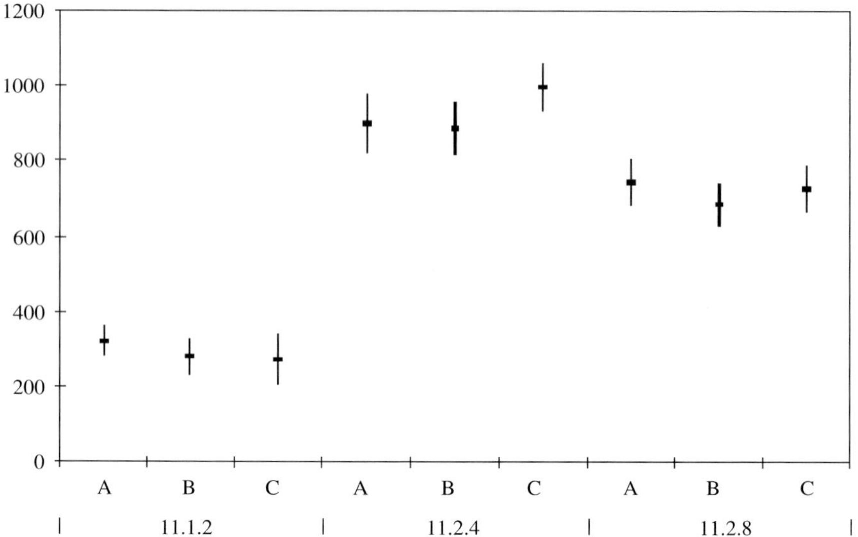

Figure 7 Verification of the suitability of used preparation method: Replicate measurements of ^{241}Am in three subsamples (A, B, C) of one sample after milling and homogenisation (IAEA, 1998c).

which might be helpful during method validation and verification, while Figure 7 provides an example of replicate analysis of several subsamples of one sample.

Spiking to determine the completeness and quality of the sample dissolution, radionuclide separation and sample enrichment may be critical because the chemical form and binding of the radionuclide is often different in the liquid spike compared to the solid sample, for example a refractory particle or different oxidation stages. Also, the use of surrogate measurements and of isotopic ratios to evaluate the activity of the radionuclide of interest needs to be validated. Both are commonly used practices. After the Chernobyl accident, ^{90}Sr assessment was often based on Cs measurements using a conversion factor derived from Cs/Sr ratio (IAEA, 1991). Also, Pu activity is often derived from the ratio of ^{239}Pu and ^{240}Pu to the 59.5 keV gamma ray decay product ^{241}Am of ^{241}Pu (IAEA, 1998b; MARSSIM, 2000). Since the ratio of americium to plutonium may vary from site to site and is reported to change even in one trial area (Burns et al., 1994), it is essential to determine the actual ratio of the sampling area. All ratios used for site characterisation and assessment, and the accuracy of the related assumptions, shall be verified experimentally in the laboratory. The uncertainties of the ratios need to be quantified and their significance on the analysis accuracy evaluated.

6.4.2. Uncertainty of survey data

The evaluation of the uncertainty related to sampling, sample processing, preparation and measurements is essential for the interpretation of the survey data, and the maximum acceptable uncertainty of data should be agreed upon in the survey planning. The uncertainty is a parameter, which describes the range of possible values based on the measurement result. The measurement uncertainty estimate takes account of all recognised effects operating on the result (EURACHEM/CITAC, 1998). Uncertainties related to sampling and processing need to be added using established procedures (Lyn et al., 2003).

Several publications discuss the uncertainty related to sampling (Ramsey and Ellison, 2007; Ramsey and Argyraki, 1997; Nielsen et al., 1992; Ingamells and Switzer, 1973), but a clear separation of sampling-method-related uncertainty from other factors like survey area heterogeneity, measurement and sample processing and preparation uncertainties is often very difficult. De Zorzi et al. (2002) described some attempt to access the sampling-method-related uncertainty by using a well-defined homogeneous sampling area for soil sampling.

The calculation of measurement uncertainty is described in detail in several guidebooks and publications (ISO, 1995a; EURACHEM/CITAC, 1998; Barwick et al., 2003; IAEA, 2004b; MARLAP, 2004; Makarewicz, 2005; Moser et al., 2003). Efforts made for the evaluation of data uncertainty shall be commensurate with the data quality needed for the survey and the risk originated from poor data quality and related wrong decisions.

6.4.3. Instrument performance tests

The verification of an adequate instrument performance before the measurement of samples is one of the requirements in a quality system conforming to ISO 17025. Instrument performance tests should be independent of the sample measurement itself. Instrument software of many companies already includes QA/QC packages with performance checks. MARLAP (2004) gives advice on instrumentation performance indicators and related tests, root cause analysis if they are out of control and possible corrective actions. Table 18 provides some examples of instrument performance tests.

The performance data produced during these QA tests are often used for control charts to evaluate long-term trends and precision of the instruments.

Table 18 Measures to control instrument performance.

Analysis technique	Performance verification methods
Gamma	Measurements of standard gamma sources: • Peak position, energy calibration • Resolution at full width half maximum • Peak detection efficiency *Background measurements*
LSC	Measurement of counting efficiency of unquenched standard sources (H-3, C-14 and background)
Alpha spectrometry	Measurement of counting efficiency of certified standard sources and background Resolution at full width half maximum
ICP-MS	Measurement of a mixed standard to determine: • Oxide ratio • Doubly charged ion ratio • Isotopic ratios at low masses /at high masses • Signal-to-noise ratio • Short-term repeatability of intensity readings *Background intensity measurements*
X-ray	Measurement of: • Intensity and peak position using control samples • Background intensity

6.4.4. Quality Control

While method validation and instrument performance checks are completed, the QC tests monitor preparation and measurement quality for the sample itself (Rius et al., 1999). The problems leading to loss of analytical control are numerous (see Table 19), and the method validation and the QC applied during the laboratory process shall help to identify and avoid them. Additional information can be found through the Internet, many publications and QA books (e.g. EPA IV, 2008; MARLAP, 2004; IAEA, 2004c, 2005a).

QC checks shall be performed randomly or systematically during sample preparation and measurement. Results of QC samples should always be seen and evaluated in comparison to the method validation results, and may help to further improve the robustness, precision and accuracy of the method. Different methods are:

- duplicate sample preparation;
- duplicate measurements;
- reagent blanks;

Table 19 Problems leading to the loss of analytical control (MARLAP, 2004; Figure 18.1).

Radiochemical processing	Source preparation	Instrumentation	Others
Processing difficulty Questionable reagent purity Low tracer/carrier recovery Excessive tracer/carrier recovery Inaccurate aliquanting of tracer/carrier Sample aliquanting inaccuracy Cross-contamination Inadequate dissolution of sample Complex matrix Sample heterogeneity Ineffective chemical isolation or separation: • chemical/radionuclide interferences • improper carrier yield • uncompensated quench • improper/inaccurate ingrowth factors • variable blank and analytical bias Laboratory blunder	Poor mounting Poor plating Improper geometry Incorrect thin plastic film thickness Improper plating on the planchet Excessive source mass Uncorrected self-absorption Quenching Recoil contamination Laboratory blunder	Electronic malfunction • preamplifier • power supply • guard • analog-to-digital convertor • amplifier gain • high voltage • discriminator • pole zero • shape constant Improper source or sample geometry Poor counting statistics Poor detector resolution Detector contamination Recoil contamination Inappropriate/outdated efficiency, background or calibration factor Background shift Improper cross-talk factors Incorrect nuclear transformation data or other constants Peak/calibration shift Counting gas • pressure too high, too low, or variable • gas impurity Loss of vacuum/coolant Temperature and humidity fluctuation Laboratory blunder	Data transcription error Incorrect units Calculation error Software limitation Inadequate/no removal of peak interferences Computer problem Loss of electrical power Electrical power fluctuations Mislabelling Loss of sample Insufficient sample information Data processing problem Interfering radionuclides Laboratory blunder

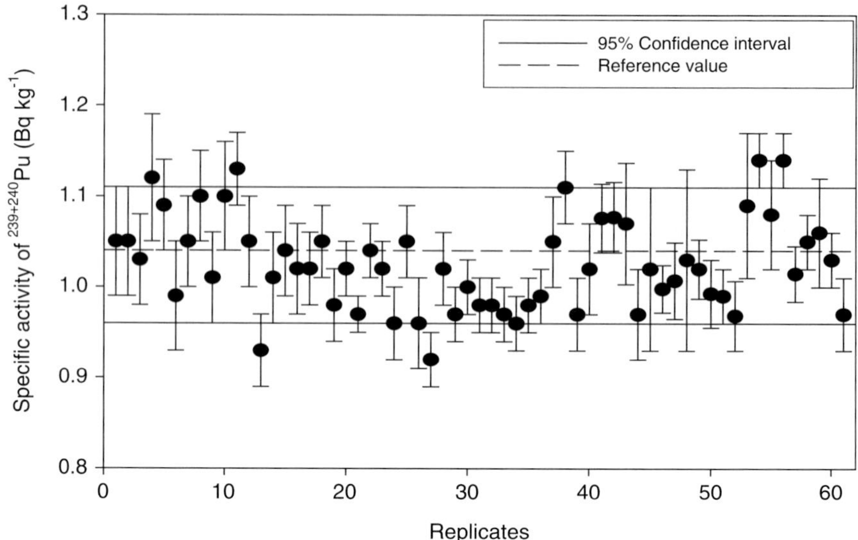

Figure 8 Control charts of analytical results of ^{239}Pu in IAEA Soil-6.

- QC samples;
- blind samples;
- RMs;
- matrix RMs and
- certified RMs.

Control charts can be prepared for all QA/QC information. Examples can be found in most statistical and QA books (Besterfield, 2008; Miller and Miller, 1993; Prichard, 1997). Figure 8 shows an example of a control chart of Pu measurements of IAEA Soil-6 Reference Material (IAEA, 2004d) prepared in the IAEA Laboratories to monitor long-term performance of sample digestion and measurement (Fajgelj et al., 1999).

6.4.5. Proficiency tests and inter-laboratory comparisons

All laboratories providing data for surveys shall routinely participate in proficiency tests and inter-laboratory exercises. This will allow objectively proofing the competence of the analyst and assures the comparability of the data used for decisions and thereby help to identify the analytical deficiencies and the accuracy of the analysis.

6.4.6. Documentation

Documentation is a QA requirement of all modern guides and standards (e.g. ISO 17025; ISO 14001; MARLAP, 2004; MARSSIM, 2000; IAEA,

2005a). For a survey, following documentation should be available for the final evaluation:

- standard operation procedures or working instructions for all methods used;
- results of validation and verification of these methods;
- records of all processing and preparation steps, including information on QA/QC measures performed, the methods used, the name of analyst, and all unusual observations which might influence the quality and/or interpretation of the results;
- records of sample trackability (see also Section 5.5);
- records of instrument calibration, links to measurement raw data, final activity calculation and description of any correction procedure implemented by the analyst and
- analysis reports, including information on calibration traceability and QC measures.

The details of the records should be agreed upon in the project design and shall be harmonised for all participants.

6.4.7. Training

Training is important at all stages of the survey project. All personnel involved in the project need to be aware of the QA requirements laid down in the project planning which are related to their tasks within the project. The level and topic of training shall be commensurate with its importance to achieve the required data quality; thus, training on measurement uncertainty evaluation is essential for the comparability of the measurement data produced in different laboratories.

6.5. Internal verification of survey data

The verification of the survey data shall be done before these data are used for classification of the site and decisions on remediation.

For this, the related QC data (e.g. blanks, laboratory replicates, results of RMs, spiked or other control samples), control charts and other QA information such as instrument performance data and recoveries mentioned in the earlier section are used. The process of data verification is described in various publications (MARLAP, 2004; MARSSIM, 2000; EPA, 2002b; IAEA, 2005a). Some evaluations are based on statistical tests; MARLAP (2004) describes the evaluation of QC samples and performance parameters using control charts and validation data as basis for comparison, but expert judgement based on QA/QC information is also valid. An external, independent auditing could further help to improve trust in the survey data and in the appropriateness of the planned remediation strategy.

7. Safety

Especially in activities related to radioactivity, the disparity between actual and perceived risks is high. This needs to be also considered when planning the application of safety and health protection requirements during sampling, sample processing and preparation and measurements. For radioactivity safety, the ALARA principle shall be applied which says that the risk shall be 'as low as reasonably achievable'. All types of hazards, also related to animals (snakes, ticks, etc.) or special environmental conditions (heat), shall be addressed during health and safety training and shall be specific for the site to be surveyed or for the laboratory work to be performed. Requirements and procedures to eliminate and minimise such potential hazards are addressed in some books and Internet websites (ISO, 2001; Luxon, 1992; COSHH IV, 2008; ICRP, 2007).

8. Post-Remediation Survey and Compliance Verification

8.1. General issues

Post-remediation surveys of sites and compliance verification to environmental and health standards are addressed in many publications (IAEA, 1999a, 2005a; MARSSIM, 2000). It shall characterise the radiological condition after the remediation and verify that the remediation was effective, the residual contamination satisfied remediation criteria for release as set in the planning process, and is in compliance with the national or international environmental standards for the intended site use, and hence confirms that the site presents no significant risk for the health of the inhabitants, the environment and the biota.

8.2. Post-remediation survey

Post-remediation surveys need a similar approach and planning as the characterisation survey before the remediation. Selection of sampling strategy, considerations on screening and individual samples, sample preparation and measurements, and QA and data quality requirements needed for the final compliance verification are comparable to the ones described in the previous sections. In most post-remediation surveys, similar methods as for the characterisation surveys were used as outlined in the previous chapters. This improves the comparability of the environmental data before and after remediation since similar data quality and uncertainty are expected.

It is advisable to use independent teams for the post-remediation survey especially if the site had a 'sensitive' history or will be used for housing (IAEA, 1999a). This will improve the trust in the 'compliance' verification. If this is not possible QA principals involving independent laboratories shall be applied e.g. by performing external quality audits, duplicate analysis or duplicate measurements at critical, previously high contaminated locations.

8.3. Interpretation of survey results and compliance verification

In most countries, compliance verification involves officials from regulatory bodies and health and safety specialists. The final step of the characterisation and compliance surveys described in this chapter is the evaluation and interpretation of the measurement data of the remediation survey and the conclusion if the site may be released for the use identified in the remediation project – in most cases, for unrestricted use. The final verification consists of following actions:

- review of post-remediation plan;
- review of field and laboratory data quality (see previous QA chapters);
- summary of data and graphical presentation;
- judgemental and statistical evaluation of data quality objectives versus measurement data and their quality;
- calculation of internal and external doses;
- comparison of measurement data and calculated dose with legal limits for the intended site use;
- interpretation of survey data in view of legal health and safety requirements and
- preparation of summary reports.

The comparison of measurement data and calculated doses with reference limits and criteria for public exposure needs high level of expertise and experience.

Statistical approaches, as described in statistical books and publications (MARSSIM, 2000; Keith, 1996; MARLAP, 2004), need to be reviewed for their suitability. MARSSIM (2000) described two statistical tests to evaluate data for the final compliance verification. The Wilcoxon Rank Sum test is used for contaminants that are present in the background while, when contaminants are not present in the background, the Sign test is used. Statistical tests are often sensitive and cannot assure compliance within an acceptable confidence if sufficient measurement data are not available for evaluation, for example in large sampling areas. Also results, below the detection limit, with large uncertainties because measurement techniques are not sensitive enough, or which are close to the limit of quantification, give problems with statistical evaluations. The compliance verification needs to consider site-specific situations, natural background levels, main exposure pathways, critical groups and the uncertainties of monitoring data and modelling. Examples and guidance on compliance monitoring are given elsewhere (IAEA, 1999a, 2005a; MARSSIM, 2000).

8.4. Long-term monitoring of remediated areas

In many countries, confirmation and verification of compliance is an ongoing process and long-term monitoring after remediation is a regulatory requirement. It shall verify that the results of the remediation survey remain valid and are not deteriorated by environmental influences (e.g. earthquakes, floods) or that the long-term public exposure conditions are slowly changing. Environmental long-term monitoring might detect unknown contamination sources in the remediated area, changes in the radionuclide mobility caused (e.g. by weathering), but can also confirm the results of post-remediation surveys.

For long-term monitoring, the sampling strategy is normally judgemental, taking into account possible pathways of radionuclide transfer and the role the radionuclide plays for the dose accumulation. Air, water and food monitoring is done in many countries as part of their environmental radioactivity monitoring programme. Type, frequency and number of measurements are often regulated in national laws. While dose rate, gross alpha/beta and gamma measurements are routinely performed in these samples, radiochemical separation followed by alpha and beta measurements is done often only on randomly selected samples or if the other methods give indication for elevated contamination levels.

Individual monitoring of a selected number of individuals (whole body dose, external dose for a period) is often limited to members of critical groups or persons living in areas which did not fulfil the remediation requirement for unrestricted use (IAEA, 2005a).

All long-term monitoring programmes are essential to maintain the trust in the effectiveness and efficiency of the remediation.

9. Conclusion

Site characterisation, the classification of contamination and the final verification of compliance with criteria for public exposure depend all on the information and data collected and measured during the site survey. To achieve the data quality and quantity needed for the evaluation and classification, the planning and performance of the survey is essential.

The processes are complex and need multidisciplinary knowledge and experience. The survey team requires expertise in:

- nuclear safety, safety standards and legislation;
- radioecology to implement information on pathways and transfer;
- field monitoring and screening techniques;
- sampling to identify sample type, sampling equipment and methodology, and to decide on the sampling strategy;
- gamma spectrometry and radiochemical methods, followed by alpha and beta spectrometry and liquid scintillation counting;
- other specific analysis techniques (e.g. radon measurements);
- dose assessment and
- QA and QC.

Only a combination of expertise, trained personnel and QA and control of measurements, validated methods as well as good documentation will assure that appropriate data quality be used as the basis for decisions and actions.

This chapter has described topics of importance for any site evaluation providing informed decision for remediation. It demonstrated the need to characterise the different release scenarios and to determine the radio-ecological sensibility/vulnerability of a contaminated site. It has also given very practical advice on planning, field and laboratory measurements, verification of data, and the role of QA and QC within the whole survey and remediation process, and has listed points for consideration. This information should help to improve the survey planning, the survey and the pathway analysis for optimised remediation strategy.

REFERENCES

Aarkrog, A. (2001). Manmade radioactivity. In: *Radioecology: Radioactivity and Ecosystems* (Eds E. Van der Stricht and R. Kirchman). Fortemps, Liege, pp. 55–78.

ACS. (2008). Data Quality Objective (DQO-PRO-) Calculator, Internet version available from the American Chemical Society: Environmental Chemistry Division, http://www.envirofacs.org/

Akleev, A. V., and M. F. Kiselev. (Eds) (2001). Ecological and medical impacts of the 1957 accident at the PP "Mayak". RF Ministry of Health, Moscow, 294pp, (in Russian).

Alexakhin, R. M., L. A. Buldakov, V. A. Gubanov, Ye. G. Drozhko, L. A. Ilyin, I. I. Kryshev, I. I. Linge, G. N. Romanov, M. N. Savkin, M. M. Saurov, F. A. Tikhomirov, and Y. B. Kholina. (2004). Large radiation accidents: Consequences and protective countermeasures. (Eds L. A. Ilyin and V. A. Gubanov). IzdAT Publisher, Moscow, 555pp.

Alexakhin, R. M., and N. A. Korneyev. (Eds) (1991). *Agricultural Radiology*. Ecologia, Moscow.

Anspaugh, L. R., and B. W. Church. (1986). Historical estimates of external ψ exposure and collective external ψ exposure from testing at the Nevada Test Site. I. Test series through Hardtack II, 1958. *Health Physics*, 51, 35–51.

Anspaugh, L. R., Y. E. Ricker, S. C. Black, et al. (1990). Historical estimates of external ψ exposure and collective external ψ exposure from testing at the Nevada Test Site. II. Test series after Hardtack II, 1958, and summary. *Health Physics*, 59, 525–532.

APHA. (1998). *Standard Methods for the Examination of Water and Wastewater*, 20th edition. American Water Works Association, Water Environment Federation, American Public Health Association (APHA), Washington, DC, ISBN 0-875553-235-7.

ASTM. (2000). ASTM C999-90(2000): Standard Practice for Soil Sample Preparation for the Determination of Radionuclides, American Society for Testing and Materials, West Conshohocken.

ASTM. (2003a). ASTM D5283-92: Standard Practice for Generation of Environmental Data Related to Waste Management Activities: Quality Assurance and Quality Control Planning and Implementation, American Society for Testing and Materials, West Conshohocken.

ASTM. (2003b). ASTM D6311-98: Standard Guide for Generation of Environmental Data Related to Waste Management Activities: Selection and Optimization of Sampling Design, 1998, American Society for Testing and Materials, West Conshohocken.

ASTM. (2006a). ASTM D5792-02 (approved 2006): Standard Practice for Generation of Environmental Data Related to Waste Management Activities: Development of Data Quality Objectives, American Society for Testing and Materials, West Conshohocken.

ASTM. (2006b). ASTM D 5956-96 (re-approved 2006): Standard Guide for Sampling Strategies for Heterogeneous Wastes, American Society for Testing and Materials, West Conshohocken.

Baetsle, L. H. (1993). Study of the radionuclides contained in wastes produced by the phosphate industry and their impact on the environment, European Commission Doc. XI-5027/94. *Proceedings of Symposium on Remediation and Restoration of Radioactive-contaminated Sites in Europe*, Antwerp, pp. 233–252.

Ball, L., K. W. Sims, and J. Schwieters. (2008). An inter-laboratory assessment of the thorium isotopic composition of synthetic and rock reference materials. *Geostandards and Geoanalytical Research*, 32(1), 65–91.

Balonov, M. I. (1993). Overview of dose to the Soviet population from the Chernobyl accident and protective actions applied. In: *The Chernobyl Papers, I, Doses to the Soviet Population and Early Health Effects Studies* (Eds S. Merwin and M. Balonov). Research Enterprises, Richland, pp. 23–45.

Balonov, M. I. (Ed.). (1994). *Guide-Book on Radiation Situation and Radiation Doses.* Received in 1991 by Population of Areas of Russian Federation Subjected to Radioactive Contamination Caused by the Accident at the Chernobyl NPP. Ariadna-Arcadiya Publishers, Saint-Petersburg (in Russian).

Barwick, V. J., S. L. R. Ellison, and B. Fairman. (2003). Estimation of uncertainties in ICP-MS analysis: A practical methodology. *Analytica Chimica Acta*, 394(2), 281–291.

Bate, L. C., and J. R. Stokely. (1982). Iodine-129 separation and determination by neutron activation analysis. *Journal of Radioanalytical Chemistry*, 72, 557–570.

Besterfield, D. H. (2008). *Quality Control*, 8th edition. Pearson Education, Upper Saddle River, NJ, ISBN-13:978-0-13-5000095-3.

Bock, R. A. (1979). *Handbook of Decomposition of Methods in Analytical Chemistry*. Wiley, New York, ISBN 0-470-26501-9.

Boson, J., L. Kenneth, T. Nylen, G. Agren, and L. Johansson. (2006). In-situ gamma spectrometry for environmental monitoring: A semiempirical calibration method. *Radiation Protection Dosimetry*, 121, 310–316.

Bunzl, K., W. Schimmack, L. Zelles, and B. P. Albers. (2000). Spatial variability of the vertical migration of fallout ^{137}Cs in the soil of a pasture, and consequences for long-term predictions. *Radiation and Environmental Biophysics*, 39, 197–205.

Burnazyan, A. I. (Ed.) (1990). Results of the studies and experience of mitigation of consequences of the accidental contamination of the territory by uranium fission products. Energoatomizdat, Moscow, (in Russian).

Burnett, W. C., G. Kim, and E. P. Horwitz. (2000). Efficient pre-concentration and separation of actinide elements from large soil and sediment samples. *Analytical Chemistry*, 72, 4882–4887.

Burns, P. A., M. B. Cooper, P. N. Johnston, L. J. Martin, and G. A. Williams. (1994). Determination of the ratios of ^{239}Pu and ^{240}Pu to ^{241}Am from nuclear weapons test sites in Australia. *Health Physics*, 67, 226–232.

Burrough, P. H., M. Van der Perk, B. J. Howard, B. S. Prister, U. Sansone, and O. Voitsekhovitsch. (1999). Environmental mobility of radiocaesium in the Pripyat catchment, Ukraine/Belarus. *Water, Air and Soil Pollution*, 110, 35–55.

Cancio, D. J., C. Gutierrez, and A. S. Rujz. (1993). Radiological considerations related with the restoration of a phosphogypsum disposal site in Spain, European Commission Doc. XI-5027/94. *Proceedings of Symposium on Remediation and Restoration of Radioactive-contaminated Sites in Europe*, Antwerp, pp. 645–652.

Carlson, G. (1970). *Losses of Radionuclides Related to High Temperature Ashing*. Swedish State Power Board, Ringhalsverket. (INIS-mf-10045). Vattenfall, Sweden.

Carlsen, T. M., L. E. Peterson, B. A. Ulsh, C. A. Werner, K. L. Purvis, and A. C. Sharber. (2001). *Radionuclide Contamination at Kazakhstan's Semipalatinsk Test Site: Implications on Human and Ecological Health*. Lawrence Livermore National Laboratory, Livermore, CA, UCRL-JC-143920.

Chao, J. H., C. L. Tsers, C. J. Lee, and C. C. Hsia. (1999). Analysis of I-129 in radwastes by neutron activation analysis. *Applied Radiation Isotopes*, 51, 137–143.

Chiappini, R., J.-M. Taillade, and S. Brébion. (1996). Development of a high-sensitivity inductively coupled plasma mass spectrometer for actinide measurement in the femtogram range. *Journal of Analytical Atomic Spectrometry*, 11, 497–503.

CITAC. (2000). CITAC 04.02/2000 Traceability in chemical measurements, http://www.citac.cc/traceability.pdf

Cizdziel, J. V., M. E. Ketterer, D. Farmer, S. H. Faller, and V. F. Hodge. (2008). Pu-239, 240, 241 fingerprinting of plutonium in western US soils using ICP-MS: Solution and laser ablation measurements. *Analytical and Bioanalytical Chemistry*, 390, 521–530.

COSHH IV. (2008). Health and Safety Executive. Control of Substances Hazardous to Health, http://www.hse.gov.uk/pubns/leaflets.htm

Da Silva, C. J., J. U. Delgado, M. T. B. Luiz, P. G. Cunha, and P. D. de Barros. (1991). Considerations related to the decontamination of houses in Goiania: Limitation and implications. *Health Physics*, 60, 87–90.

Dale, P., I. Robertsson, and M. Toner. (2008). Review: Radioactive particles in dose assessments. *Journal of Environmental Radioactivity*, 99, 1589–1595.

De Zorzi, P., M. Belli, S. Barbizzi, S. Menegon, and A. Deluisa. (2002). A practical approach to assessment of sampling uncertainty. *Accreditation and Quality Assurance*, 7, 182–188.

DOE. (1992). *EML Procedure Manual 1990* (revised 1992), HASL-300, 27th edition, Vol. 1. Environmental Measurement Laboratory, US DOE, New York.

DOE. (1997). *HASL-300 Procedure Manual of the Environmental Measurements Laboratory*. US Department of Energy, New York, February 1997, http://www.eml.st.dhs.gov/publications/procman/

Donaldson, L. R., A. H. Seytnour, and A. E. Nevissi'k. (1997). University of Washington's radioecological studies in the Marshall islands, 1946–1977. *Health Physics*, 73, 214–222.

Drozdovitch, V. V., G. M. Goulko, V. F. Minenko, H. G. Paretzke, G. Voigt, and Ya. I. Kenigsberg. (1997). Thyroid dose reconstruction for the population of Belarus after the Chernobyl accident. *Radiation and Environmental Biophysics*, 36, 17–23.

Eisenbud, M., and T. Gesell. (1997). *Environmental Radioactivity from Natural, Industrial and Military Sources*. Academic Press, Elsevier, Amsterdam, 4th edition, ISBN 0-12-235154-1.

Ellis, K. M., and J. N. Smith. (1987). Dynamic model for radionuclide uptake in lichen. *Journal of Environmental Radioactivity*, 5, 185–208.

Emons, H., A. Fajgelj, A. van deer Veen, and R. Watters. (2006). New definitions on reference materials. *Accreditation and Quality Assurance*, 11, 576–578.

EPA. (1990). US Environmental Protection Agency. Manual for the Certification of Laboratories Involved in Analysing Public Drinking Water Supplies. EPA-570/9-90-008, Washington, DC.

EPA. (2002a). US Environmental Protection Agency. Guidance on Choosing a Sampling Design for Environmental Data Collection (QA/G-5S) EPA/240/R-02/005, December 2002, http://www.epa.gov/quality/qa_docs.html#guidance

EPA. (2002b). US Environmental Protection Agency. Guidance on Environmental Data Verification and Data Validation (EPA QA/G-8) EPA/240/R-02/004, Office of Enviromental Information, Washington, DC, http://www.epa.gov/quality/qa_docs.html

EPA. (2006). Guidance on Systematic Planning using the Data Quality Objectives Process (QA/G-4), EPA/240/B-06/001, February 2006, http://www.epa.gov/quality/qs-docs/g4-final.pdf

EPA IV. (2008). Internet version 2008. EPA Quality System: Agency-Wide Quality System Documents, http://www.epa.gov/quality/qa_docs.html#EPArqts

ERAMS. (2001). Environmental Radiation Ambient Monitoring System Quality Manual, EPA 2001, Montgomery, http://www.epa.gov/narel/radnet/pdf/eramsqamfinal.pdf

EURACHEM/CITAC. (1998). *EURACHEM/CITAC Guide: The Fitness for the Purpose of Analytical Methods — A Laboratory Guide to Method Validation and Related Topics*. EURACHEM Working Group, LGC, Teddington.

Fajgelj, A., Z. Radecki, K. I. Burns, J. Moreno Bermudez, P. P. De Regge, P. R. Danesi, R. Bojanowski, and J. LaRosa (1999). Intended use of the IAEA reference materials, Part I: Examples on reference materials for the determination of radionuclides or trace elements. *Proceedings of the Workshop on Proper Use of Environment Matrix Reference Materials*. The Royal Chemical Society, Berlin, pp. 22–23, April, ISBN 0-85404-739-5.

Fesenko, S. V., R. M. Alexakhin, M. I. Balonov, I. M. Bogdevitch, B. J. Howard, V. A. Kashparov, N. I. Sanzharova, A. V. Panov, G. Voigt, and Y. M. Zhuchenka. (2007). An extended critical review of twenty years of countermeasures usedc in agriculture after the Chernobyl accident. *Science of the Total Environment*, 383, 1–24.

Fesenko, S. V., R. M. Alexakhin, and N. I. Sanzharova. (2000). Site characterisation techniques used in restoration of agricultural areas on the territory of the Russian Federation contaminated after the accident at the Chernobyl NPP. In: *Site Characterisation Techniques Used in Restoration Activities*. IAEA, Vienna, Austria, pp. 129–152.

Fesenko, S. V., A. Colgan, et al. (1997). The dynamics of the transfer of caesium-137 to animal fodder in areas of Russia affected by the Chernobyl accident and doses resulting from the consumption of milk and milk products. *Radiation Protection Dosimetry*, 69, 289–299.

Fesenko, S. V., N. I. Sanzharova, S. I. Spiridonov, and R. M. Alexakhin. (1996). Dynamics of ^{137}Cs bioavailability in a soil–plant system in areas of the Chernobyl Nuclear Power Plant Accident Zone with a different physico-chemical composition of radioactive fallout. *Journal of Environmental Radioactivity*, 34, 287–313.

Fetisov, V. I., G. N. Romanov, and E. G. Drozhko. (1993). Practice and problems of environment restoration at the location of the industrial association Mayak, 1994, European Commission Doc. XI-5027/94. *Proceedings of Symposium on Remediation and Restoration of Radioactive-contaminated Sites in Europe*, Antwerp, pp. 507–521.

Geringer, T., J. Feichtinger, K. Horrak, E. Lovranich, M. Kraus, M. Blaickner, and M. Schwaiger. (2007). Internationale Messtechnikuebung ISIS 2007. *Strahlenschutz aktuell* **41**(2) (in German).

Gilmore, G., and J. Hemingway. (1995). *Practical gamma-ray spectrometry*. Wiley, Chichester, ISBN 0-471-95150-1.

Gorsuch, T. T. (1962). Losses of trace elements during oxidation of organic materials: The formation of volatile chlorides during dry ashing in presence of inorganic chlorides. *Analyst*, 1031, 112–115.

Grindler, J. E. (1962). *The Radiochemistry of Uranium*. National Academy of Sciences National Research Council, NAS-NS 3050, Washington, DC.

Heilgeist, M. (2000). Use of extraction chromatography, ion chromatography and liquid scintillation spectrometry for rapid determination of strontium-89 and strontium-90 in food in cases of increased release of radionuclides. *Journal of Radioanalytical and Nuclear Chemistry*, 245, 249.

Herranz, M., R. Idoeta, A. Abelairas, and A. Legarda. (2006). Radon fixation for determination of ^{224}Ra, ^{226}Ra and ^{228}Ra via gamma-ray spectrometry. *Radiation Measurements*, 41(4), 486–491.

Hou, X., X. Feng, Q. Qian, and Ch. Chai. (1998). A study of iodine loss during the preparation and analysis of samples using ^{131}I tracer and neutron activation analysis. *The Analyst*, 123, 2209–2213.

Hou, X., and P. Roos. (2008). Review article: Critical comparison of radiometric and mass spectrometric methods for the determination of radionuclides in environmental, biological and nuclear waste samples. *Analytica Chimica Acta*, 608(2), 105–139.

Hutter, A., P. Wang, and S. Weeks. (2001). Innovative characterization, monitoring and sensor technologies for environmental radioactivity at US DOE sites. *Journal of Radioanalytical and Nuclear Chemistry*, 249(1), 209–214.

IAEA. (1973). *Control of Iodine in the Nuclear Industry*. IAEA Technical Report Series No. 148. International Atomic Energy Agency, Vienna.

IAEA. (1981). *Low Level Tritium Measurements*. IAEA-TECDOC-246. International Atomic Energy Agency, Vienna.

IAEA. (1989). *Measurement of Radionuclides in Food and the Environment – A Guidebook*. IAEA Technical Report Series No. 295. International Atomic Energy Agency, Vienna.

IAEA. (1990). *The Environmental Behaviour of Radium*. IAEA Technical Report Series No. 310 (2). International Atomic Energy Agency, Vienna.

IAEA. (1991). *Technical Report: The International Chernobyl Project, Part D*. Report by the International Advisory Committee. International Atomic Energy Agency, Vienna.

IAEA. (1994). *Handbook of Parameter Values for the Prediction of Radionuclide Transfer in Temperate Environments*. IAEA Technical Report Series No. 364. International Atomic Energy Agency, Vienna.

IAEA. (1995a). *Planning and Management of Uranium Mine and Mill Closures*. IAEA-TECDOC-824. International Atomic Energy Agency, Vienna.

IAEA. (1995b). *Decommissioning of Facilities of Mining and Milling of Radioactive Ores and Closeout of Residues*. IAEA Technical Reports Series No. 362. International Atomic Energy Agency, Vienna.

IAEA. (1996). *International Basic Safety Standards for Protection against Ionizing Radiation and for the Safety of Radiation Sources*. IAEA Safety Series 115. International Atomic Energy Agency, Vienna.

IAEA. (1997a). *Application of Radiation Protection Principles to the Cleanup of Contaminated Areas — Interim Report for Comment*. IAEA-TECDOC-987. International Atomic Energy Agency, Vienna.

IAEA. (1997b). *Closeout of Uranium Mines and Mills: A Review of Current Practices*. IAEA-TECDOC-939. International Atomic Energy Agency, Vienna.

IAEA. (1997c). *IAEA Laboratory Manual on Sample Preparation Procedures for X-Ray Micro Analysis*. IAEA/AL/109. International Atomic Energy Agency, Vienna.

IAEA. (1997d). Biomonitoring of Atmospheric Pollution (with Emphasis on Trace Elements) — BioMAP, *Proceedings of an International Workshop Held in Lisbon, Portugal,* 21–24 September 1997, IAEA-TECDOC-1152, International Atomic Energy Agency, Vienna.

IAEA. (1998a). *Factors for Formulating Strategies for Environmental Restoration.* IAEA-TECDOC-1032. International Atomic Energy Agency, Vienna.

IAEA. (1998b). *Characterization of Radioactively Contaminated Sites for Remediation Purpose.* IAEA TECDOC-1017. International Atomic Energy Agency, Vienna.

IAEA. (1998c). *The Radiological Situation at the Atolls of Mururoa and Fangataufa.* Technical Report Volume 1: Radionuclide Concentration Measured in the Terrestrial Environment of the Atolls, International Advisory Committee. International Atomic Energy Agency, Vienna.

IAEA. (1998d). The radiological situation at the atolls of Mururoa and Fangataufa, *Volume 6: Doses due to radioactive materials present in the environment or released from the atolls, International Advisory Committee,* International Atomic Energy Agency, Vienna.

IAEA. (1998e). The radiological situation at the atolls of Mururoa and Fangataufa, *Technical Report 3, International Advisory Committee (Working Group 3).* International Atomic Energy Agency, Vienna.

IAEA. (1999a). *Compliance Monitoring for Remediated Sites.* IAEA-TECDOC-1118. International Atomic Energy Agency, Vienna.

IAEA. (1999b). *Technologies for the Remediation of Radioactively Contaminated Sites.* IAEA-TECDOC-1086. International Atomic Energy Agency, Vienna.

IAEA. (1999c). *Technical Options for the Remediation of Contaminated Groundwater.* IAEA-TECDOC-1088. International Atomic Energy Agency, Vienna.

IAEA. (2000a). Biomonitoring of Atmospheric Pollution (with Emphasis on Trace Elements) — BioMAP II, *Proceedings of an International Workshop Held in Praiia da Vitória, Azores Islands, Portugal,* 28 August–3 September 2000, IAEA-TECDOC-1338, International Atomic Energy Agency, Vienna.

IAEA. (2000b). *Regulations for the Safe Transport of Radioactive Material* (1996 edition). Safety Standard Series No. TS-R-1(ST-1 revised). International Atomic Energy Agency, Vienna.

IAEA. (2000c). *Site Characterisation Techniques Used in Environmental Restoration Activities.* IAEA-TECDOC-1148. International Atomic Energy Agency, Vienna.

IAEA. (2001). *Design Criteria for a Worldwide Directory of Radioactively Contaminated Sites (DRCS).* IAEA-TECDOC-1251. International Atomic Energy Agency, Vienna.

IAEA. (2002). *Advisory Material for the IAEA Regulations for the Safe Transport of Radioactive Material.* Safety Standard Series No. TS-G-1.1 (ST-2). International Atomic Energy Agency, Vienna.

IAEA. (2003). *Collection and Preparation of Bottom Sediment Samples for Analysis of Radionuclides and Trace Elements.* IAEA-TECDOC-1360. International Atomic Energy Agency, Vienna.

IAEA. (2004b). *Quantifying Uncertainty in Nuclear Analytical Measurements.* IAEA-TECDOC-1401. International Atomic Energy Agency, Vienna.

IAEA. (2004c). *Soil Sampling for Environmental Contamination.* IAEA TECDOC-1415. International Atomic Energy Agency, Vienna.

IAEA. (2004d). *IAEA Analytical Quality Control Services.* Reference Materials Catalogue. 1st edition, International Atomic Energy Agency, Vienna.

IAEA. (2005a). *Environmental and Source Monitoring for Purposes of Radiation Protection.* Safety Standards Series No. RS-G-1.8. International Atomic Energy Agency, Vienna.

IAEA. (2005b). *Report on a "Radiological Survey of the Arak and Kura Rivers, Azerbaijan.* IAEA/AL/161. International Atomic Energy Agency, Vienna.

IAEA. (2006a). *Environmental Consequences of the Chernobyl Accident and their Remediation: Twenty Years of Experience*. Report of the Chernobyl Forum Expert Group Environment, International Atomic Energy Agency, Vienna.

IAEA. (2006b). *Report on the Second ALMERA Network Meeting and the Soil Sampling Intercomparison Exercise*. IAEA/AL/164. International Atomic Energy Agency, Vienna.

IAEA. (2006c). *Applicability of Monitored Natural Attenuation at Radioactively Contaminated Sites*. IAEA Technical Reports Series 445. International Atomic Energy Agency, Vienna.

IAEA. (2009). *Quantification of radionuclide transfer in terrestrial and freshwater environments for radiological assessments*. IAEA TECDOC. International Atomic Energy Agency, Vienna, in press.

ICRP. (2007). Recommendations of the International Commission on Radiological Protection. ICRP Publication 103, *Annals ICRP* 37(2–4).

ICRU. (2006). Sampling for Radionuclides in the Environment. Report 76. *Journal of the ICRU* 6(1).

Ingamells, C. O., and P. Switzer. (1973). A proposed sampling constant for use in geochemical analysis. *Talanta*, 20, 547–568.

Iranzo, E., and C. R. Richmond. (1987). *Plutonium contamination twenty years after the nuclear accident in Spain*. Oak Ridge National Laboratory, Tennessee.

ISO. (1995a). *Guide to the Expression of Uncertainty in Measurements*. Produced by BIPM, IEC, IFCC, ISO, IUPAC, IUPAP and OIML. ISO, International Organization for Standardization, ISBN 92-67-10188-9.

ISO. (1995b). Water quality – Sampling, Part 12: Guidance on sampling of bottom sediments, International Organization for Standardization, ISO 5667-12:1995.

ISO. (2001). Soil quality – Sampling, Part 3: Guidance on safety, International Organization for Standardization, ISO 10381-3.

ISO. (2002a) Soil quality – Sampling, Part 1: Guidance on the design of sampling programmes, International Organization for Standardization, ISO 10381-1.

ISO. (2002b). Soil quality – Sampling, Part 2: Guidance on sampling techniques, International Organization for Standardization, ISO 10381-2.

ISO. (2002c). Soil quality – Sampling, Part 4: Guidance on the procedure for investigation of natural, near natural and cultivated sites, International Organization for Standardization, ISO 10381-4.

ISO. (2002d). Soil quality – Sampling, Part 5: Guidance on investigation of soil contamination of urban and industrial sites, International Organization for Standardization, ISO 10381-5.

ISO. (2004). Environmental management systems — Requirements with guidance for use, International Organization for Standardization, ISO 14001.

ISO. (2005). General requirements for the competence of testing and calibration laboratories, 2nd edition, 2005-05-15, International Organization for Standardization, ISO/IEC 17025:2005(E).

ISO. (2007). Measurement of radioactivity in the environment – Soil, Part 4: Measurement of plutonium isotopes (plutonium 238 and plutonium 239+240) by alpha spectrometry, International Organization for Standardization, ISO/DIS 18589-4:2007.

Izrael, Yu. A., E. V. Kvasnikova, I. M. Nazarov, and Sh. D. Fridman. (1994). Global and regional contamination of the territory of the European part of the former USSR by ^{137}Cs. *Meteorology and Hydrology*, 5, 5–9, (in Russian).

Jia, G., and G. Torri. (2007). Estimation of radiation doses to members of the public in Italy from intakes of some important naturally occurring radionuclides (^{238}U, ^{234}U, ^{235}U, ^{226}Ra, ^{228}Ra, ^{224}Ra and ^{210}Po) in drinking water. *Applied Radiation and Isotopes*, 65, 849–857.

KAEA. (1998). Atomic Energy Agency of the Republic of Kazakhstan. Communication to the UNSCEAR Secretariat from T. Zhantikin.

Keith, L. H. (1996). *Principles of Environmental Sampling*, 2nd edition. ACS Professional Reference Book, American Chemical Society, Washington, DC, ISBN 0-8412-3152-4.

Ketterer, M., and S. Szechenyi. (2008). Review: Determination of plutonium and other transuranic elements by inductively coupled plasma spectrometry: A historical perspective and new frontiers in the environmental sciences. *Spectrochimica Acta Part B*, 63, 719–737.

Kettunen, M., T. Heininen, and M. Pulakka. (1997). Finland's preparedness to make airoborne gamma measurement in emergency situation, www.srp-uk.org/srpcdrum/p10-7.doc

Kim, C. S., C. K. Kim, P. Martin, and U. Sansone. (2007). Determination of Pu isotope concentrations and isotope ratio by ICP-MS: A review of analytical methodology. *Journal of Analytical Atomic Spectrometry*, 22, 827–841.

Krey, P. W., and D. C. Bogen. (1987). Determination of acid leachable and total plutonium in large soil samples. *Journal of Radioanalytical and Nuclear Chemistry*, 115, 335–355.

Krouglov, S. V., R. M. Alexakhin, N. A. Vasilyeva, A. D. Kurinov, and A. N. Ratnikov. (1990). About the radionuclide composition formation in soils after the Chernobyl accident. *Pochvovedeniye*, 3, 26–34, (in Russian).

Lee, M. H., M. Douglas, and S. B. Clark. (2005). Development of in situ fission track analysis for detecting fissile nuclides in contaminated solid particles. *Radiation Measurements*, 40, 37–42.

Li, D.-M., L.-X. Zhang, X.-H. Wang, and L. B. Liu. (2003). Ashing and microwave digestion of aerosol samples with a polypropylene fibrous filter matrix. *Analytical Chimica Acta*, 482, 129–135.

Lide, D. R. (Ed.) (2006). *CRC Handbook of Chemistry and Physics, 2006–2007*, 87th edition. CRC Press, Taylor and Francis, Boca Raton.

Long, S., L. Martin, S. Baylis, and J. Fry. (2004). Development and performance of the vehicle-mounted radiation monitoring equipment used in the Maralinga rehabilitation project. *Journal of Environmental Radioactivity*, 76, 207–223.

Luxon, S. G. (Ed.) (1992). *Hazards in the Chemical Laboratory*, 5th edition, Royal Society of Chemistry, Cambridge, 675pp, ISBN 0-85186-229-2.

Lyn, J. A., M. H. Ramsey, R. J. Fussel, and R. Wood. (2003). Measurement uncertainty from physical sample preparation: Estimation including systematic error. *The Analyst*, 128, 1391–1398.

Makarewicz, M. (2005). Estimation of the uncertainty components associated with the measurement of radionuclides in air filters using ray spectrometry. *Accreditation and Quality Assurance*, 10, 269–276.

Makishima, A., T. A. Chekol, and E. Nakamura. (2008). Precise measurement of $^{228/226}Ra$ for ^{226}Ra determination employing total integration and simultaneous ^{228}Th correction by multicollector ICP-MS using multiple ion counters. *Journal of Analytical Atomic Spectrometry*, 23, 1102–1107.

Malherbe, L. (2002). Designing a contaminated soil sampling strategy for human health risk assessment. *Accreditation and Quality Assurance*, 7, 189–194.

MARLAP. (2004). Multi-Agency Radiological Laboratory Analytical Protocol Manual, NUREG-1576, EPA 402-B-04-001A, NTIS PB2004-105421, http://www.epa.gov/radiation/marlap/manual.html

MARSAME. (2006). Multi-Agency Radiological Assessment of Material and Equipments, NUREG-1575, Suppl.1 EPA 402-R-06-002, DOE/EH-707. Draft report for comments, December 2006 http://www.epa.gov/radiation/marssim/publicpreview.html#obtain

MARSSIM. (2000). Multi-Agency Radiation Surveys and Site Investigation Manual, NUREG-1575, Rev.1; EPA 402-R-97-016, Rev.1; DOE/EH-0624 Rev.1, August 2000. http://www.epa.gov/radiation/marssim/index.html

Maxwell, S. L., and S. T. Nichols. (2000). Actinide recovery method for large soil samples. *Radioactivity and Radiochemistry*, 11(4), 46–54.

Miller, J. C., and J. N. Miller. (1993). *Statistics for Analytical Chemistry*. Ellis Horwood, Chichester, 3rd edn., 233pp.

Moreno, J., J. Larosa, P. R. Danesi, N. Vajda, K. Burns, P. De Regge, and M. Sinojmeri. (1998). Determination of ^{241}Pu by LSC in combined procedure for Pu radionuclides, ^{241}Am and ^{90}Sr analysis in environmental samples. *Radioactivity and Radiochemistry*, 9, 35–44.

Moser, J., W. Wegscheider, T. Meisel, and N. Fellner. (2003). An uncertainty budget for trace analysis by isotope-dilution ICP-MS with proper consideration of correlation. *Analytical and Bioanalytical Chemistry*, 377(1), 97–110.

Muramatsu, Y., Y. Takada, H. Matsuzaki, and S. Yoshida. (2008). AMS analysis of ^{129}I in Japanese soil samples collected from background areas far from nuclear facilities. *Quaternary Geochronology*, 3(3), 291–297.

Nielsen, S. P., A. Aarkrog, P. A. Colgan, E. McGee, H. J. Synnott, K. J. Johansson, A. D. Horrill, V. H. Kennedy, and N. Barbayiannis. (1992). *An intercomparison of sampling techniques among five European laboratories for measurements of radiocaesium in upland pasture and soil. Risø-R-620 (EN)*. Risø National Laboratory, Roskilde, Denmark.

Nikipelov, B. V., G. N. Romanov, L. A. Buldakov, N. S. Babaev, Yu. B. Kholina, and E. I. Mikerin. (1987). The radiation accident in the South Urals in 1957. *Atomnaya Energia*, 67(2), 74–80, (in Russian).

NUREG. (2002). Radiological Surveys for Controlling Release of Solid Materials (NUREG-1761). US Nuclear Research Commission, http://www.nrc.gov/reading-rm/doc-collections/nuregs/staff/sr1761/

Oliveira, J. M., and F. P. Carvalho. (2006). Sequential extraction procedure for determination of uranium, thorium, radium, lead and polonium radionuclides by alpha spectrometry in environmental samples. *Czechoslovak Journal Physics*, 56(Suppl. D), 545–555.

Paatero, J., T. Jaakkola, and S. Kulmala. (1998). Lichen as a deposition indicator for transuranium elements investigated with the Chernobyl fallout. *Journal of Environmental Radioactivity*, 38, 223–247.

Paulus, L. R., D. W. Walker, and K. C. Thompson. (2003). A field test of electret ion chambers for environmental remediation verification. *Health Physics*, 85, 371–376.

Pointurier, F., A. Hubert, N. Baglan, and P. J. Hemet. (2008). Evaluation of a new generation quadrupole-based ICP-MS for uranium isotopic measurements in environmental samples. *Journal of Radioanalytical and Nuclear Chemistry*, 276, 505–511.

Prichard, F. E. (Co-ordinating author). (1997). *Quality in the Analytical Chemistry Laboratory*. Wiley, Chichester, 307pp.

RADREM. (1989). A report by the methodology sub-group to the Radioactivity Research and Environmental Monitoring Committee (RADREM) on: *Sampling and Measurement of Radionuclides in the Environment*. Her Majesty's Stationery Office, London, ISBN 0 11 752261 9.

Ramsey, M. H. (2002). Appropriate rather than representative sampling, based on acceptable levels of uncertainty. *Accreditation and Quality Assurance*, 7, 274–280.

Ramsey, M. H., and A. Argyraki. (1997). Estimation of measurement uncertainty from field sampling: Implication for the classification of contaminated land. *Science of the Total Environment*, 198, 243–257.

Ramsey, M. H., and S. L. R. Ellison. (2007). *EURACHEM/EUROLAB/CITAC/Nordtest/AMC Guide: Measurement Uncertainty Arising from Sampling: A Guide to Methods and*

Approaches. Eurachem, ISBN 978 0 948926 26 6, http://www.eurachem.org/guides/UfS_2007.pdf

Ramzaev, V., A. Mishin, V. Golikov, T. Argunova, V. Ushnitsi, A. Zhuravskaya, P. Sobakin, J. Brown, and P. Strand. (2008). Radioecological studies at the Kraton-3 underground nuclear explosion site in 1978–2007: A review. *Proceedings of the International Conference on Radioecology and Environmental Radioactivity* (Eds P. Strand, J. Brown, and T. Jølle), NRPA, Norway, pp. 448–451.

Reiman, R. T. (1994). Mobile in situ gamma-ray spectrometry system. *Transactions of the American Nuclear Society*, 70, 47.

Retsch. (2008). http://www.retsch.com/support/applications/product-selection-milling/

Ritter, R. (1964). Ueber die Fluechtigkeit vin Caesiumsalzen bei der Veraschung von Lebensmitteln. *Die Naturwissenschaften*, 5, 104–105, (in German).

Rius, F. X., R. Boqué, J. Riu, and A. Maroto. (1999). Estimating uncertainties of analytical results using information from the validation process. *Analytic Chimica Acta*, 391, 173–185.

Robinson, W. L., and V. E. Noshkin. (1999). Radionuclide characterization and associated dose from long-live radionuclides in close-in fallout delivered to the marine environment at Bikini and Enewetak Atolls. *Science of the Total Environment*, 237/238, 311–327.

Rozental, J. J., P. O. Cunha, and C. A. Oliveira. (1989). Aspects of the initial and recovery phases of the radiological accident in Goiania. *Proceedings of the International Symposium on Recovery Operations in the Event of a Nuclear Accident or Radiological Emergency*, IAEA, Vienna, pp. 593–597.

Rudy, C. (1993). Identification and Radiological Characterization of Contaminated Environment in Ukraine, European Commission Doc. XI-5027/94. *Proceedings International Symposium on Remediation and Restoration of Radioactive-contaminated Sites in Europe*, Antwerp, pp. 324–348.

Salbu, B. (1990). Hot Particles. Special session: Environmental contamination following a major nuclear accident. *Proceedings of a Symposium*, 16–20 October 1989, Vol. 2, IAEA, Vienna.

Salbu, B., and O. C. Lind. (2005). Radioactive particles released from various nuclear sources. *Radioprotection*, 40, 527–532.

Sanzharova, N. I., S. V. Fesenko, R. M. Alexakhin, V. S. Anisimov, V. K. Kuznetsov, and L. G. Chernyayeva. (1994). Changes in the forms of ^{137}Cs and its availability in plants as dependent on properties of fallout after the Chernobyl Nuclear Power Plant accident. *Science of the Total Environment*, 154, 9–22.

SCKCEN. (2006). Centre detude de l'energie nuclear, Johan Paridaens: Intergration of a gamma measurement equipment into an Augusta A109BA Helicopter of the Belgian Armed Force for use in emergency situations, Operation office: Boeretang 200, BE-2400 MOL, Scientific Report 2006, www.sckcen.be

Shaw, G. (2007). Introduction. In: *Radioactivity in the Terrestrial Environment, Series: Radioactivity in the Environment* (Ed. G. Shaw). Elsevier, Amsterdam, pp. 1–19.

Sill, C. W. (1988). Volatility of caesium and strontium from a synthetic basalt. *Nuclear and Chemical Waste Management*, 8, 97–105.

Sill, C. W., K. W. Puphal, and F. D. Hindman. (1974). Simultaneous determination of alpha emitting nuclides of radium through californium in soil. *Analytical Chemistry*, 46, 1725–1737.

Sill, C. W., and D. S. Sill. (1995). Sample dissolution. *Radioactivity and Radiochemistry*, 6(2), 1–28.

Sloof, J. E., and B. Th. Wolterbeck. (1992). Lichens as biomonitors for radiuocaesium following the Chernobyl accident. *Journal of Environmental Radioactivity*, 16, 229–242.

Smith, L. S., F. Markun, and T. TenKate. (1992). Comparison of acid leachate and fusion methods to determine plutonium and Americium in environmental samples.

Analytical Chemistry Laboratory ANL/ACL-92/2 Argonne National Laboratory, Argonne, IL, June.

Suomela, J., L. Wallberg, and J. Melin. (1993). Method for the determination of Sr-90 in food and environmental samples by Cherenkov counting. Swedish Radiation Protection Institute. SSI-Rapport 93-1, p. 19.

Theocharpoulos, S. P., G. Wagner, J. Sprengart, M.-E. Mohr, A. Desaules, H. Muntau, M. Christou, and P. Quevauviller. (2001). European soil sampling guidelines for soil pollution studies. *Science of the Total Environment*, 264, 51–62.

Thomé, D. J. and B. W. Hurley. (1996).Federal Radiological Response in the United States, DOE/NV/11718-036 UC-700, http://www.nv.doe.gov/library/publications/Environmen-tal/DOENV_11718_036.pdf

Trapeznikov, A. (1993). Radiological investigation of the Techa River (The Urals) and of the soil and vegetation cover in its flood plain, European Commission Doc. XI-5027/94. *Proceedings of Symposium on Remediation and Restoration of Radioactive-contaminated Sites in Europe*, Antwerp, pp. 233–252.

UNSCEAR. (2000). *Sources and Effects of Ionizing Radiation.* United Nations Scientific Committee on the Effects of Atomic Radiation, Scientific Annex J within 2000 UNSCEAR Report to the General Assembly, New York.

USEPA. (2001). Public Health and Environmental Radiation Protection Standards for Yuca Mountain. Final Rule. June 13, 2001. Federal Register Part IV, 40 SFR Part 197. United States Environmental Protection Agency, Nevada.

USGS. (1977). US Geological Survey. Methods for the Determination of Radioactive Substances in Water and Fluvial Sediments. US Government Printing Office, Washington, DC.

US NRC. (1997). *Multi-Agency Radiation Survey and Site Investigation Manual (MARSSIM)*, NUREG-1575, EPA 402-R-97-016, US Nuclear Regulatory Commission, Washington, DC.

Van den Broeck, K., and C. Vandecasteele. (1998). Elimination of interferences in the determination of arsenic and determination of arsenic in percolate waters from an industrial landfill by inductively coupled plasma mass spectrometry. *Analytical Letters*, 31, 1891–1903.

Van der Veen, A., and A. Alink. (1998). A practitioner's approach to the assessment of the quality of sampling, sample preparation and sub sampling. *Accreditation and Quality Assurance*, 3, 20–26.

Villa, M., H. P. Moreno, and G. Manjón. (2005). Determination of ^{226}Ra and ^{224}Ra activity concentration by liquid scintillation counting in sediment samples. *Radiation Measurements*, 39, 543–550.

Voitsekhovitch, O., V. Kanivets, I. G. Billiy, G. Laptev, and U. Sansone. (1996). Peculiarities of particulate Cs-137 transport and sedimentation in Kiev Reservoir. *Proceedings of the First International Conference on the Radiological Consequences of the Chernobyl Accident.* European Commission and the Belarus, Russian and Ukrainian Ministries on Chernobyl Affairs, Emergency Situations and Health. Minsk, Belarus 18–22 March, 1996. EUR 16544 EN, ISBN 92-827-5248-8, 179-183.

Walling, D. E., and T. A. Quine. (1994). Use of fallout radionuclides measurements in soil erosion investigations, invited presentation. *Proceedings of an 'International Symposium on Nuclear Techniques in Soil–Plant Studies for Sustainable Agriculture and Environmental Preservation*, Vienna, IAEA-SM-334/35, pp. 597–619.

Wrenn, M. M. N. (1974). Environmental levels of plutonium and transplutonium elements, WASH 1359. Atomic Energy Commission, Washington, DC, 89pp.

Yamamoto, M., T. Tsukatani, and Y. Katayama. (1996). Residual radioactivity in the soil of the Semipalatinsk nuclear test site in the former USSR. *Health Physics*, 71, 142–148.

CHAPTER 3

Decision-Aiding Tools for Remediation Strategies

Vyacheslav Shershakov[1,*], Sergey Fesenko[2], Ivan Kryshev[1] *and* Natalya Semioshkina[3]

Contents

*Corresponding author. Tel.: +7-48439-71706; Fax: +7-48439-40910
E-mail address: vs@feerc.obninsk.org; vs@ckb.obninsk.org

[1] Research and Production Association (RPA) "Typhoon", 82 Lenin Street, Kaluga Region, 249038 Obninsk, Russia
[2] International Atomic Energy Agency, Agency's Laboratories, A-1440 Seibersdorf, Austria
[3] Helmholtz Zentrum München-Institute of Radiation Protection, 85764 Neuherberg, Germany

Radioactivity in the Environment, Volume 14
ISSN 1569-4860, DOI 10.1016/S1569-4860(08)00203-9

1. Introduction

Recently, modern decision-aiding techniques have been developed to assist in optimising the selection of countermeasures or remedial actions for contaminated settlements (Andersson et al., 2008), agricultural ecosystems (Howard et al., 2005), forests (Fesenko et al., 2005; Shaw et al., 2001) and water bodies (Monte et al., 2000; Monte, 2001). The number of protective measures to be considered can be rather large and can include up to 59–92 actions (Andersson et al., 2008; Howard et al., 2005).

The overall goal of remediation planning in contaminated areas is to suggest remedial measures in such a way that they provide, with deduction of expenses incurred, an optimum net benefit for the population and economy. The immediate result of remedial actions should be the reduction of existing radiation doses to the population and averting the potential for prolonged future exposure.

Implementation of remedial measures on contaminated areas, as a rule, is labour-intensive and costly. Apart from the benefits of reducing doses for the public, remedial actions may have undesirable impacts which need to be assessed and which were often ignored in the past (IAEA, 2002). Through careful analysis of both direct and indirect radiation impacts and social and economic impacts, the effectiveness of remedial actions aimed at ensuring radiation safety of the public is finally determined (IAEA, 2002). This requires that responsible decision making about remediation should take into account very diverse attributes. The diversity of attributes presents a particular challenge and many of them cannot be reduced to monetary or dose units alone. For remediation to be effective, large-scale organisational arrangements are normally required. In addition to technology availability and its feasibility, the moral preparedness of the public for remedial measures is equally important (IAEA, 2002).

Management decisions on remedial measures can be classified as strategic or tactical. The strategic measures, as demonstrated by the mitigation activities after the Chernobyl accident, may include restructuring the existing management system, setting up new entities and taking political decisions on complex and costly measures to reduce radioactive contamination of the area. The tactical decisions direct the performance of standard tasks for reducing the public exposure through temporary arrangements such as changing land-use patterns, using sorbents and so on.

A preliminary analysis of all the advantages and disadvantages of different remedial actions on doses and human health effects, as well as the associated costs, is indispensable for taking optimum decisions. The task of decision makers can be facilitated by using modern tools such as computer modelling that simulates the implementation of remedial measures as a function of time and space and assesses their benefits and shortcomings.

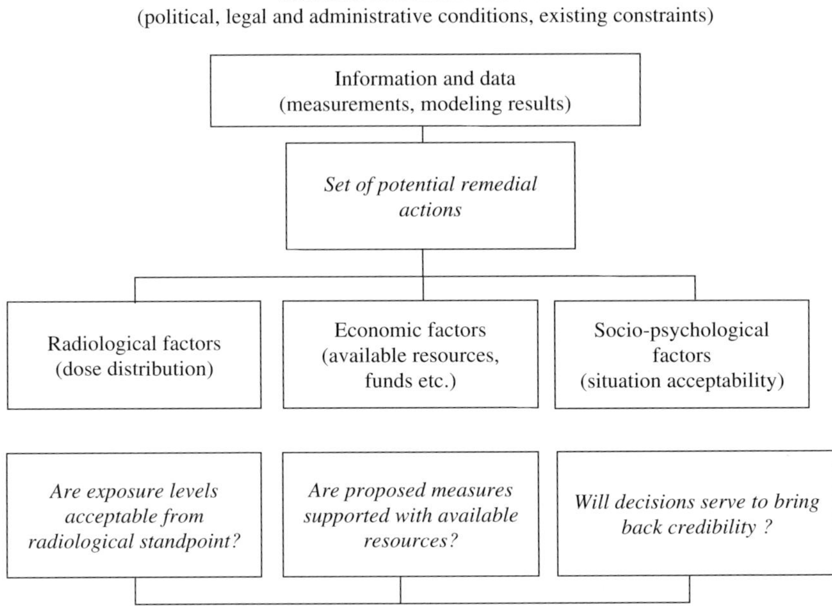

Figure 1 Key elements of the decision-making support system (DMSS) for remedial measures on contaminated areas.

The conceptual framework and basic elements of the decision-making support system (DMSS) for application of remedial measures in contaminated areas are shown in Figure 1.

2. Justification and Optimisation of Remedial Measures on Contaminated Areas: The ALARA Principle

The radiation safety system recommended by ICRP is based on three ALARA principles referred to as justification of practice, optimisation of protection and limitation of individual doses. In practice, the ALARA principles are rather difficult to implement for remediation of contaminated areas and cannot be achieved on the basis of scientific concepts alone. Scientific expert judgements about the relative importance of different kinds of risk and about balancing of risks and benefits play an important role in the process.

Decision-aiding techniques are useful tools for those who must decide on the level of protection by defining the best trade-offs between the various factors and constraints involved in the process, taking into account the inherent uncertainties and value judgements. This can only be achieved on the basis of a clear identification of available alternative protection options, factors and constraints involved in the processes followed by the quantification of as many options as possible.

The most important issue is to approach the problem systematically with the delineation of the options, factors and constraints that need quantification. Without any doubt the quantification process is the most difficult and time-consuming step because all the necessary data must be gathered or generated by models. These data should be used in combination with decision-aiding techniques to optimise the decision process.

Decision-aiding techniques have been widely used for selecting preferable remedial measures on contaminated areas. In the early 1970s, ICRP recommended the use of cost-benefit analysis. In this analysis, the cost of radiological detriment that can be averted was compared with the cost of protective measures (ICRP, 1973, 1983, 1989). The analysis of cost-effectiveness of radiation protection is another available technique. For the goals of greater complexity including consideration of large-scale remedial actions, decision-making can be improved by complementing the above-mentioned methods with multi-criteria analysis (Shershakov and Trakhtengerts, 1997; Lochard and Schneider, 2001). Each method for selecting optimum remedial actions on contaminated areas is characterised in the following sections.

2.1. Cost-effectiveness and cost-benefit analyses

In radiological protection, the usage of both cost-effectiveness and cost-benefit analyses is based on comparing the expenses for reducing the exposure levels and benefits achieved in terms of reduction of the collective radiation dose. A key element in this procedure, when applied to the selection of radiological protection options, is the use of a monetary value for the unit of collective dose. This allows expressing the benefit of protection (i.e. the reduction of dose due to the implementation of a protection option) in the same units as the protection costs.

The definition and use of the monetary value of dose has been introduced by ICRP in Publication 22 (ICRP, 1973). However, its application for practical purposes was always a matter of debate, mainly for ethical reasons. Therefore, the approach behind this concept has not been widely accepted.

Cost-effectiveness analysis is suggested not as an optimisation method but as a method to exclude from further consideration non-cost-effective remedial actions and to provide a ranking and comparison of the remaining options. The main steps of the method are as follows: (i) to characterise every protection option with its protection cost and with the associated

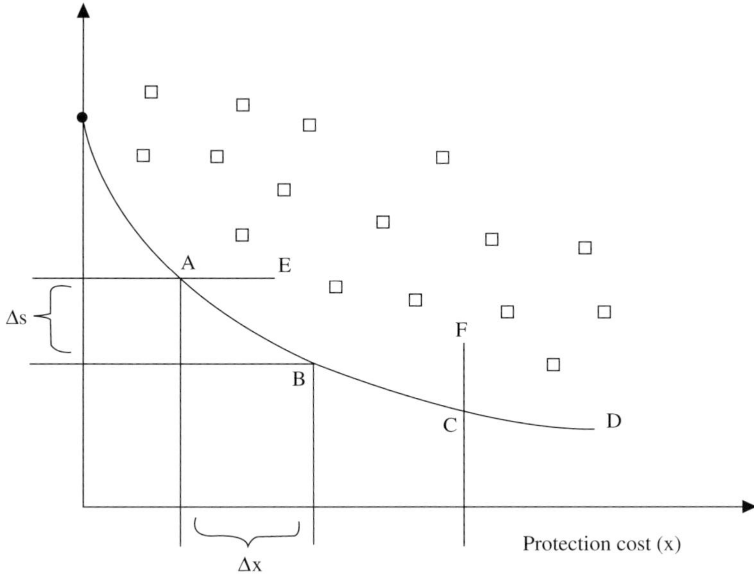

Figure 2 Cost-effectiveness analysis.

residual collective dose (shown as dots in Figure 2) and (ii) to select options which either minimise the collective dose for a fixed protection cost or minimise the protection cost for a limited collective dose (ICRP, 1989).

This process can be illustrated graphically in a simple way as shown in Figure 2. Each option is represented by a dot and all those which are cost-effective lie on the cost-effectiveness curve. For example, option A allows for a residual level of exposure at a lower cost than E, and option C gives a lower residual collective dose for the same cost as F. All the options which are above the curve are called 'totally dominated' (worse) options and must be discarded from further consideration (Lochard and Schneider, 2001).

Cost-effectiveness analysis mainly relies on the evaluation of the 'marginal cost' of every protection option, which has to be compared with the closest, less or more expensive options. If a small additional cost leads to a much higher effectiveness in terms of risk reduction, the option is non-cost-effective. Finally, each cost-effective option can be characterised by the increase in cost from one option to the next (Δx) and the corresponding decrease in collective dose (Δs). The quotient ($\Delta x/\Delta s$) is called the cost-effectiveness ratio. It provides a basis for ranking the various options. However, the determination of the cost-effectiveness curve and the corresponding cost-effectiveness ratios does not provide any basis for the selection of the optimum option. This is done by the introduction of a reference value for the cost-effectiveness ratio in the framework of the cost-benefit analysis (Lochard and Schneider, 2001).

There are several ways to carry out a cost-benefit analysis. The most obvious way is to express factors influencing the balance between costs and benefits in monetary terms and to aggregate them in order to select the option with the lowest monetary value of this aggregate (Lochard and Schneider, 2001). Then, the monetary cost (Y) of the collective dose may be expressed as follows:

$$Y = \sum \alpha_j S_j$$

where α_j is the cost of a man-Sievert (man-Sv) for the exposed population j and S_j is the collective dose of the exposed population j.

It should be realised that the value of α_j may depend on many factors, such as the category of population exposed, the spread of exposure over time and the level of individual dose.

The total cost of each option can be estimated as the sum of the costs of the protection option and of the associated value of the collective dose. The optimum solution corresponds to the minimum value of the total cost as shown in Figure 3. It should be noted that, at the optimum level of protection, the marginal cost of protection is equal to the marginal cost of the unit of collective dose avoided.

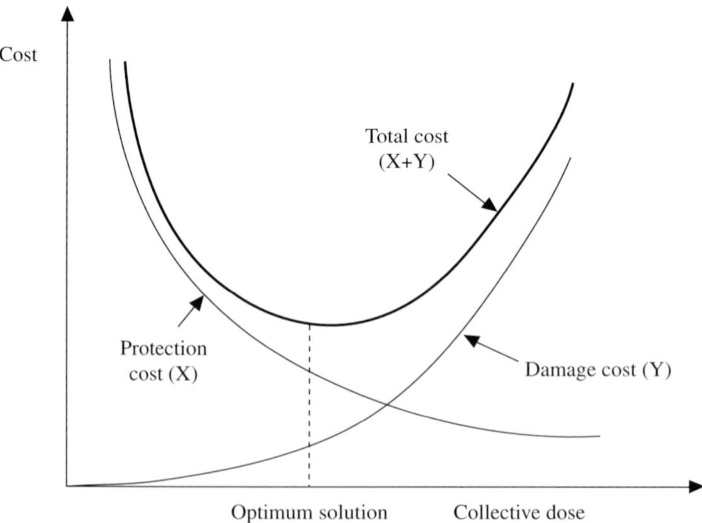

Figure 3 Cost-benefit analysis as applied to choosing measures to reduce exposure.

2.2. Multi-attribute utility analysis

2.2.1. General approach

A current trend in the use of decision-aiding technologies is based on the obligatory consideration of societal judgements (ICRP, 2007). Many such judgements are difficult or are hardly ever possible to quantify in monetary terms. Therefore, the challenge is how to find the optimum solution based on a variety of quantitative and qualitative criteria which are expressed in totally different ways. The multi-attribute utility analysis (MAUA), the technique recommended by ICRP for optimisation of radiation protection efforts (ICRP, 1989), is one of the possible options on how to approach this problem.

The technique is based on building a scoring scheme (or multi-attribute utility function) for each option on the basis of criteria to be taken into account for remediation planning (i.e. feasibility and cost of remedial actions, effectiveness in terms of reduction of collective and individual doses, perception of the remedial actions by the population and stakeholders, etc.). In this way remedial actions may be identified. However, it raises a need to define the criteria for the decision process.

The set of criteria should be identified in compliance with the specific objective to ensure the optimal choice of remediation of contaminated areas. The set of criteria used practically for all multi-criteria problems of assessing effectiveness of alternative strategies include

- economic criteria;
- human health and public safety criteria;
- socio-economic factors;
- environmental management and protection criteria;
- criteria of public opinion and interests of population groups.

Thus, each remedial action should be evaluated according to various criteria (either quantitatively or qualitatively). In further steps, weighting factors have to be assigned to each criterion to account for their relative importance. Although some techniques to identify such values are available, the selection of weighting factors should always be justified depending on the task considered.

Finally, every remedial measure must be specified based on its total utility (U), which is calculated as follows:

$$U_i = \sum k_j u_{ij}$$

where i is the remedial measure number, j the criterion number, k_j the weighting coefficient showing the relative importance of each option (normalised $\sum k_j = 1$) and u_i is the utility of measure i.

Every such option, associated with each criterion, may be defined either as a linear or as a non-linear function, reflecting the preferences of the decision makers. Finally, the remedial action that corresponds to the highest total utility values can be selected. It should also be realised that since the choice of weighting factors is mainly based on decision-makers' judgements, it is highly desirable to perform a sensitivity analysis considering different sets of weighting factors in order to test the 'robustness' of the results (Lochard and Schneider, 2001).

2.2.2. Construction of the utility function in the criteria space 'collective dose-financial costs'

An example of the construction of the utility function to illustrate the approach considers a modelling case determining the decision-maker's preferences in the criteria space 'collective dose-financial cost'. It is assumed that different measures can be applied in a region of interest according to existing regulations. There are no specific directions for the decision maker concerning protective actions and he/she is faced with the need to select a combination of measures relying on his/her available resources. The criteria applied here will be the 'collective radiation dose' and 'financial costs'. The costs to apply various combinations of protective measures are projected to range from 400 to 600 c.u. (conventional units).

The collective radiation dose, depending on the protective measures applied, is projected to vary from 10 to 40 units. The criteria space *K* in this case is two-dimensional. Figure 4 shows the criteria space *K* for this example, with the 'financial cost' taken as a baseline criterion.

On the *x*-axis the financial cost is plotted and the *y*-axis represents the collective radiation dose. The criteria space is represented as a mesh, the utility function taking the values U_{ij} in the mesh points. Clearly, the goal of reducing the collective dose implies application of more efficient protective measures, leading, as a result, to increasing costs. Thus, the criteria scales are selected in such a way that an increase in the criteria corresponds to a decrease in the utility.

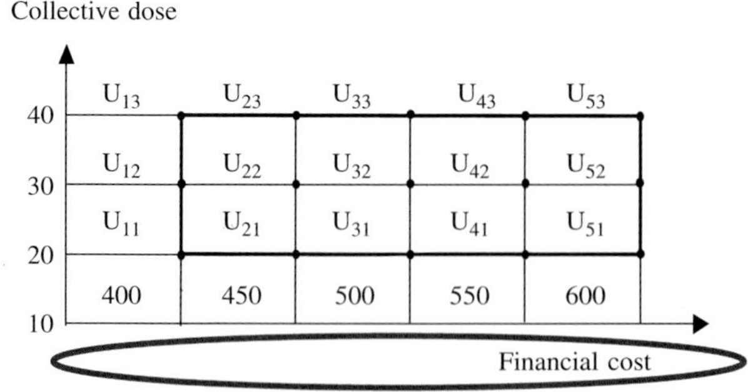

Figure 4 Criteria space for choosing protective measures.

Suppose the decision maker first considers the worst situation represented by the point (1,3) – the least efficient measure is applied and its application cost is 400 c.u. It can therefore be written as follows:

$$U_{13} = 400 \tag{1}$$

Descending from point (1,3) to point (1,2), with the financial costs retained at the level of 400 c.u., means improvement in the situation: the collective dose is reduced by 10 units. Say, the decision maker is prepared to increase the financial costs by 50–150 c.u., which will mean that, by his appraisal, the point equivalent to (1,3) is somewhere between (2,2) and (4,2). Schematically this can written as

$$U_{13} \rightarrow U_{12} \Rightarrow (U_{22}, U_{42})$$

Hence it follows:

$$\begin{cases} U_{13} > U_{22} \\ U_{13} < U_{42} \end{cases} \tag{2}$$

Likewise,

$$U_{12} \rightarrow U_{11} \Rightarrow (U_{21}, U_{31})$$

$$\begin{cases} U_{12} > U_{21} \\ U_{12} < U_{31} \end{cases} \tag{3}$$

In the case under study, it can be assumed that the difference in the utility function values, given constant collective dose value, is equal to the difference in the financial costs, that is

$$\begin{cases} U_{23} = U_{13} + 50 \\ U_{33} = U_{23} + 50 \\ U_{43} = U_{33} + 50 \\ U_{53} = U_{43} + 50 \\ \ldots \quad \ldots \quad \ldots \quad \ldots \\ \ldots \quad \ldots \quad \ldots \quad \ldots \end{cases} \tag{4}$$

This assumption means that, the collective dose being the same, the financial resources spent on implementation of protective measures may be greater than necessary.

By solving jointly the system of inequalities (2) and (3) and the system of equalities (1) and (4) one obtains an ensemble of the utility functions satisfying the system of decision-maker's judgements. The solution of this system can be sought in classes F1, F2, F3, F4 and F5.

It can be seen from inequalities (2) and (3) and equalities (4) that the utility functions are linear along the x-axis, whereas on the y-axis they are not. Hence, no solutions exist in class F1 and it needs to move to class F2 and so on.

Now we can consider the solutions in the function class F5. By the construction, the utility function in this case is determined by the values U_{12} and U_{13}. Then, the system of relationships defining the ensemble of the utility functions takes the form:

$$\begin{cases} 400 > U_{12} + 50 \\ 400 < U_{12} + 100 \\ U_{12} > U_{11} + 50 \\ U_{12} < U_{11} + 100 \\ 400 > U_{12} > U_{11} \end{cases} \tag{5}$$

Figure 5 shows, in graphic form, the solution of the system of inequalities (5) as a delineated parallelogram.

A set of points in the delineated parallelogram Pol is a geometric representation of the ensemble of the utility functions. Using the derived utility function ensemble, we compare two strategies S_{32} and S_{41}. Strategy S_{32} is represented in the criteria space by point (500,30), and strategy S_{41} by point (550,20). By the construction

$$U(S_{32}) = U(500, 30) = U_{32} = U_{12} + 100$$
$$U(S_{41}) = U(550, 20) = U_{41} = U_{11} + 150$$

where point $(U_{11}, U_{12}) \in \text{Pol}$. When comparing the strategies, we assume that all the points of the set Pol have the same 'weight', that is this set is

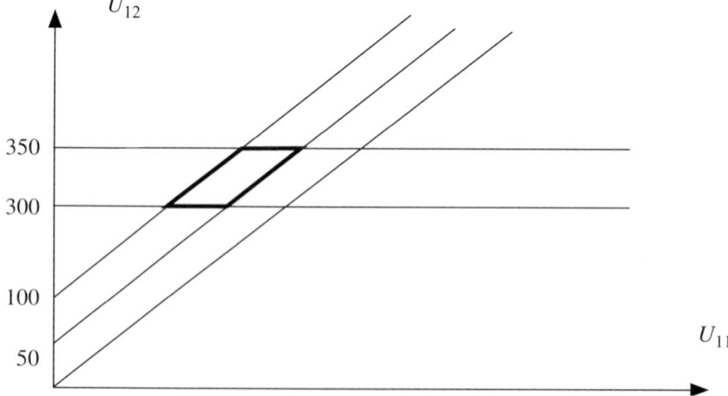

Figure 5 Geometric representation of the utility function given in inequalities (5).

characterised by a uniform distribution with the density given by $dU_{11}\, dU_{12}/2500$.

In the case of soft ranking, the strategy S_{41} is more preferable than the strategy S_{32}, since E can be calculated as

$$E(U(S_{32}) - U(S_{41})) = E(U_{12} - U_{11} - 50)$$
$$= \int_{\text{Pol}} (U_{12} - U_{11}) \frac{dU_{12}\, dU_{11}}{2500} - 50 = 75$$

Hard ranking will give the same result because

$$\text{STD}(S_{32}, S_{41}) = 25\sqrt{\frac{13}{3}} = 52$$

Finally, ranking by the principle 'it cannot be worse' demonstrates that the strategies under study have the same rank.

2.2.3. Preparation of information for constructing the preference function and decision making

The process of data preparation and decision making is illustrated by the user interface which serves as a multi-functional software medium enabling preparation of information to construct the preference function using any of the above methods (Trakhtengerts et al., 2004). Three major stages can be identified in the decision-maker operation with the interface: description of criteria, elucidation of decision-maker preferences and comparison of alternatives (options).

2.2.3.1. *Criteria description.* Using the interface components shown in Figure 6, the decision maker can introduce to the system inputs and describe the required criteria (attributes), including their dimensions and measurement range.

2.2.3.2. *Selection of representatives and elucidation of decision-maker preferences.* Using the interface component shown in Figure 7, the initial positions of alternatives in the multi-criteria space are selected and the expert inquiry procedure (trade-off) is performed for elucidating expert preferences.

2.2.3.3. *Comparison of alternatives.* In the interactive mode, the decision maker can compare the options of interest using the constructed function of decision-maker preferences. Figure 8 shows the component of the user interface enabling the comparison.

The obtained preference functions are entered into a dedicated database and can be used for ranking alternative options.

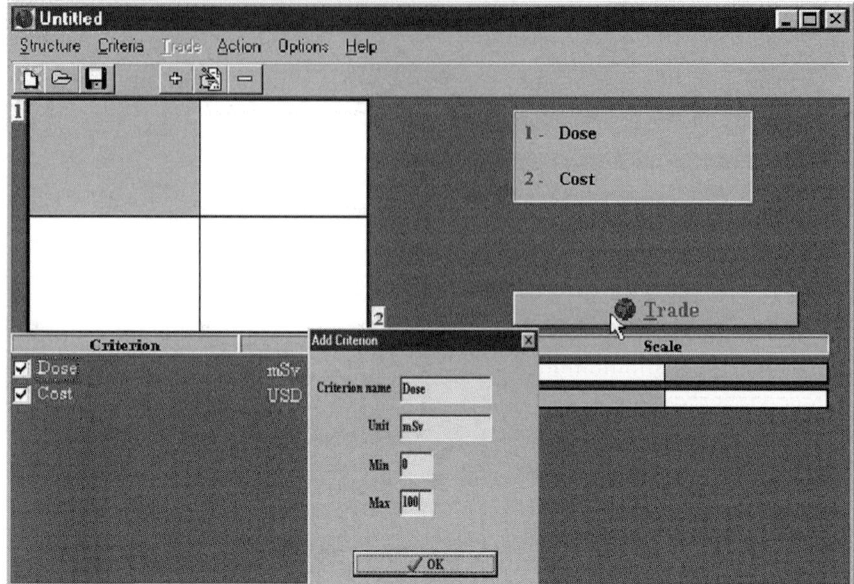

Figure 6 View of interface components for criteria entry and description.

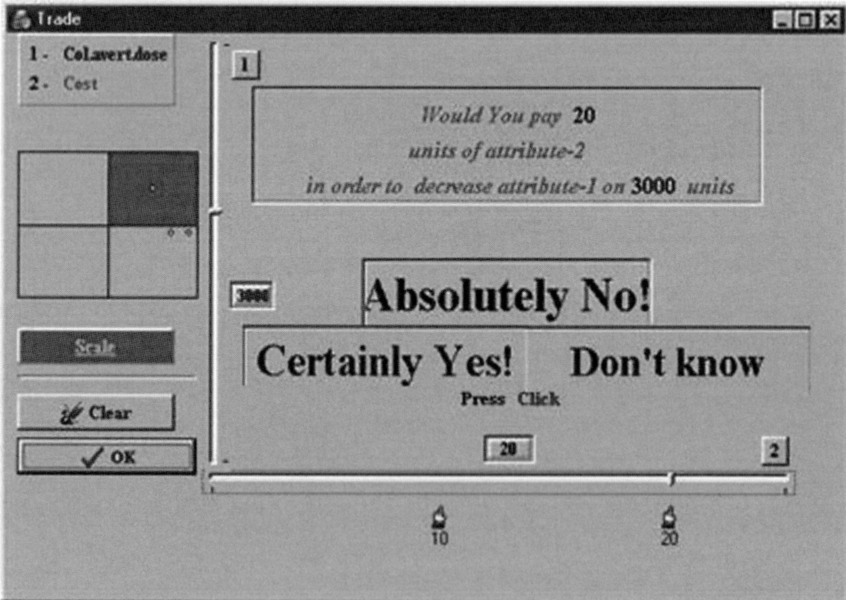

Figure 7 View of interface component for choosing initial position of alternatives in the multi-criteria space.

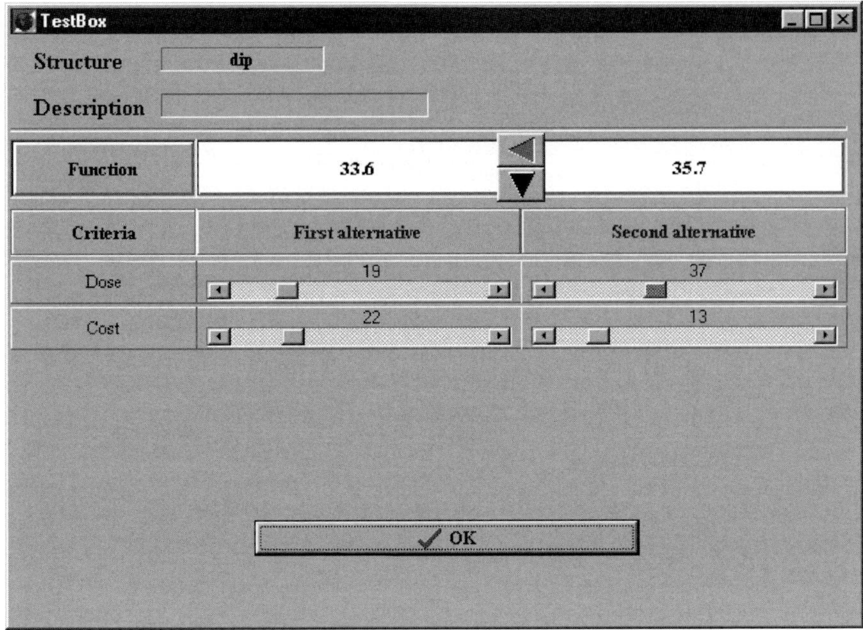

Figure 8 View of interface component for comparison of alternatives.

For ranking the possible alternative options of remedial actions prepared by the system, the module uses the preference functions generated earlier, extracting them from the database. The decision maker can interactively become familiar with the preference function features (specified criteria, range of criteria, etc.) using the interface, as shown in Figure 9. If needed, the decision maker can adjust (make more accurate) the chosen preference function.

All options of actions supplied by the system are related to the specified criteria (attributes) accounting for the outcome of their application in the situation under consideration. For ranking these actions based on the selected preference function, within the user interface, parameters of actions should be matched with the criteria used in the preference function (see Figure 10).

Figure 10 shows some outcomes of ranking 15 options of early measures to protect the public that were considered by the decision makers during the emergency management exercises. Three options were rejected as not feasible. The criteria chosen for ranking were the collective dose received by the public due to implementation of protective actions and the total cost of carrying out the plan of actions.

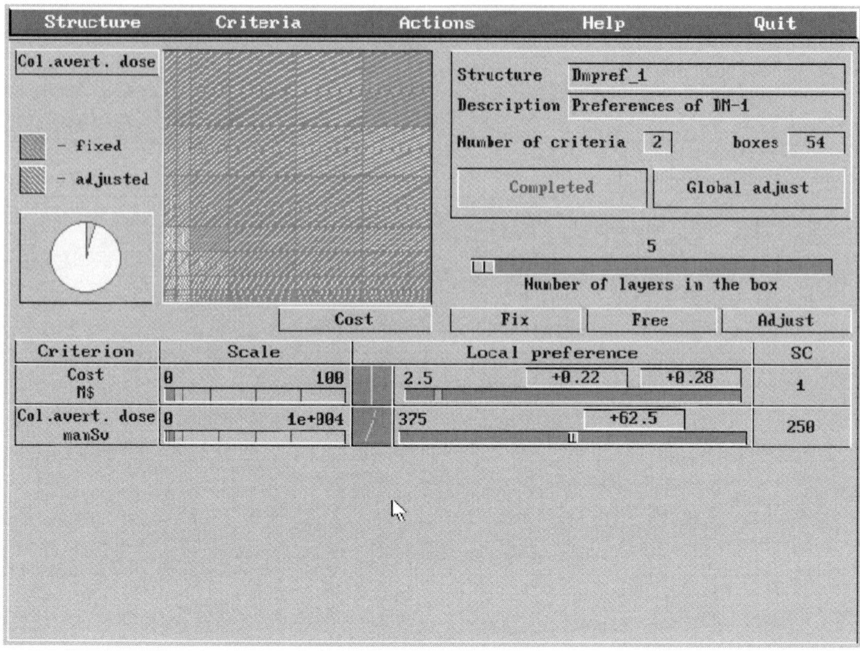

Figure 9 View of interface component showing the preference functions.

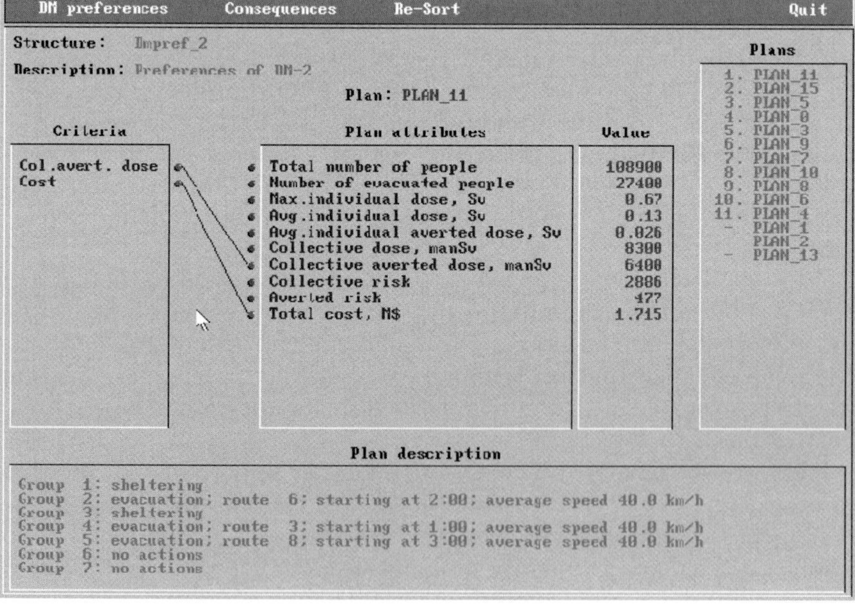

Figure 10 Results of ranking the protective options.

3. Application of the Decision-Aiding Technologies for Justification of Remediation Strategies in Different Environments

An important part of the decision-making process is predicting how the situation may evolve as a result of application or non-application of remediation. The changes may be due to natural processes (e.g. radioactive decay or radionuclide migration) or result from remedial action implementation. For predicting change in the radiological situation, radioecological models are used extensively.

In the computer-based decision-making systems used for choosing optimum remedial actions on contaminated areas, models and data are required to simulate various processes occurring during implementation of remedial measures. Simulation models can be based on complex mathematical algorithms, but the interface to interact with them should be straightforward and user-specific, with estimates delivered promptly and visually. The accuracy of prediction depends on how adequately the model reflects the process of interest and how accurately the model parameters and initial state of the system are chosen. Choice of parameter values and input data for the models can be based on direct or indirect measurements, as well as expert judgements. Often, measurement data are not feasible, and some assumptions about the magnitude of parameter values and input data have to be made. In this case, typical parameter values and input data are used in the models. Thus, along with models, decision-making systems should include expert databases with characteristics of potential remedial measures (technical, economic, social), as well as relevant regulations and standards.

Thus, to be effectively applied in practice, the multi-criteria techniques have to be supported with the following components:

- models for environmental transport of radionuclides;
- models for different exposure pathways for the public (workers) in contact with the contaminated area or foodstuffs;
- models for calculating doses and radiation risk;
- expert databases with information about the available methods to ensure radiation safety for the public and environment in the course of remediation of contaminated sites;
- databases with economic characteristics of remedial measures;
- standards and a regulatory framework for decision making.

The totality of multi-criteria analysis and its supporting components forms a system of multi-criteria analysis allowing optimal decision making and ensuring human and environmental radiation safety in the process of remediation of areas affected by radioactive contamination.

3.1. Decision-making framework for remediation of aquatic ecosystems

3.1.1. Constructing a system of criteria for choosing an optimal strategy for remediation of water bodies

An optimal strategy should be chosen on the basis of all the available information, the ultimate goal being minimisation of negative consequences of radioactive contamination, not only radiological ones but also social and economic impacts that may be triggered by a specific decision. This general goal, however, cannot always be expressed in quantitative terms and it is therefore broken up into subgoals that are possible to measure and the quantitative criteria of achieving these subgoals are defined.

Thus, when performing the analysis, general goals (higher level goals) should be first formulated, and then divided into more specific goals (lower level goals). Goals may form a hierarchy and, in turn, can be subdivided into subgoals. The subgoals can be broken up further on, until a factor is found that allows measuring achievement of the goal. This way of structuring the problems of decision selection is informal, as no formal theory exists to direct this step. Nevertheless, building a hierarchy is always useful, because complex problems are thereby broken into pieces.

In the multi-criteria analysis and whilst choosing an optimal strategy for remediation of water bodies, the higher-level goal of minimising the adverse impacts of contamination can be divided into three subgoals: minimisation of social damage, economic damage and detriment to the environment.

The social impacts are associated with two subgoals: minimising the impacts on human health and living conditions. Obviously, radiation dose is an important factor affecting human health. In ensuring the radiation safety of water bodies, the goal of minimising the health effects can be subdivided into more specific subgoals: minimising *average doses* for the public from the use of a water body, minimising *doses for a critical group* and minimising the *collective dose* from the use of a water body. The general health status of people can be affected by *specific stress factors* (the possibility of an accident, changes in living conditions, relocation, etc.). ‘Life restrictions’ are understood to include measures to limit drinking water consumption, fish intake, water use for irrigation and recreational use of water bodies. In this case, the criterion for achieving the goal is reducing, as much as possible, the doses associated with the use of a contaminated water body.

Economic impacts from actions ensuring radiation safety in the process of remediation of a contaminated water body can be subdivided into direct costs that can easily be described in quantitative terms and indirect effects such as loss of the prestige value of the area surrounding the water body and negative market response in the spheres related to the water body. Direct costs can be subdivided into the cost of recovery measures, economic losses from termination of water use, cost of delivery of clean foodstuffs in case of banning the use of local food and cost of setting up clean water supply systems.

Another possible task is economic and radiological assessment of the generation of secondary waste in the course of remedial action implementation, specifically the costs of waste transportation and disposal, exposure of personnel handling waste and so on.

Detriment to the environment has only recently become part of decision making related to radiation safety actions on contaminated areas. If the contaminated water body is a unique natural ecosystem, the detriment from radiation exposure and measures should be considered in detail, whereas in other cases, as a first approximation, the sole criterion of 'the degree of the water body disturbance due to countermeasure implementation' can be used (e.g. in the case of filling or draining, water body destruction is 100%).

3.1.2. Assessing individual criteria and constructing the utility function

The major challenge in solving multi-criteria problems is non-uniformity of individual criteria that are initially measured by different units (monetary, dose, volumetric, etc). The values of the lowest level criteria become the basis for determining higher level criteria (integrated criteria).

Because of this, in the multi-criteria techniques, measurements of criteria in physical units are replaced by criteria in conventional units. All integrated criteria are measured in relative units in the interval [0, 1] or [0, 100]. The values close to 0 represent the low utility of the countermeasure with respect to the integrated criteria of interest, whereas values close to 1 denote high utility. Changing from physical units to relative units is done with the use of the conversion function. To define the conversion function, the upper and lower boundaries of the criterion variation due to countermeasure implementation should be specified. The shape of the conversion function is normally defined by a linear dependence. If necessary, a monotonically decreasing conversion function can also be used. If the criterion changes non-monotonically in some parts of the interval [0, 1], piecewise-linear conversion functions, as well as the exponential, Gaussian and other functions, can be utilised (Eltarenko, 1995).

The problem of the criteria space uniformity, however, cannot be fully resolved by changing over to relative units for individual criteria, since comparison of diverse objects is rather difficult in the multi-criteria space. Hence, the coefficients of relative importance of criteria (W_j) need to be defined. Besides, the criteria should be weighted to determine the desirable outcome for each alternative comparison and such weighting factors are determined by experts with the involvement of decision makers.

When defining the relative importance coefficients, it should be ensured that the criterion is informative from the standpoint of decision choice. If the criterion remains the same or varies insignificantly in all countermeasure variants, it may be regarded to be of little informative value for selecting preferences and to have low importance value W_j.

As a result of assessments, each action S_q will be described by a set of criteria $(K_i^q, \ldots, K_m^q)$, where K_i^q is a factor at the specific level i for alternative S_q. The next step will be assessing each of these measures using the multi-factor model in order to reveal the best of them. The aggregating operator should satisfy the following requirements:

(i) the input parameters are normalised values of aggregated criteria u_1, ..., u_m (m is the number of the lower level criteria);
(ii) the aggregating operator should be continuous and monotonous with respect to $u_1, \ldots, u_m$;
(iii) the operator is commutative with respect to the normalised criteria;
(iv) the parameters of the aggregating operator are the relative importance coefficients (criteria weights) W_j; the singular values of the aggregating operator being as follows: $U(1, 1, \ldots, 1) = 1$; $U(0, 0, \ldots, 0) = 0$; $U(u_1, 0, \ldots, 0) > 0$ with $u_1 > 0$; $U(u_1, 0, \ldots, 0) < 1$ with $u_1 < 1$.

The simplest and most frequently used utility function is the additive aggregation operator, that is the function summing up the normalised values of criteria with their relative importance coefficients.

Following the aggregation and the derivation of the utility function, the strategies should be ranked by the utility function value. The most preferable strategy will be the one for which the utility function value is the highest.

As a whole, solving the multi-criteria problem is an iteration process and after assessing the integrated criteria it may become necessary to add new ones to the criteria structure (or to exclude some of the non-informative criteria), to modify the conversion functions or parameters of the aggregation operators.

3.1.3. Example of assessing effectiveness of the remedial actions in freshwater ecosystems

A decision-making process for remediation of a water body contaminated with radionuclides is illustrated here by an example based on application of the MAUA (Sazykina and Kryshev, 2006).

It has been assumed that the water body (lake) has a surface area of $5\,km^2$ and a depth of 5 m and is contaminated with ^{137}Cs and ^{90}Sr. The contamination levels of the environmental compartments are as follows:

(i) specific activity concentration of ^{137}Cs in bottom sediments is $2.0 \times 10^4\,Bq\,kg^{-1}$ and of ^{90}Sr is $10^4\,Bq\,kg^{-1}$;

(ii) concentration of ^{137}Cs in water is 20 Bq l^{-1} and of ^{90}Sr is 100 Bq l^{-1};
(iii) radionuclide activity concentrations in fish and water fowl are 20 kBq kg^{-1} for ^{137}Cs and 3.0 kBq kg^{-1} for ^{90}Sr.

There are 500 people living in the vicinity of the lake, some of whom are engaged in fishery, others raise fowl or work in agriculture. The annual fish catch is 10 tons. The annual fish consumption is 80 kg year^{-1} per person for the critical group (fishermen) and the average fish consumption for people living on the lake banks is 20 kg year^{-1} per person. The rural population consumes 10 kg year^{-1} of fowl, the fowl production being 5 tons year^{-1}.

The alternative strategies under consideration include

- Strategy I: Doing nothing, relying on natural self-cleaning of the water body.
- Strategy II: Undertaking radical clean-up of the water body by removing the upper layer of bottom sediments (5 cm layer).
- Strategy III: Relocating people living on the lake banks and setting up a reserve.
- Strategy IV: Banning water use, fishing and raising fowl for 30 years, fencing the lake perimeter and guarding the territory. As a replacement, establishing alternative sources of water supply (ponds, wells) and importing clean fish and meat for the local population.

After a list of alternative strategies and a system of criteria are put together, the lower level criteria are calculated for each strategy in natural units, that is doses for the population, cost of measures, damage to the local economy and so on. As a next step, the criteria values are converted from natural into conventional units adjusting to a scale from 0 to 1 and Table 1 shows the score for the example under consideration. The estimates of the aggregated utility function for each strategy with allowance for normalised

Table 1 MCA criteria in normalized units representing the relative acceptability of measures with respect to particular criteria (Sazykina and Kryshev, 2006).

Remediation strategy	Dose for critical group (mSv)	Average dose (mSv)	Remed-iation cost	Economic damage	Social damage	Environ-mental detriment
Doing nothing	0.25	0.52	1	0.98	0.7	1
Removal of bottom sediment	0.85	0.95	0.05	0.9	1	0.5
Setting up a reserve	0.92	0.98	0.93	0.8	0.6	1
Banning use of water body	0.92	0.98	0.98	0.97	0.9	1

Table 2 The utility function values in MCA for different countermeasure strategies applied to the contaminated water body (Sazykina and Kryshev, 2006).

Countermeasure strategy	Aggregated utility function	Ranking by applicability
Banning the use of water body	0.95	1st
Setting up a reserve	0.86	2nd
No actions	0.70	3rd
Removal of bottom sediments	0.68	4th

Note: The maximum value of the aggregated utility function is 1.

criteria u_i (Table 1) and the estimated weighting coefficients W_i are shown in Table 2. The strategies are also ranked by the priority (preference) of application.

3.2. Decision-making framework for remediation of forest ecosystems

Several studies were carried out mainly after the Chernobyl accident to provide optimisation of remediation strategies in contaminated forest. In these studies, forest remedial actions were considered under various aspects: radiological, economic, environmental or ecological and social. Economic and social consequences are connected with the restrictions of access, which should be applied in the contaminated forests, as well as with extra expenses for the remediation implementation. These restrictions result in losses or reduction of wood and other forest products such as mushrooms, berries and medical herbs in which radionuclide contents exceed the temporary permissible levels (TPLs). This has negative economic consequences and leads to negative psychological effects on the population which now considers a forest as a source of risk. Besides, some measures that could be potentially effective have negative secondary ecological effects (Belli, 2000) and their implementation can decrease the ecological stability of a forest.

In particular, a good example of the application of a comprehensive cost-benefit analysis of long-term management options for contaminated forests was given by Shaw et al. (2001). It has been found that in general only few forest countermeasure options can be potentially effective even for high contamination levels. However, the significances of various exposure pathways as well as the cost of implementation are clearly very dependant on local conditions. Accordingly, the cost-effectiveness of certain forest remedial actions can be much higher and some of the options can be recommended for implementation even in the long term after the Chernobyl Nuclear Power Plant (ChNPP) accident.

Besides, the experience gained in the long term after the ChNPP accident has clearly shown a need to consider not only radiological and economic factors but also the perception of remedial measures by the local population as well as the assessment of secondary ecological effects. These factors are very difficult to identify and sometimes these are hardly possible to quantify in monetary terms. Therefore, a set of different criteria including social acceptability, provided in an MAUA, has to be applied to justify the optimised selection. Such an approach was considered to elucidate the relative importance of health, economic and social factors that influenced the relocation policy adopted in the Soviet Union after the ChNPP accident (French, 1992) and partially the agricultural management (Salt et al., 1999).

The Novozybkov district of the Bryansk region of Russia located 180 km northeast of the ChNPP was selected as a case study area for the application of MAUA approach for a justification of remediation strategies. The average deposition density of ^{137}Cs on the territory of this district was about 750 kBq m^{-2}, whereas contamination of forest soils varied in a range from 150 to 2,500 kBq m^{-2}. The data obtained for the study area were united within zones with different levels of contamination (zone A, above 1,480 kBq m^{-2}; zone B, between 555 and 1,480 kBq m^{-2}; and zone C, between 185 and 555 kBq m^{-2}) and, consequently, with different scales of application of measures. The more detailed description of this area is given elsewhere (Fesenko et al., 2000).

According to the Russian radiation safety regulations (RSS, 1999), countermeasures or any remedial actions can be optionally applied in areas where dose to the population exceeds 1 mSv year^{-1}; countermeasures are necessary when doses exceed 5 mSv year^{-1}. In the dose range 1-5 mSv, measures should be applied on an optimised basis.

Information on internal and external doses to inhabitants of the settlements located on the contaminated territory is available for several years after the accident and was published in recent years. Mean data of annual effective doses in the settlements of different zones calculated on the basis of information available from dose catalogues as well as on results of the FORESTLAND model (Fesenko et al., 1999) predictions are shown in Figure 11.

The results indicate that the application of remediation is of importance up to 2025, 2015 and 2005 in zones A, B and C, respectively; however, measures on restriction or optimisation of mushroom and berry consumption can be considered even outside these time periods. Evaluated on the basis of the results given in Fesenko et al. (2000), the contributions of different pathways to the exposure of critical population group members and the rest of the population are given in Table 3.

In terms of individual exposure pathways, the highest contribution to the total dose to the population is the external dose within (or near) settlements, followed by internal doses via milk and meat products, mushrooms and berries and external dose associated with the forest contamination.

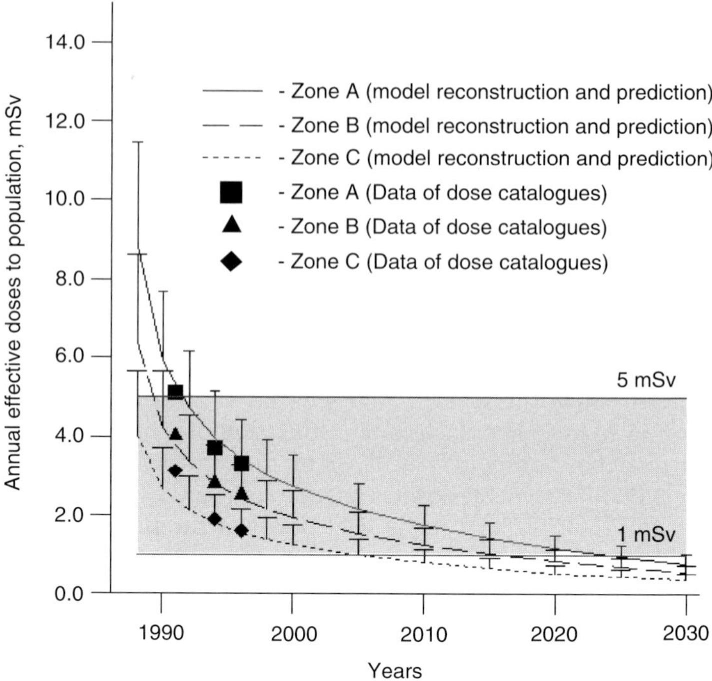

Figure 11 Variation with time of average effective doses to the population of the study area. Bars reflect dose variability across zones A, B and C dependant on exposure of the population in individual settlements. The dose range where remedial actions should be applied on an optimised basis is shown by shadow.

For critical population group members, the forest pathways play a dominant role (exceeding 70% of the total dose) and it might be expected that application of forest remedial actions is the most effective way of decreasing the contamination impact. The most important pathway is related to the consumption of 'forest milk'. However, even when using alternative feedstuff, this pathway is excluded; the contribution of other forest pathways (such as mushroom consumption and external irradiation of foresters within the contaminated forest) is rather important, varying in the range from 23% (only mushrooms and berries) to 57% (all forest exposure pathways).

Based on these results, the potential effectiveness of forest remedial actions in terms of dose to critical population group reduction can be ranked as follows: 'forest milk consumption' > 'mushroom consumption' > 'external dose in the forest (to foresters)' > 'berry consumption' > 'visiting forest for recreation and mushroom (berry) gathering'.

Table 3 Contributions of exposure pathways in zones with different contamination levels (%).

	Zone A		Zone C	
	Population	Critical group	Population	Critical group
External dose within the forest	3	1–18	4	2–17
External dose within the settlement	41	11–17	44	17–24
Internal dose from milk and meat consumption	33	58–66	35	54–60
Internal dose from mushroom consumption	19	11–14	13	1–12
Internal dose from berry consumption	2	1–2	1	1–2
Internal dose from consumption of other products	2	1–2	4	1–2
Contribution of internal dose to total dose	56	71–83	52	66–75
Contribution of forest pathways to total dose	24	83–88	18	75–83

The effectiveness of forest measures in terms of percentage of total dose reduction estimated based on the data of Table 3 as well as the related cost per hectare of forested area, their acceptability and cost of 1 man-Sv averted are given in Table 4. Collective doses which can be averted due to forest measures application were calculated based on the approach presented elsewhere (Fesenko et al., 2000) and the cost of the measures was taken from studies carried out in this region and includes the necessary resources, manpower, equipment and consumables (Belli, 2000).

It can be seen from the data of Table 4 that the most effective way to reduce the impact of contaminated forests on the population is the application of options for reducing exposure from mushroom consumption. As for the critical population group, this pathway is also of importance; however, as outlined earlier, a more effective option is to decrease the 'forest milk' exposure pathway. The lowest effectiveness in terms of dose reduction as well as the highest cost and low acceptability are typical for soil-based countermeasures, which excludes these options from further analysis.

A decision analytic tool based on a *PRIME* (preference ratios in multi-attribute evaluations) technique was applied to justify optimal countermeasure strategies (Gustafsson, 1999). In the framework of this technique,

Table 4 Effectiveness of forest measures in terms of annual effective dose reduction, cost, acceptability and cost of 1 man-Sv being averted after their application in 2003 (thousand USD).

Actions	Decrease in effective dose (%)		Cost ($ per ha)	Acceptability	Cost of 1 man-Sv
	Critical group	Population			
Restrictive measures					
A. Abandonment	83–88	18–24	105–107	Very low	297–866
B. No foresters access	79–84	–	109–111	Low	231–770
C. No public access		18–24	103–104	Low	295–861
D. Restriction on grazing of domestic animals or using forest grass for fodder	54–58	–	4.3	Moderate	16.5–53.7
E. Restriction on mushroom collection	9–11	13–19	24–26	Low	14.2–53.4
F. Restriction on berries collection	0.9–1.8	1–2	14–15	Low	87.8–305
Optimisation in forest management					
G. Limitation of tree harvesting to the areas with low doses	2–5	–	0.48	Very high	246–785
H. Limitation on mushroom collections to the species with low accumulation	3–4	3–4	0.64	Moderate	5.8–21.6
I. Mushroom processing before consumption	3–4	3–4	0.46	Moderate	6.6–24.6
Application of Cs binders					
J. Decrease of contamination of "forest milk"	30–49	–	0.14	High	1.5–5.2
Soil-based measures (only for berries)					
K. Liming	0.5–1.0	0.6–1.2	12.7	Low	>1000
L. Application of potassium	0.4–0.9	0.5–1.0	130	Low	>1000

the following steps allowing the identification of a reasonable decision are considered:

- identification of the alternatives and their attributes;
- preference assessment;
- determination of the best alternative.

The objectives of the first step are identification and description of possible alternatives. Within this step, each countermeasure listed in Table 4 (or their combinations) has been considered as an alternative, and the above-mentioned criteria were taken as their attributes which were defined for each alternative from Table 4. Levels of acceptability were transformed into a numeric scale: very low, 5; low, 25; moderate, 50; 'high, 75; and very high, 90.

The objective of the preference assessment is to identify priorities in the selection of optimal forest remediation strategies. The first phase in preference assessment is to assess the weights of the attributes. The weight, ranging from 0 to 100, is a subjective ranking based on expert judgement, taking into account general preferences related to the forest remedial actions.

Two main possible strategies of forest countermeasure application in the long term after the ChNPP accident can be considered: the first (dose reduction strategy) is based on the preference of maximal reduction of effective dose and the second (acceptability strategy) is aimed at a maximal acceptability of the measures. Taking into account these preferences, attributes were weighted in the following two ways: (1) reduction of effective dose (100) > acceptability (70–90) > cost of 1 man-Sv averted (60–80) and cost for countermeasure application (40–50), and (2) acceptability (100) > reduction of the effective dose (70–90) > cost of 1 man-Sv averted (60–80) and cost for countermeasure application (40–50).

The third step directly concentrates on the selection of optimal alternatives and includes analysis of 'value intervals' and decision rules, which allow the selection of the optimal alternative (actions). The notion of the value interval of an alternative refers to the alternative's value interval of the main goal, since it contains the total value of an alternative (Figure 12).

It can be seen on the basis of these results that the restriction on using forest grass for fodder (alternative D) or using Prussian blue to decrease 'forest milk' contamination (J) in combination with limitation of tree harvesting to the area with low doses (alternatives D&G and J&G) are likely dominant in the selection of optimal actions for the critical population group. However, this fact does not allow a final decision to be made.

As for the rest of the population, the results have shown that a combination of forest countermeasures such as 'limitation on mushroom

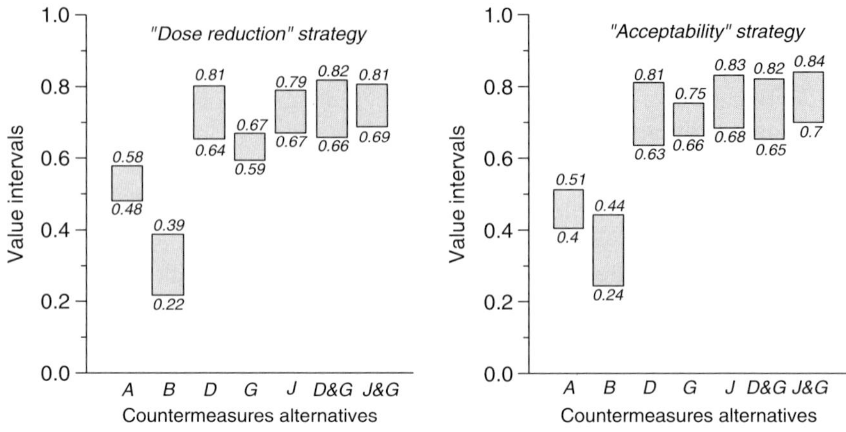

Figure 12 Value intervals of alternatives in forest countermeasures applications (critical population group).

and berry collections to the species with low accumulation in the combination' and 'mushroom processing before consumption' (alternative H&I) has likely a good chance of selection as the best option.

Decision rules assist the decision maker in the determination of the best alternative. *PRIME* decision provides four decision rules that apply to different situations. The available rules are maxi-max, maxi-min, central values and mini-max regret (French, 1986).

Mini-max regret takes a different approach and calculates the possible loss of value for each alternative by using dominance data. Thus, this technique selects the alternative with the minimal possible loss of value. The results of such calculations have shown that, for both 'dose reduction' and 'acceptability' strategies, alternatives J&G for the critical population group and H&I for the rest of the population are marked as the most appropriate practically and the possible loss of value calculated based on the mini-max regret technique is shown in Figure 13 to prove this conclusion.

The results obtained allow the conclusions that application of forest remedial measures in the most contaminated areas of Russia are of importance for at least several further decades and an application of the decision-making approach based on combined consideration of radio-ecological, social and economic features of forest countermeasures is useful for the selection of optimal long-term actions. It also allows a conclusion that restrictive options as well as soil-based forest measures are not applicable in the long term after the Chernobyl accident and more attention should then be given to the optimisation of forest and forest product usage.

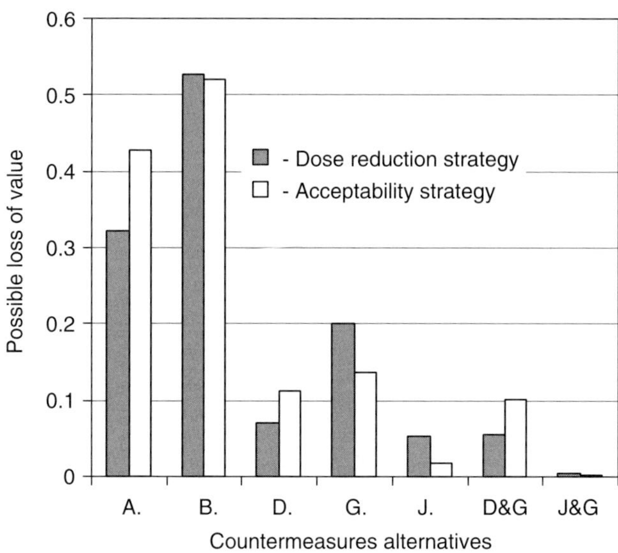

Figure 13 Loss of value for alternatives in application of forest countermeasures (critical population group).

3.3. Decision-making framework for remediation of rural areas contaminated after the Chernobyl accident

After the accident at the ChNPP, a lot of remedial measures were developed and adopted in agriculture for use in the event of radioactive contamination of the environment. Within the 20 years since the accident, various measures have been implemented and a vast amount of data on their effectiveness has been generated, together with information on ancillary factors such as the required resources and costs. These measures vary considerably in their effectiveness, cost and practicability in actual situations. The efficiency of application thereof depends on many factors, including soil and climatic conditions and specific features of the agricultural production management. Because of this, implementation of these measures on the basis of general expert estimations often resulted in inadequate decisions. Therefore, application of the decision-aiding technologies was widespread to provide advice on remediation strategies at different levels of decision-making process.

In particular, an example of the application of an MAUA to justify the optimised selection of remedial actions in agriculture (which is similar to the approach described in Section 3.2 for contaminated forests) is given elsewhere (Panov et al., 2006).

The major focus of environmental decision support systems (EDSSs) intended for optimising remediation strategies in agriculture was to derive holistic approaches to reliably predict radionuclide transfer in different ecosystems making use of tools such as Geographical Information Systems (GIS) and geostatistics to directly implement cost-effective actions and assess their potential environmental and social side effects. Following the Chernobyl accident, different EDSSs have been produced, such as FORCON (Fesenko et al., 1996), AGRO (Recommendations, 1998) and RESTORE (Van der Perk et al., 1998; Gillett et al., 2001) which have enhanced the ability to optimise remediation strategies in contaminated areas.

One such system, which has been given the acronym ReSCA, has been developed within an IAEA Technical Cooperation Project (IAEA, 2006; Fesenko et al., 2007), the approach used by the system being different from that given above and representing another way of selecting agricultural remediation strategies on affected territories. The remediation tool considers seven different remediation actions: radical improvement (RI) with or without drainage, surface improvement of meadows (SI), ferrocyn application to cows (FA), clean feed for pigs (FP), mineral fertilizers for potato fields (MF), an information campaign on mushroom consumption (IM) and removal of contaminated soil (i.e. decontamination of the settlement, RS). The tool also allows the simulation of the application of the above options to a part of the settlement or its population.

It has been assumed that decontamination is effective for the entire time period after its application, radical and surface improvement being assumed to be effective for 4 years, whereas other actions being effective for only 1 year. For the case where information on food contamination is not readily available, contamination of the foodstuffs can be obtained by using data on soil ^{137}Cs contamination density and generic transfer factors (T_{ag}) to crops evaluated by the project experts for the long term after the accident.

The social acceptability of the above remediation actions were evaluated based on questionnaires completed in the framework of the project in 2003 (Fesenko et al., 2007). However, for practical application, and in the framework of an algorithm, these levels of applicability were transformed into a numerical scale via the application of a score from 0.1 to 1.

Remediation options are considered by the tool in relation to three aspects: radiological, economic and social. This provides an opportunity to make flexible decisions within the limitations on funds allocated for remediation purposes. The following expression is used for the prioritisation of the remediation actions:

$$\beta \frac{\min(\mathrm{CD}_r)}{\mathrm{CD}_r} + (1-\beta)\mathrm{DA}_r$$

where CD_r is the cost of 1 man-Sv averted as a result of application of the remediation option r and DA_r is the degree of acceptability of the

Table 5 Effectiveness, costs and degrees of acceptability of the remediation actions accounted for by the ReSCA tool.

Remediation action	Reduction factor	Degree of acceptability	Cost (Euros)		
			Belarus	Russia	Ukraine
RI[a]	1.7-8	1	350	390	450
SI	2	1	300	340	400
FA	2-3	0.75	30	60	40
FP	3	0.6	6	7	20
MF	3	1	0.8	2.5	1
IM	1.5	0.5	3	3	3
RS[b]	1.2/1.3/1.5	0.1	109/163/325	109/163/325	109/163/325

Note: Costs for RI/SI/FA are per cow, others are per inhabitant.
[a]The tool allows the simulation of various RI options: the first application, repeated application and RI consisting of drainage with different effectiveness and cost.
[b]It is assumed by the tool that one third, a half and the whole settlement is decontaminated.

corresponding action. Parameter β allows the user to give preference either to economic or to social aspects of the remediation planning. Thus, for value β equal to 1, the remedial actions are ranked according to the costs per averted dose, whereas for minimum β (0.01), the ranking is based mainly on acceptability of remediation actions. The remediation strategy is being built as a list of separate remediation actions until the total cost becomes higher than the total amount of funds allocated for remediation purposes. Thus, for given input and model parameters, several strategies can be generated varying by the amount of available funds and/or user priorities. The tool provides a variety of output results – from individual fields to all affected settlements taken for the evaluation. One of the tasks of implementation of remedial actions in affected areas is a reduction in the number of settlements with annual effective dose above 1 mSv. Table 5 and Figure 14 illustrate the results obtained for 105 settlements in Zhitomir and Rivno oblasts of Ukraine (Fesenko, 2007).

Two remediation strategies with differing priorities are shown in Figure 14. The first strategy (radiological strategy) is based on radiological and economic factors (cost of 1 man-Sv averted), whereas the second strategy (social strategy) is focused on remedial actions which had highest applicability by the population. The main characteristics of these strategies in terms of total averted dose and cost of remediation are given in Table 6.

Table 6 data conclude that both strategies are quite cost-effective. In spite of the fact that the 'radiological strategy' is characterised by a twofold lesser cost of collective dose averted and is much cheaper, the second strategy provides higher reduction in collective dose to the population and can be better supported by the public.

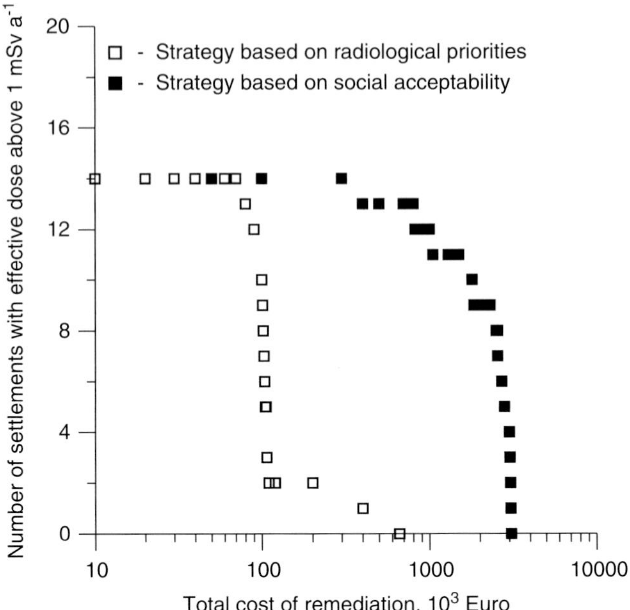

Figure 14 Dependence of the number of settlements with annual effective dose exceeding 1 mSv on the total cost of the remediation for strategies based on the radiological effectiveness and social acceptability of the remedial actions.

Table 6 Some characteristics of remediation strategies calculated by the ReSCA remediation tool.

Parameters	Social strategy	Radiological strategy
Total averted dose (man-Sv)	113	47
Total cost of the strategy (thousand Euros)	3050	671
Cost per averted dose (thousand Euros per man-Sv)	26.8	14.3

Making such calculations, the ReSCA system allows a balancing between these two options that provides opportunity for making justifiable decisions on remediation planning.

It can be concluded, based on these results, that the total investments to achieve doses below 1 mSv in all affected settlements may be in a range of 670–3,050 thousand Euros and, depending on availability of funds, optimal management actions can be suggested for implementation in the long term after the Chernobyl accident.

It can also be seen from Figure 14 that the increase in funds does not always provide a proportional decrease in the number of settlements where the effective dose is above 1 mSv year^{-1}. It is unlikely that funds for remediation of less than 100,000 Euros for a 'radiological strategy' and of 700,000 Euros for a social strategy would provide a reduction in the number of settlements with an annual effective dose above 1 mSv.

The highest effectiveness of remediation in terms of the reduction in the number of such settlements can be achieved for funds between 90,000 and 120,000 Euros for a 'radiological strategy' and between 900,000 and 3,000,000 Euros for a strategy based on social acceptability. Thus, a user can make a well-justified choice depending on the current financial constraints.

An example of the tool application at the level of a single settlement is shown in Figure 15.

The case study selected for this demonstration was the settlement of Veprin in the Bryansk region, Russia. The settlement is located at a distance of around 190 km from the ChNPP and is surrounded by forest. At the time of the last survey (2004), the settlement comprised 54 homesteads with 254 inhabitants who kept 20–28 dairy cows and 30–40 goats. The density of ^{137}Cs contamination of the settlement area in Veprin is about 600 kBq m^{-2} and the ^{137}Cs activity per unit settlement area differs by up to 10% from the contamination of the nearest fields used as pastures for private cows (670 kBq m^{-2}).

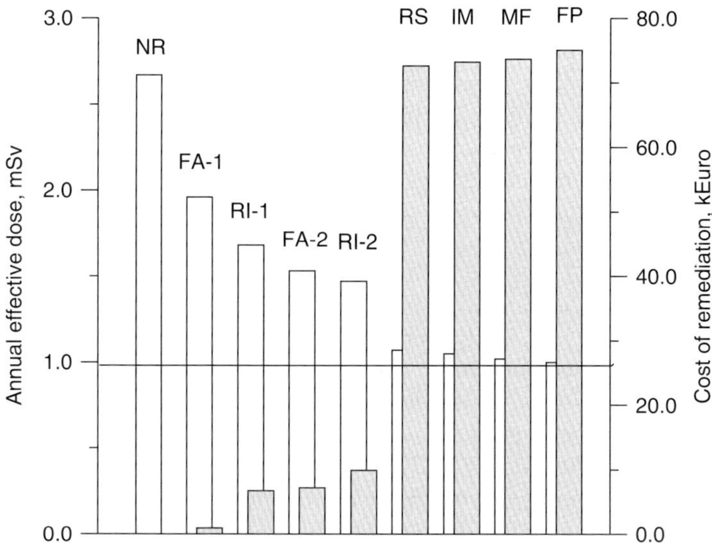

Figure 15 Evaluation of the effectiveness of the remediation actions in an affected settlement. Effective dose values are shown by empty bars and associated costs by shaded bars, NR means no remedial actions are applied, the other notifications are the same as in Table 5.

The mean ^{137}Cs specific activity in the cow milk of Veprin for the indoor period was $260 \pm 190\,Bq\,l^{-1}$, whereas in the grazing period it reached $400–450\,Bq\,l^{-1}$; the ^{137}Cs concentration in goat milk ranged from 212 to $573\,Bq\,l^{-1}$. The ^{137}Cs concentrations in beef and pork were at levels of 600–800 and $100–200\,Bk\,kg^{-1}$, respectively, whereas in vegetables, in particular in potato, they ranged at $20–50\,Bk\,kg^{-1}$. Rather high ^{137}Cs concentrations (mean of $2{,}100\,Bk\,kg^{-1}$, dry weight) were observed in mushrooms collected in the surrounding forests. For grazing the cows and making hay (the main feed in the winter period), inhabitants of the settlement mainly use pastures and haylands located on peat-boggy soils with a high level of grass contamination and less contaminated pastures located near the settlement. This fact shows a need to formulate a stepwise strategy when considering the effect of remediation of each fodderland. The total annual effective dose before remediation was estimated to be 2.67 mSv. The contribution of internal exposure is slightly higher (55%) than that of the external dose.

The sequence of the necessary remedial actions, for the case in which the acceptability of the remedial actions was ignored, is shown in Figure 15. The strategy is based on the application of the main remedial actions to decrease internal dose to the population by the application of ferrocyn to cows grazing more contaminated grass, followed by radical improvement of more contaminated meadow, and after that the system recommends taking the same actions for cows grazing less contaminated fields near the settlement.

The cost of 1 man-Sv averted for these actions ranges from 3.5 to 14.0 thousand Euros. The next action is against external exposure – to decontaminate the territory of the settlement with the cost of averted dose ranging from 35 to 90 thousand Euros and finally application of rather costly actions: clean feed for pigs and mineral fertilizers for potato fields or less effective actions such as an information campaign on mushroom consumption. The total cost of this strategy, which allows averting 2.7 man-Sv, is 75.1 thousand Euros.

As can also be seen from Figure 15, the reduction of the effective dose below $1\,mSv\,year^{-1}$ could be achieved only by application of the majority of possible options, including decontamination of the settlement area. However, it can also be concluded from these data that such a decrease can be achieved by the application of quite expensive options with relatively high cost per averted dose (28.2 thousand Euros per man-Sv).

At the same time, the results achieved allow the conclusion that the application of remediation actions 20 years after the Chernobyl accident is still effective and allows considerable reduction in the exposure of the rural population.

Similar results can be achieved for different priorities in remedial action selection, forming a basis for a reasonable justification of remediation strategies in the long term after the Chernobyl accident. Thus, the results presented here demonstrate the high effectiveness of a flexible decision support system (DSS) capable of providing practical advice on agricultural countermeasure strategy and show a need in application of such decision support technologies for justification of remediation strategies.

4. Review of Presently Available Decision Support Systems

4.1. Introduction

Within the last 15–20 years, decision-aiding technologies based on application of user-friendly computer programmes have received considerable attention in remediation planning in areas affected by the Chernobyl accident and have made a substantial contribution to the improvement of the existing emergency planning. The effectiveness of remediation strategies as well as necessary resources and costs to implement remedial actions varied depending on many factors, such as deposition characteristics, soil and climatic conditions and agricultural production management. Because of these multiple factors affecting the feasibility of remedial actions, generalised recommendations which do not account for the diversity often result in inadequate decisions when applied at a local scale. These considerations have led to a need for development of practical EDSSs capable of providing advice on remediation strategies.[1]

This chapter provides an overview of the 'state-of-the-art' for the developed EDSSs and considers mainly the EDSSs developed either in the framework of the European projects (Schulte et al., 2002) or in the framework of the IAEA international projects: FORM (Frissel et al., 1996; IAEA, 1996) and RESCA (Ulanovsky et al., 2009). In both cases, the EDSSs are the products of international cooperation where experience in the development of national DSSs was assimilated and harmonised across the countries involved.

Terrestrial ecosystems can be roughly subdivided into three main categories: rural environment, semi-natural environment and urban environment. Such a categorisation has been used to classify the EDSSs considered (Table 7), and the environments covered by these EDSSs are shown schematically in Table 8.

As stated earlier, the majority of the EDSSs mentioned in Tables 7 and 8 were developed within EC framework programmes and these are mainly

[1]Description of the EDSSs given in this chapter is based on the contributions to the EC EVANET Project and, in particular, EDSS SAVE/SAVE-IT (N. Crout), EDSS RESTORE(G. Voigt), EDSS STRESS (M. van der Perk), EDSS RIFE (B. Rafferty), EDSS Landscape (L. Moberg), EDSS CESER (C. Salt) and EDSS TEMAS(C. Vazquez).

Table 7 List of Environmental Decision-Support Systems intended for justification of remediation strategies in the terrestrial environments (Semioshkina et al., 2004).

EDSS	Deposition scenario		Transfer parameters		Spatial resolution	Soil characteristics	Effect of remediation
	Chernobyl	Others	T_{ag}	TF			
RESTORE	✓		✓		✓	✓	✓
CESER		✓		✓			✓
SAVE	✓	✓	✓	✓	✓	✓	✓
TEMAS		✓	✓				✓
STRESS	✓				✓		
RECLAIM	✓		✓			✓	✓
RESCA/IAEA	✓		✓		✓	✓	✓
FORESTLAND		✓		✓		✓	
FORIA	✓						✓
RIFE2	✓		✓	✓			✓
FORM/IAEA	✓	✓	✓				✓

Table 8 Environmental domains of the EDSSs considered.

	Environments			
	Rural	Semi-natural	Urban	Freshwater
RESTORE	✓	✓	✓	✓
CESER	✓			
SAVE	✓	✓		
TEMAS	✓	✓	✓	
STRESS	✓	✓		
RECLAIM	✓			
ReSCA/IAEA	✓			
FORESTLAND		✓		
FORIA		✓		
RIFE		✓		
FORM/IAEA		✓		

based on the approaches and data accumulated in the aftermath of the Chernobyl accident (Semioshkina et al., 2004).

The ReSCA EDSS is a product of the IAEA regional Technical Cooperation Project dealing with remediation of territories affected by the Chernobyl accident. The specific tasks of the project were to develop the software tool ReSCA – *Remediation strategies after the Chernobyl accident* – and to evaluate the effectiveness of remedial actions in the long term after the Chernobyl accident (Fesenko et al., 2007).

As mentioned earlier, the current trend in the development of the EDSSs consists of the combined use of several components: radioecological and dosimetric models; databases on the effectiveness of possible remedial options; databases on the parameters of the environments which influence the effectiveness of remedial actions and radionuclide transfers; and databases on the parameters of remedial actions, which are to be considered in identifying optimum remediation strategies, especially a decision-aiding module, providing analyses of the alternatives and support of the decisions (see Figure 16).

Ideally, an EDSS can contain all these modules and is coupled with a GIS. Figure 17 demonstrates the user interface of the RESTORE EDSS with the open model dialog box.

The majority of the EDSSs have a user-friendly interface facilitating application of the software for the evaluation of effectiveness of the remedial actions, representing a variety of different options for spatial analysis of the consequences of the radioactive contamination (Tarsitano et al., 2005).

Graphical interfaces for most EDSSs (RESTORE, SAVE, TEMAS, STRESS, RECLAIM, etc.) were developed based on standard algorithms

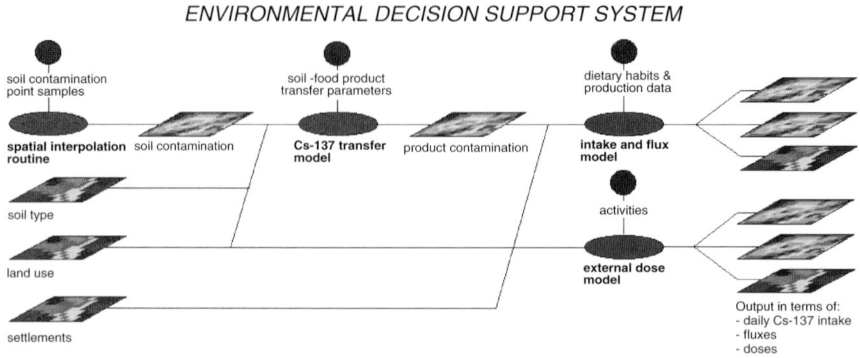

Figure 16 Schematic overview of the data processing and radionuclide transfer modelling with the EDSS.

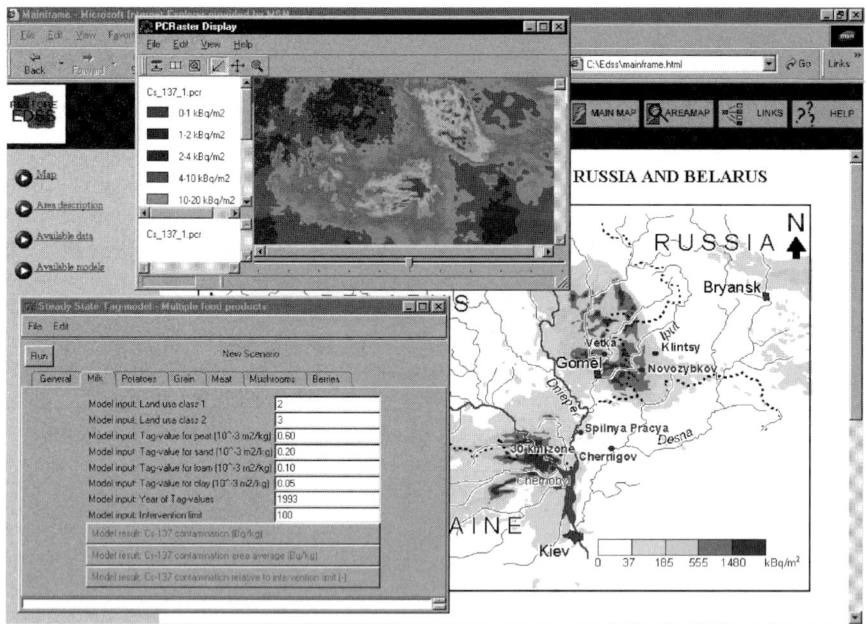

Figure 17 User interface of the RESTORE EDSS. Examples of the model dialog box and the PCRaster display are also shown (Van der Perk et al., 1998).

and software (mainly PCRaster), providing uniformity and information exchange among different systems. The interface and service programmes have been written in Delphy language.

At the same time, in many cases, current EDSSs are intended only for the provision of information (such as doses related to some specific exposure pathways or dose reduction following the remediation) just to support some decisions; therefore, other more specific software is required for making final conclusions (Semioshkina et al., 2004).

4.2. Models used for characterisation of radionuclide transfer in the environment

Mathematical simulation models can serve as a general tool for assessing various environmental impacts of changes in land use and land management, such as erosion and nutrient losses. The selection of the models to be used depends on the purpose of the exercise, on the data availability and accessibility and on the scale of the assessment.

The typical feature of the impact assessment of remedial actions is to compare the effects of different management practices, such as different ploughing methods and manure applications, and effects of land use and crop rotation changes (Voigt and Semioshkina, 2000). Therefore, management-oriented models (instead of research-oriented models) or their extensions are most suited for this purpose.

The scale of the model (soil profile-field parcel-drainage basin) depends on the modelling scale. In the case of predicting the changes in a single watershed, a drainage basin scale model might be the best selection. For handling larger areas, or for making nation-wide assessments, the use of field scale models is often a more versatile solution. Often the selection of the model scale depends also on the availability and accessibility of the spatial data (Howard et al., 2005).

Six of the current EDSSs (namely, CESAR, TEMAS, RECLAIM, SAVE, RESTORE and ReSCA) have been designed to estimate the behaviour of radionuclides with an overall aim of identifying proper management options in rural environment (Semioshkina et al., 2004). Some of them are based on relatively simple models associated with the databases on site-specific information; some of them implement semi-mechanistic models for radionuclide transfer to plants and the rest use dynamic models that allow time-dependent estimates.

The radionuclide behaviour in the environment, with respect to activity concentration in food products, is considered by SAVE, RESTORE and TEMAS DSSs (Semioshkina et al., 2004).

The model implemented in SAVE (Absalom et al., 2001) is similar to that used in RESTORE, which is a semi-mechanistic model for the prediction of the migration of ^{137}Cs from soil to plants. The primary difference between these two models is that the Absalom model estimates the soil-plant transfer factor values considering chemical characteristics of the soil, such as pH, exchangeable K^+, clay and organic matter content, whereas the second one evaluates them by taking into account physical characteristics of the soil, such as soil texture and bulk density.

The STRATEGY DSS transfer model is a direct evolution of the SAVE transfer model (Howard et al., 2005). It also utilises the Absalom model to estimate the activity concentration in food products. The animal-plant pathway has been uniformly modelled adopting the T_{ag} approach and none of the models considers the physiological aspect of radionuclide uptake.

The ReSCA utilises a simple approach based on both site-specific and generic data. The model parameters comprise consumption habits, dose conversion coefficients, effectiveness and costs of the remediation actions.

The application of the tool requires settlement-specific input information such as contamination levels of areas used in farming, data on the consumption habits and ^{137}Cs concentrations of agricultural and forest products. A special interface between national databases on contamination of soil and foodstuffs in affected regions was established to provide its implementation in Belarus, Russia and Ukraine. At the same time, a set of generic T_{ag} values were given for the four major soil groups: sand, loam, clay and peat (including wet peat). These generic T_{ag} values are not used in the software tool; a user can apply these values to calculate ^{137}Cs concentrations in foodstuffs if there are no measured data to input into the software tool (Ulanovsky et al., 2009).

Some of the above models are directly linked with the module providing evaluation of alternatives and provide information for making decisions in a user-friendly way.

Five models which can be used for estimating radionuclide transfer in forest ecosystems, namely RIFE1, LANDSCAPE, FORM, SAVE and FORIA, have been developed to facilitate selection of the optimal management options in contaminated forests (Figure 18).

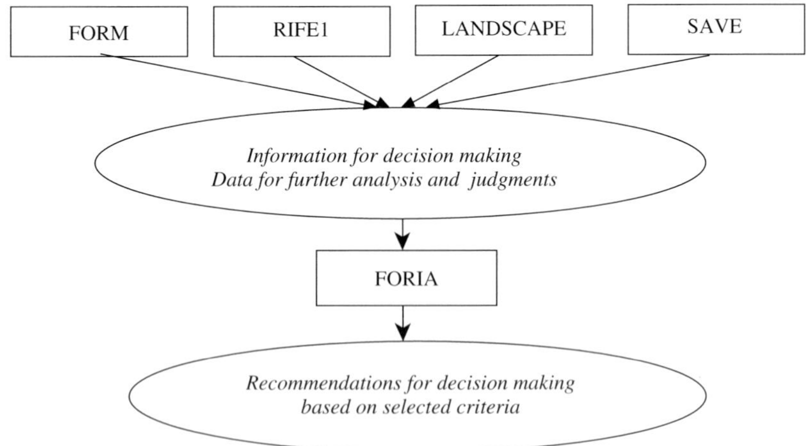

Figure 18 Conceptual schema of interactions between forest EDSSs for identification of optimum management roots in contaminated forests.

The RIFE1 and FORM (which is the IAEA forest model (Frissel et al., 1996; Semioshkina et al., 2004)) are dynamic compartment models, and they are characterised by similar model design but with minor differences. Thus, the RIFE1 model estimates the tree contamination, and therefore the timber contamination, as a function of the tree biomass, using a logistic equation to simulate the plant growth, whereas FORM does not present such a feature.

LANDSCAPE is a more complex model, comprising six sub-models, giving information on radionuclide activity concentrations in basic forest products and on internal and external doses associated with the forest contamination (Moberg et al., 1999; Riesen et al., 1999). SAVE adopts a T_{ag} model to estimate the transfer of ^{137}Cs from soil to plants. Two main approaches have been applied for the long-term changes of ^{137}Cs activity concentrations in wild animals: the T_{ag} approach with associated ecological half-lives was implemented in the SAVE, RIFE1 and FORM EDSSs (Frissel et al., 1996; Semioshkina et al., 2004), and dynamic compartment models were used in the LANDSCAPE EDSSs (Moberg et al., 1999).

Independent of the models used, all the above EDSSs except FORIA did not provide direct information on possible decisions (alternatives) for contaminated forest management. Therefore, it is assumed that the aim of these system applications is to deliver information to the FORIA EDSS which is intended for analysis of the different alternatives for potentially useful management actions (Figure 17). And conversely, based on the objectives of the study and the constraints associated with the models, a companion EDSS is to be selected to provide information for further processing by the FORIA system.

4.3. Evaluation of remediation strategies

With regards to remediation, three main approaches have been developed and implemented:

- The CESER-TEMAS approach which considers the remedial action side effects and is intended to identify the most suitable remedial actions for the given scenario.
- The SAVE-RESTORE approach which estimates the effectiveness of the selected remedial actions from the dose reduction point-of-view and evaluates a suitable strategy for remedial actions for a given scenario.
- The ReSCA approach which is based on the assessment of the environmental, social and economic side effects of each remedial action and on direct involvement of the stakeholders in the remediation planning at different levels.

It should be mentioned that only SAVE, TEMAS and FORIA directly estimate the effects of remedial action implementation, the rest of the EDSSs just provide supportive information for decision making (Semioshkina et al., 2004). Unlike SAVE (which estimates the effect of selected remedial actions in respect of the dose reduction), TEMAS and FORIA also consider the remedial action side effects. FORIA was not intended to estimate the forest contamination, but its peculiarity is in the provision of not only the possible remedial actions which can be applied for a given scenario, but also the side effects for each such action (Semioshkina et al., 2004).

Significant economic and social disruptions arise after radioactive contamination as a result of the contamination and measures carried out to minimise radiation doses to the public. Implementation of remedial actions is needed to reduce exposure of population to contamination, and simultaneously economic and social impacts should be minimised. It is clearly desirable that remedial actions are implemented in the most appropriate way, targeting expensive resources to areas and/or food products for which they are most required. As a consequence of nuclear accidents, in particular the Chernobyl accident, many studies have been performed to develop appropriate restoration strategies for contaminated areas. The majority of models developed and applied in Europe and in the NIS describing pathways of radionuclides through the human food chains have been based on information relevant to intensive agricultural systems (Semioshkina et al., 2004). An understanding of the processes controlling transfer of radionuclides to the population is required to provide a sound basis for restoration strategies.

The following examples provide details of typical approaches implemented by the most widespread EDSSs, illustrating the processes considered and the modern practice of remediation planning.

4.3.1. CESER EDSS

The CESER methodology has been developed to assess the potential side effects of long-term remedial actions in agricultural systems following a radionuclide deposition event (Salt et al., 1999).

The remedial action selection process aims to identify suitable measures for a given contamination scenario which will ensure that food products do not exceed the CEC intervention limits for radiocaesium and radiostrontium. A large number of possible remedial actions which could be potentially applied are not realistic for use. During the development of the DSS, the first step was to eliminate remedial actions that are unlikely to be used in practice. This was achieved through a screening of the

Table 9 Radionuclide deposition scenario (kBq m^{-2}) (Salt et al., 1999).

	^{137}Cs	^{90}Sr	Pu	Example of contamination scenario
Scenario 1	100	2	0.02	Remote area of the Chernobyl-like source term
Scenario 2	100	100	0.02	Remote area of source term with higher Sr fraction
Scenario 3	1000	200	0.2	Area close to site of accident
Scenario 4	5000	500	1	Area very close to site of accident

literature on a wide range of remedial actions using the following general criteria (Salt et al., 1999):

- radiological effectiveness (relative reduction in human dose or soil-plant-animal transfer);
- direct monetary costs (e.g. labour, materials);
- practicability (ease of execution);
- acceptability (e.g. animal welfare, toxicity).

The groups of remedial actions reviewed were soil-plant-based measures (application of fertilisers and chemical binders, mechanical/physical treatment, crop and land use change), animal-based measures (chemical treatment, feeding regime, animal management) and change of land use.

The EDSS was focused on environmental contamination being caused by atmospheric release and subsequent deposition of radionuclides. Of the various radionuclides which may be emitted in the course of a nuclear accident, only few show high transfer rates to human via food chains and pose a long-term radiation problem due to long half-lives. Therefore, the radionuclide deposition scenarios (Table 9) developed within the project focus on ^{137}Cs and ^{90}Sr (Salt et al., 1999). They reflect different source-terms and variable distances from the nuclide release.

Following on from the initial screening, the choice of remedial actions for inclusion in the DSS was further restricted. Remedial actions have to be tailored as closely as possible to fit in with the local farming conditions (in this case in Scotland) to ensure maximum effectiveness. Detailed knowledge of the agricultural management and environmental conditions along with knowledge of the magnitude and composition of the fallout is required to predict which food products are most likely to exceed intervention limits, and thus identify which production systems most urgently require application of remedial actions.

The selection of remedial actions for different types of Scottish agricultural production systems in the DSS was based on the following steps (Salt et al., 1999):

Step 1	For each deposition scenario, contamination levels in food products were predicted using 95% confidence intervals of transfer factors from IAEA. Where Community Food Intervention Levels (CFILs) are likely to be exceeded, remedial actions are necessary. The calculations agreed well with the post-Chernobyl experience regarding which production systems were affected.
Step 2	Only those remedial actions were selected that were feasible under the prevailing local farming conditions. This is particularly important for animal production systems where the feeding regime and farm management (e.g. stock movement; time spent indoors and outdoors) may determine whether a remedial action is feasible.
Step 3	Some remedial actions were found to be too expensive or drastic under certain deposition scenarios and are therefore not always recommended (e.g. cease agricultural production).
Step 4	The additional dose to farmers executing the remedial actions was calculated for each deposition scenario. Calculations were based on average times required to execute remedial actions. If these seemed to be unreliable, they were modified by location-specific information. Dose conversion factors are taken from BMU.[a]

[a]Dosiskoeffizienten zur Berechnung der Strahlenexposition, Bundesanzeiger Nr.160a und b (Beilage) von 28.8.2001.

In the deposition Scenario 4, the external dose to the population will exceed 1 mSv year^{-1} and the only option in this situation is evacuation of the population and termination of agriculture. The area may be left fallow or be converted to forestry. The same approach was used for both crop and animal management systems (Salt et al., 1999).

The prioritised impacts were quantified using a combination of mathematical modelling, experimentation, expert judgement and contingent valuation (Table 10).

CESER EDSS utilised two models which were not specifically developed for the conditions and purposes required to simulate remedial action impacts. However, a lot of efforts were made to adapt these models to the unique conditions of the study sites and remedial action scenarios (Salt et al., 1999). As a result of the prioritisation of impacts, it was determined that in order to quantify key non-radiological effects mechanistic models must be considered, including

- a soil hydrology model to estimate erosion and nutrient losses (ICECREAM),

Table 10 The combination of mathematical modelling, experimentation, expert judgement and contingent valuation.

Modelling	Soil erosion and sedimentation Soil nutrient transport to water Soil pollutant transport to water
Calculations	Ammonia emissions Product quantity
Experimentation	Soil nutrient transport to water Soil pollutant transport to water
Expert judgement	Soil organic matter Animal welfare Product quality Product quantity Biodiversity Landscape quality
Contingent valuation	Landscape quality

- plant growth which influences soil hydrology and a nutrient cycling model (ICECREAM),
- a soil chemistry model to evaluate the effects of fertilising and liming (PHREEQC),
- agricultural management operations (based on Multiple Criteria Decision Making (MCDM) techniques).

ICECREAM is a versatile catchment management model, which includes sub-models to simulate soil hydrology, soil erosion, surface loss and subsurface transport of nutrients and trace substances, plant growth and the impact of agricultural management operations (e.g. ploughing, application of manure or pesticides) (Salt et al., 1999). ICECREAM is a Finnish adaptation of the CREAMS/GLEAMS model (Salt et al., 1999). The basis for its selection was mainly its adaptations made to the conditions where the model is used in this context, and its user interface that allows a series of model runs over wide climate-soil-crop-management combinations. ICECREAM was used to model soil erosion and losses of phosphorus and nitrogen in response to ploughing remedial actions, changes in animal feeding regimes, changes in the number of livestock and cessation of crop/animal production. Simulations performed for case study areas in Scotland were used as a basis for selecting appropriate impact scores for the CESER DSS (Salt et al., 1999).

The PHREEQC model simulates the equilibrium chemical composition of multi-species systems taking into account a variety of chemical reactions. PHREEQC was used to model the influence of potassium fertilisation and liming on soil chemistry, including pH and concentrations of major and trace nutrients and of toxic substances in soil solution and consequently their plant availability. Simulations performed over a wide range of soil properties for Scottish soils were used to set thresholds for soil suitability for liming and K application in the CESER DSS. The results illustrate that in soils where these remedial actions are effective against radiostrontium and radiocaesium, they may promote leaching of nutrients (e.g. magnesium) and trace pollutants (e.g. cadmium). The impact scores for the criteria 'nutrient transport to water' and 'pollutant transport to water' were adjusted accordingly in the DSS (Salt et al., 1999).

At the basis of all MCDM techniques is the evaluation of a two-dimensional matrix in which one dimension is made up of alternatives and the other consists of criteria. In the context of the CESER project, the alternatives are the different possible remedial actions from which the decision maker must select. Criteria are the means by which the remedial action alternatives are assessed. In the CESER DSS, the criteria consist of a mixture of environmental and agricultural considerations. A similar approach has been used in the RESTRAT project to evaluate restoration options for small but highly contaminated areas, such as waste disposal facilities (Jackson et al., 1999).

A range of specific MCDM algorithms was tested for incorporation into the CESER DSS. Ideal point analysis was selected because it proved to be easy and intuitive for the user to apply without being overly simplistic in its analysis. This method not only allows the user to weight criteria based on their own agenda, but it also allows the user to specify the 'ideal' score (also referred to as the criteria objective) and the compensatory level. If a method is said to be compensatory, it signifies that a poor performance by a particular alternative on one or more criteria can be 'compensated for' by a good performance on other criteria. The final outcome on using such an approach is largely dependant on the structure of the weighting and preferences that are imposed on the system by the decision maker. Non-compensatory methods, on the other hand, involve a criterion-by-criterion evaluation of the alternatives in which the strengths and deficiencies are taken at face value and evaluated as such. Therefore, if an alternative does not achieve good results on a particularly important criterion, that alternative would be excluded from further consideration (Salt et al., 1999). This is despite the fact that it might perform extremely well on the subsequent criteria. The compensatory level can be adjusted by manipulating the p parameter. In the CESER DSS, it can be set to equal any number ranging from 1, which causes the assessment to be fully compensatory, to 10, which makes it non-compensatory. Ideal point analysis (also called goal programming) measures the deviation between the scores for each set of 'alternative' solutions and the 'ideal' set of solutions.

The alternative, which minimises the distance between itself and the ideal case, is deemed the optimal solution. In the equation used to calculate the distance, d, between the actual and the ideal set of solutions, the variable h_j represents the standardised set of ideal point values and q_{ji} is the standardised set of alternative scores.

The variable p in this equation symbolises the metric parameter. It is used to set the compensatory level of the assessment, with 1 being compensatory and numbers greater than 1 increasingly less compensatory. The weighting function varies between 0 and 10. The weights follow the 'rating system' in which the number of points allocated to each objective is representative of that objective's relative importance in the decision-making process.

$$\min d = \{^{j}\gamma_j^p(|h_j - q_{ji}|)^p\}^{1/p}, \quad j = 1$$

$$q_{ji} = \frac{\max \rho_{ji} - \rho_{ji}}{\max \rho_{ji} - \min \rho_{ji}}$$

where d is the distance score to be minimised, h_j the standardised ideal point value for criteria j, q_{ji} the standardised value, ρ_{ji} the weight for criteria j and p the metric parameter (usually 1, 2 or ∞).

The ability to make 'trade-offs' in criteria performance, within the bounds of certain thresholds, is viewed as a key component of the CESER assessment methodology. It accurately simulates the real-world decision-making environment in which losses in one arena can be justified by the gains made in another (Salt et al., 1994, 1999; Salt and Culligan Dunsmore, 2000; Salt and Rafferty, 2001).

4.3.2. RECLAIM EDSS

The overall aim of RECLAIM EDSS development was to describe time-dependent and spatial factors determining external and internal radiation doses, so that optimised generic remediation strategies for reducing dose and reclamation of abandoned land can be developed. An emphasis was placed on comparing and contrasting areas in Russia (Mayak and Chernobyl), Ukraine (Chernobyl) and Kazakhstan (weapons test site), which have received radioactive contamination with different fallout characteristics and where the population is still subjected to increased internal and external radiation exposure (Strand et al., 1999; Voigt and Semioshkina, 2000).

An emphasis in the approach to determine dose in the framework of the RECLAIM EDSS from these different contamination sources was to estimate the comparative significance of intensive agricultural production, private agricultural production and wild foodstuffs from forests for different population groups such as urban and rural dwellers (Strand et al., 1999; Voigt and Semioshkina, 2000). Work has been focused on studies of key parameters determining transfer of radiocaesium to human, and on development of methods to estimate fluxes of radiocaesium (or radiostrontium in the case of Kyshtym area) via different food production systems.

Levels, trends and distribution of external doses in different population groups and at different sites have been assessed, and the importance of external and internal dose contributions to the total dose compared. Consideration and implementation of remedial actions to reduce both external and internal dose need to take into account the availability of resources and materials to implement remedial strategies, the efficiency of remedial actions and also the acceptability to the local population. Important here was to provide the affected people with appropriate information about restoration strategies, for the possibility of self-help (Strand et al., 1999).

A major focus of the RECLAIM EDSS development has been to provide a basis for self-help advice to reduce dietary radiocaesium intake via the consumption of forest berries, fungi and privately produced milk (Voigt and Semioshkina, 2000). Providing people with the ability to manage their own radiological situation will be of increasing importance as government spending on remedial actions declines and there is a move from collective systems to private farming enterprises (collective farms within the Ukraine were abolished as of March 2000). If successful, the provision of advice on how individuals can restrict their own ^{137}Cs intake represents a cost-effective alternative by which their radiation dose can be considerably reduced. Indeed, for private milk production, if following the advice presented led to ^{137}Cs activity concentrations in milk falling to below national intervention limits, a portion of the milk could be sold for financial gain. In the case of forest berries and fungi, the reduction in dose would be achieved with a negligible effect on the nutritional content of the diet. There is also the considerable psychological benefit associated with adopting such approaches as they will enable local populations to understand more clearly their own situation and to influence it.

It is required to consider the likely contamination levels of foodstuffs produced in areas excluded from productive use for at least two reasons:

(i) people have illegally returned to them and are producing foods for self-consumption;
(ii) people are exploiting the potential economic productivity of these areas.

Both internal and external doses within abandoned areas have been estimated using models developed during the course of collaborating projects. The external dose rate (predicted by spatially applying the model developed within RESTORE) in much of the abandoned areas within Belarus is estimated to be sufficiently high to exclude rehabilitation for some time. Effective remedial actions to reduce external exposure have been reviewed and those which could be employed by people living within abandoned areas, such as ploughing or digging and turning over of soil in gardens and orchards, have been identified (Voigt and Semioshkina, 2000).

4.3.3. TEMAS EDSS

The assessment of applicable remedial actions is an important step in analysing intervention strategies as a part of restoration management. For agricultural systems, the potential remedial actions considered were of two types: (1) those modifying the availability of radionuclides in the root zone (mechanical remedial actions) and (2) those causing modification of the soil-to-crop transfer (agrochemical remedial actions). In particular, remedial actions such as ploughing (deep and normal/shallow), crop removal, turf harvesting and the application of potassium and lime or other calcium-containing treatments were considered (Gutierrez and Vazquez, 2000; Semioshkina et al., 2004).

Different situations can be envisaged conditioning the applicability of remedial actions. The first relates to soils disturbed during the period between the deposition and the intervention. In these cases, only remedial actions causing modification of the transfer factors, known as agrochemical remedial actions, are expected to be feasible and, therefore, are considered in TEMAS (Gutierrez and Vazquez, 2000). In the contrary case (soils not disturbed during the period between deposition and intervention), one can distinguish whether there are crops directly contaminated by the deposition or not. If yes, mechanical remedial action (crop removal, ploughing) would be applicable: crop removal will be applicable only if the existing biomass is enough to allow such action from the technological point of view. The removed material will be disposed of as waste according to a corresponding analysis. Since a fraction of soil is also removed in the operation, crop removal involves a reduction of the soil contamination. In case there are no crops directly contaminated by the deposition, or not enough amount of biomass exists to allow the removal, ploughing as well as agrochemical remedial actions are applicable.

Special consideration is required for using crop removal as an efficient remedial action for directly contaminated crops. If the activity concentrations in the mature crop are in excess of the relevant predetermined intervention level, a decision on removing the standing crops needs to be made quickly in order to minimise the wastes arising. To assist in this decision, a robust methodology has been developed to predict the activity concentrations in immature crops following wet and dry deposition. If removal is decided upon, the removed crops must be treated as wastes. In order to allow for the selection of the most adequate disposal route, it is required to estimate the volume of generated wastes and their activity concentration. The biomass to be removed is estimated by applying an adequate crop growth model, taking into account their real sowing date and efficiency factors dependant on the machinery used (Gutierrez and Vazquez, 2000; Semioshkina et al., 2004).

In general, the effectiveness of remedial actions can be expressed in terms of reductions either in external dose rates or in radionuclide transfer to crops. According to the methodology used for the estimation of transfer factors, on poor soils, recommended values should be applicable after fertilisation in order to estimate effectiveness (Gutierrez and Vazquez, 2000).

Potassium addition reduces uptake of radiocaesium in low-fertility soils, while lime has a similar effect on uptake of radiostrontium. The model describing these effects supposes an inverse linear relationship between the exchangeable K and Ca content in the soil and the soil-to-plant transfer factor (TF) for caesium and strontium, respectively. This model calculates the TF's for the real conditions of K and Ca in the soil under intervention. If the exchangeable K or Ca in the soil is lower than the optimum, fertilisation can reduce the TF up to the recommended TF for the corresponding type of soil. The first year fertilisation is addressed to make the level of K or Ca in the soil adequate and the following fertilisation must compensate for the losses of K or Ca.

The practicability of the remedial actions was assessed in terms of their technical feasibility, cost, effectiveness and secondary effects. A remedial action is technically feasible if the required equipment, techniques and resources are readily available within the affected area. Physical factors affecting feasibility also need to be identified. For example, a shallow soil would not be amenable to deep ploughing; a set of physical constraints has been proposed. Costs depend on the resources required and the associated unit costs; for example, labour has been expressed in terms of man-hours. Secondly, effects can be very difficult to quantify, but may be expressed in terms of changing yield or quality.

The IAEA model (IAEA, 2001) has a module to evaluate the most appropriate actions to be applied in forests. Nevertheless, the constraints due to the long cycling time and the barrier formed by the trees for mechanical actions reduce the feasible intervention options to those related to industries and to restrictions of use by the public. The model concludes that remedial actions in the forest are often not feasible if the cost per saved man-Sv exceeds 17,000 Euros. An exhaustive study of the potential remedial actions in forests was made and Table 11 shows different remedial actions considered by the system, their benefits and costs.

4.3.4. SAVE/SAVE-EC EDSS

Only those remedial actions and management strategies which fit readily within the semi-mechanistic framework of the underlying models have

Table 11 Comparison of the benefits and costs of the remedial action used.

Remedial action	*Caveats*	Benefit	Cost
1. Normal operation	–	No loss of production	No dose reduction
2. Limit access for the public	–	Lower doses from ingestion and external radiation	No cost
3. Early cutting	Access to machinery and work force. Seasonal dependant (seasonally dependant?)	Less contaminated wood	Loss in revenue
4. Delay cutting	Limited time	Radionuclide decay. Better planning possible	Delay in revenue
5. Removal of Cs at industrial process	Available techniques	Dose reduction	Cost for the industry
6. Limit occupational time at ash deposit	Available techniques	Dose reduction	Machinery costs
7. Blending of raw material	Selection of sites	Dose to workers below maximum permissible level	Transport costs
8. Optimise use of raw material	Must be combined with 5	Lower dose	Loss in revenue

been included in the software (Absalom et al., 2001). The remedial actions considered by the EDSS are grouped into four categories:

- administration of ferrocyn to animals,
- soil management,
- food chain management and
- animal feeding.

The animal-based remedial actions are limited to the daily intake (mg day^{-1}) of ferrocyn or NH_4-iron-hexacyanoferrates (assumed to be derived from either a rumen boli, salt-lick or within concentrate). A dose reduction factor response curve for goat milk has been implemented, with a 10-fold reduction and increase in ferrocyn effectiveness assumed for dairy/beef cow and chickens, respectively. Pigs and sheep are assumed to behave in the same way as goats. The user defines the period over which the ferrocyn is administered and can vary the administration rate by animal type (Howard and Wright, 1999).

Soil remedial actions include the effect of redistribution of radiocaesium throughout the plough layer during normal ploughing (for all arable crops), the reduction in radiocaesium uptake as a consequence of increased K^+ concentration following application of K_2O and change in soil pH associated with application of lime-based fertilizers. For the case of K-based fertilizers, the rate, fraction of K^+ and date of annual application are defined by the user on a regional basis. It is assumed that the soil potassium status returns to the initial (data base) value after 12 months at a constant rate to approximately simulate the effect of plant uptake. The effect of liming is considered by allowing the user to define a 'target' pH (a function of the soil type and quantity of lime applied) and the annual date of application. The pH is then assumed to return to the original (database) level over the course of a year at a constant rate (Howard and Wright, 1999; Voigt and Semioshkina, 2000).

Animal feeding and food chain management can also be manipulated in accordance with current regulations from the Council of the European Communities for radionuclide activity concentration intervention levels in human foodstuffs (CFILs) and maximum permitted levels (MPLs) of radiocaesium in animal feedstuffs. Two categories of CFILs can be defined by the user: normal and emergency. The former are the levels set in the long term after a nuclear incident, whereas the later override the former for a short-term period set by the user. If the CFILs are activated, then any product with activity levels greater than the CFIL are automatically excluded from the calculation of human intake. The CFILs fall into three broad categories: dairy/liquid (dairy and liquid foods), agricultural (meat and arable) and semi-natural (minor foodstuffs consumed at generally low quantities). Animal intakes are similarly 'corrected' using appropriate MPLs, which fall into three categories of animals: dairy/beef, poultry (sheep, goats and chickens) and other (pigs). The default values for CFILs and MPLs are as given by Nisbet et al. (1998). For the case of maize and grass silage, there is also the facility to destroy the first, second or third grass cuts and first maize harvest (Nisbet et al., 1998).

4.3.5. ReSCA EDSS

Here remediation of the contaminated territories is considered to be done through implementing certain remedial actions, that is agrotechnical, social, administrative and other actions, which result in reduction of population exposure in the given settlement. In total, seven different remedial actions have been found to be relevant and considered (Fesenko, 2007; Ulanovsky et al., 2009):

1. radical improvement of grassland (RI), for areas with wet peat; this action also includes drainage (RI+D);
2. surface improvement of grassland (SI);
3. ferrocyn application to cows (FA);
4. clean feed for pigs (FP);
5. mineral fertilizers for potato fields (MF);
6. information campaign on mushroom consumption (IM);
7. removal of contaminated soil (RS).

The effect of a remedial action *r* is expressed by the so-called reduction factor for a pathway *f*, that is a factor being numerically equal to the ratio of the annual dose after application of the remedial action, D^f_r, and before the remedial action, D^f_0. Remedial actions result in reduction of the population dose, which is expressed by the averted collective dose. The latter represents the dose averted due to application of a remedial action in the given settlement. The attitude of the population to remedial actions can vary considerably – from complete repulsion to enthusiasm. Rehabilitation practice has shown (see e.g. Jacob et al., 2001) that remedial actions aimed at improving quality of pastures are generally very well accepted, whereas a decontamination of the settlement accompanied by upper soil removal is generally strongly disliked. Situations can easily be foreseen when the attitude of the population to certain remedial actions has to be accounted for in planning a strategy of remediation.

A sequence of remedial actions undertaken in the settlements of a region or a country is called remediation strategy. The total cost of the strategy is the sum of the costs of individual remedial actions. The total effect of the remediation strategy is a result of all the remedial actions applied. Theoretically, the cost per averted dose criterion would result in the creation of the most cost-effective strategy. Nevertheless, it can be favourable to take into account also the degree of acceptability of remedial actions in the population. In the present work, the criterion used for the selection of remedial actions to be included in the remediation strategy combines cost-per-averted-dose, that is a ratio of cost and averted dose for single remedial actions, and the degree of public acceptance of a remedial action. All remedial actions are arranged according to this combined criterion and sequentially included in the remediation strategy under construction.

5. Conclusion

In recent years, various practical remedial actions have been developed for use in the event of radioactive contamination of the environment. In particular, following the major radiation accidents in Kyshtym and Chernobyl, various measures have been implemented and a vast amount of data on their effectiveness have been generated together with information on ancillary factors such as the required resources and costs. These measures vary considerably in effectiveness, cost and practicability in actual situations. The efficiency of application thereof depends on many factors, including soil and climatic conditions and specific features of the agricultural production management. Because of this, implementation of action on the basis of general expert estimations often results in inadequate decisions. The above facts show a need for the development of EDSS and multi-criteria analysis systems capable of providing advice on remediation strategies at a different level of decision making. Such systems would be of importance in the case of an emergency, and their further development is necessary to expand existing capabilities to provide decision support on optimised management practices in areas contaminated by radioactive materials.

REFERENCES

Absalom, J. P., S. D. Young, N. M. J. Crout, A. Sanchez, S. M. Wright, E. Smolders, A. F. Nisbet, and A. G. Gillett. (2001). Predicting the transfer of radiocaesium from organic soils to plants using soil characteristics. *Journal of Environmental Radioactivity*, 52, 31–43.

Andersson, K. G., J. Brown, K. Mortimer, J. A. Jones, T. Charnock, S. Thykier-Niellsen, J. C. Kaiser, G. Proehl, and S. P. Nielsen. (2008). New developments to support decision-making in contaminated inhabited areas following incidents involving a release of radioactivity to the environment. *Journal of Environmental Radioactivity*, 99, 439–454.

Belli, M. (Ed.). (2000). *FORECO. Forest Ecosystems: Classification of Restoration Options, Considering Dose Reduction, Long-Term Ecological Quality and Economic Factors*, Final report. Agenzia Nazionale per la Protezione dell' Ambiente (ANPA), Rome.

Eltarenko, E. A. (1995). *Assessment and Decision Making on Many Criteria.* Moscow Engineering and Physical Institute, Moscow.

Fesenko, S. V., R. M. Alexakhin, M. I. Balonov, I. M. Bogdevich, B. J. Howard, V. A. Kashparov, N. I. Sanzharova, A. V. Panov, G. Voigt, and Y. M. Zhuchenko. (2007). An extended review of twenty years of countermeasures used in agriculture after the Chernobyl accident. *Science of the Total Environment*, 383, 1–24.

Fesenko, S. V., N. I. Sanzharova, B. T. Wilkins, and A. F. Nisbet. (1996). FORCON: Local decision support system for the provision of advice in agriculture – Methodology and experience of practical implementation. *Radiation and Protection Dosimetry*, 64, 157–164.

Fesenko, S. V., S. Spiridonov, and R. Avila. (1999). Modelling of ^{137}Cs behaviour in forest game food chains. In: *Contaminated Forests. Recent Developments. Risk Identification and Future Perspectives. NATO Science Series. 2. Environmental Security, 58* (Eds I. Linkov and W. Schell). Kluwer Academic Publishers, Amsterdam, pp. 239–247.

Fesenko, S. V., G. Voigt, S. I. Spiridonov, and I. A. Gontarenko. (2005). Decision making framework for application of forest countermeasures in the long term after the Chernobyl accident. *Journal of Environmental Radioactivity*, 82, 143–166.

Fesenko, S. V., G. Voigt, S. I. Spiridonov, N. I. Sanzharova, I. A. Gontarenko, M. Belli, and U. Sansone. (2000). Analysis of the contribution of forest pathways to the radiation exposure of different population groups in Bryansk region of Russia. *Radiation and Environmental Biophysics*, 39, 291–300.

French, S. (1986). *Decision Theory: An Introduction to the Mathematics of Rationality.* Ellis Horwood, Chichester.

French, S. (1992). *The Decision Conference in the USSR.* In International Chernobyl Project. Input from the CEC to the Evaluation of the Relocation Policy Adopted by the Former Soviet Union. CEC report EUR 14543 EN, Brussels.

Frissel, M. J., G. Shaw, C. Robinson, E. Holm, and M. Crick. (1996). Model for the evaluation of long term countermeasures in forests. In: *Radioecology and the Restoration of Radioactive Contaminated Sites, IUR/NATO Handbook. NATO Advanced Science Institute Series* (Eds F. E. Luykx and M. J. Frissel). Kluwer Academic Publishers, Amsterdam, ISBN 0-7923-4136-8.

Gillett, A. G., N. M. J. Crout, J. P. Absalom, S. M. Wright, S. D. Young, B. J. Howard, C. L. Barnett, S. P. McGrath, N. A. Beresford, and G. Voigt. (2001). Temporal and spatial prediction of radiocaesium transfer to food products. *Radiation and Environmental Biophysics*, 40, 227–235.

Gustafsson, J. (1999). *PRIME: An Introduction and Assessment*. Helsinki University of Technology, Helsinki.

Gutierrez, J., and C. Vazquez. (Eds.). (2000). *Techniques and Management Strategies for Environmental Restoration and Their Ecological Consequences*. Final Report of TEMAS Project. Editorial CIEMAT, ISBN 84-7834-378-4.

Howard, B. J., N. A. Beresford, A. Nisbet, G. Cox, D. H. Oughton, J. Hunt, B. Alvarez, K. G. Andersson, A. Liland, and G. Voigt. (2005). The STRATEGY project: Decision tools to aid sustainable restoration and long-term management of contaminated agricultural ecosystems. *Journal of Environmental Radioactivity*, 83, 275–295.

Howard, B. J., and S. M. Wright. (Eds). (1999). *Spatial Analysis of Vulnerable Ecosystems in Europe: Spatial and Dynamic Prediction of Radiocaesium Fluxes into European Foods (SAVE)*. Final Report. Commission of European Communities.

IAEA. (1996). *The IAEA Model for Aiding Decisions on Contaminated Forests and Forestry Products: Application to Intervention/Cleanup Criteria*. Working Material, International Atomic Energy Agency, Vienna.

IAEA. (2001). *Modelling the Migration and Accumulation of Radionuclides in Forest Ecosystem*. IAEA-TECDOC-XXX (BIOMASS/T3FTM/WD02). IAEA, Vienna.

IAEA. (2002). *Non-Technical Factors Impacting on the Decision-Making Processes in Environmental Remediation*. IAEA-TECDOC-1279. IAEA, Vienna.

IAEA. (2006). *Environmental Consequences of the Chernobyl Accident and Their Remediation: Twenty Years of Experience*. Report of the Chernobyl Forum Expert Group 'Environment'. IAEA, Vienna.

ICRP. (1973). *Implication of Commission Recommendations that Doses be Kept as Low as Readily Achievable*. ICRP Publication 22. Pergamon Press, Oxford, 18pp.

ICRP. (1983). *Cost-Benefit Analysis in the Optimisation of Radiation Protection*. ICRP Publication 37. *Annals of the ICRP*, 10 (2/3). Pergamon Press, Oxford, 75pp.

ICRP. (1989). *Optimisation and Decision-Making in Radiological Protection*. ICRP Publication 55. *Annals of the ICRP*, 20 (1). Pergamon Press, Oxford, 60pp.

ICRP. (2007). *The 2007 Recommendations of the International Commission on Radiological Protection*. ICRP Publication 103. *Annals of the ICRP*, 37 (2–4). Elsevier, Amsterdam.

Jackson, D., S. K. Wragg, A. Bousher, T. Zeevaert, Y. Stiglund, V. Brendler, J. P. Hedemann, and S. Nordlinder. (1999). Establishing a method for assessing and ranking restoration strategies for radioactively contaminated sites and their close surroundings. *Nuclear Energy*, 38, 223–231.

Jacob, P., S. Fesenko, S. K. Firsakova, I. A. Likhtarev, C. Schotola, R. M. Alexakhin, Y. M. Zhuchenko, L. Kovgan, N. I. Sanzharova, and V. Ageyets. (2001). Remediation strategies for rural territories contaminated by the Chernobyl accident. *Journal of Environmental Radioactivity*, 56, 51–76.

Lochard, J., and T. Schneider. (2001). Application of decision aiding technologies for relocation strategies. In: *Radioecology: Radioactivity and Ecosystems* (Eds. E. Van der Stricht, R. Kirchman). Fortemps, Liege, pp. 55–78.

Moberg, L., L. Hubbard, R. Avila, L. Wallberg, E. Feoli, M. Scimone, C. Milesi, B. Mayes, G. Iason, A. Rantavaara, V. Vetikko, R. Bergman, T. Nylern, T. Palo, N. White, H. Raitio, L. Aro, S. Kaunisto, and O. Guillitte. (1999). *An Integrated Approach to Radionuclide Flow in Semi-Natural Ecosystems Underlying Exposure Pathways to Man*. Final Report of the LANDSCAPE Project. SSI Report 99:19, Stockholm, Sweden.

Monte, L. (2001). A generic model for assessing the effects of countermeasures to reduce the radionuclide contamination levels in abiotic components of fresh water systems and complex catchments. *Environmental Modelling & Software*, 16, 669–690.

Monte, L., J. Van deer Steen, U. Bergström, E. Gallego Díaz, L. Håkanson, and J. Brittain. (Eds.). (2000). *The Project MOIRA: A Model-Based Computerised System for Management Support to Identify Optimal Remedial Strategies for Restoring Radionuclide Contaminated Aquatic Ecosystems and Drainage Areas*. Final Report. ENEA RT/AMB/2000/13.

Nisbet, A. F., R. F. M. Woodman, J. Brown, J. G. Smith, and B. T. Wilkins. (1998). *Derivation of Working Levels for Animal Feedstuffs for Use in the Event of a Future Nuclear Accident*. NRPB Report (NRPB-R299). National Radiological Protection Board, Chilton, Didcot.

Panov, A. V., S. V. Fesenko, and R. M. Alexakhin. (2006). Methodology for assessing the effectiveness of countermeasures in rural settlements in the long term after the Chernobyl accident on the multi-attribute analysis basis. *Second European IRPA Congress on Radiation Protection. Book of Abstracts*, IRPA, Paris, France, 173pp.

Recommendations. (1998). *Guide on Agriculture Management on Contaminated Territories*. Ukrainian Institute of Agricultural Radiology, Kiev, 102pp. (in Ukrainian).

Riesen, T., R. Avila, L. Moberg, and L. Hubbard. (1999). Review of forest models developed after the Chernobyl NPP accident. In: *Contaminated Forests. Recent Developments in Risk Identification and Future Perspectives. NATO Science Series II, Environmental Security* (Eds I. Linkov and W. Schell)Contaminated Forests. Recent Developments in Risk Identification and Future Perspectives. NATO Science Series II, Environmental SecurityKluwer Academic Publishers, Dordrecht, pp. 151–161.

RSS. (1999). *Radiation Safety Standards - 1999. Hygienic Standards*. Russian Federation Health Ministry, Moscow, (in Russian).

Salt, C. A., and M. Culligan Dunsmore. (2000). Development of a spatial decision support system for post-emergency management of radioactively contaminated land. *Journal of Environmental Management*, 58(3), 169–178.

Salt, C. A., H. S. Hansen, G. Kirchner, H. Lettner, S. Rekolainen, M. Culligan Dunsmore. (1999). *Impact Assessment Methodology for Side-Effects of Countermeasures Against Radionuclide Contamination in Food Products*. Research Report No. 1, Nord-Troendelag College, Steinkjer, Norway, ISBN 82-7456- 119-8.

Salt, C. A., R. W. Mayes, P. M. Colgrove, and C. S. Lamb. (1994). The effects of season and diet composition on the radiocaesium intake by sheep grazing on heather moorland. *The Journal of Applied Ecology*, 31(1), 125–136.

Salt, C. A., and B. Rafferty. (2001). Assessing potential secondary effects of countermeasures in agricultural systems: A review. *Journal of Environmental Radioactivity*, 56, 99–114.

Sazykina, T. G., and I. I. Kryshev. (2006). *Multi-Criteria Analysis for Evaluating the Radiological and Ecological Safety Measures in Radioactive Waste Management. Communications of Higher School*. Nuclear Power Engineering (Izvestia vissikh uchebnikh zavedenij. Yadernaya energetica), Obninsk, 1, pp. 39–46 (in Russian).

Schulte, E. H., G. N. Kelly, and C. A. Jackson (Eds.) (2002). *Decision Support for Emergency Management and Environmental Restoration*. European Communities, Luxembourg, ISBN 92-894-1744-7.

Semioshkina, N., G. Voigt, and D. Tarsitano. (2004). *Evaluation and Network of EC-Decision Support Systems in the field of Terrestrial Radioecological Research*. Final Report of the EVANET Project.

Shaw, G., C. Robinson, E. Holm, M. J. Frissel, and M. Crick. (2001). A cost-benefit analysis of long-term management options for forest following contamination with ^{137}Cs. *Journal of Environmental Radioactivity*, 56, 185–208.

Shershakov, V. M., and E. A. Trakhtengerts. (1997). Decision-making in an emergency by using computer analysis with dynamically changeable rules. *Radiation Protection Dosimetry*, 73, 141–142.

Strand, P., M. Balonov, I. Travnikova, K. Hove, L. Skuterud, B. Prister, and B. J. Howard. (1999). Fluxes of radiocaesium in selected rural study sites in Russia and Ukraine. *Science of the Total Environment*, 231, 159–171.

Tarsitano, D., N. Semioshkina, and G. Voigt. (2005). EVANET-TERRA – Evaluation and network of SC-decision support systems in the field of terrestrial radioecological research. *Radioprotection*, 40, 261–268.

Trakhtengerts, E. A., V. M. Shershakov, and D. A. Kamaev. (2004). *Computer-Based Support on Management of Elimination of Consequences of Radiation Impact*. SINTEG, Moscow.

Ulanovsky, A., P. Jacob, S. Fesenko, I. Bogdevitch, V. Kashparov, N. Sanzharova. (2009). ReSCA – Decision support tool for remediation planning in the long term after the Chernobyl accident. *Environmental Modelling & Software*, in press.

Van der Perk, M., P. A. Burrough, and G. Voigt. (1998). GIS based modelling to identify regions of Ukraine, Belarus, and Russia affected by residues of the Chernobyl nuclear power plant accident. *Journal of Hazardous Materials*, 61, 85–90.

Voigt, G., and N. Semioshkina (Eds). (2000). *Restoration Strategies for Radioactive Contaminated Ecosystems*. GSF Bericht. ISSN 0721-1694.

CHAPTER 4

Remediation of Areas Contaminated after Radiation Accidents

Rudolph M. Alexakhin*

Contents

*Corresponding author. Tel.: +7 4843964802; Fax: +7 4843968066
E-mail address: alexakhin@riar.obninsk.org; alexakhin@ya.ru

Russian Institute of Agricultural Radiology and Agroecology, Kievskoe shosse, 109 km, Obninsk, Kaluga region 249032, Russia

Radioactivity in the Environment, Volume 14
ISSN 1569-4860, DOI 10.1016/S1569-4860(08)00204-0

1. The Main Features of Remediation after Radiation Accidents and Incidents

1.1. Radiation accidents and incidents requiring remediation

Remediation of areas affected by radioactive contamination after radiation accidents and incidents is an interdependent combination of different recovery measures to mitigate the adverse effects of the release of radioactive materials into the human environment. In specific terms, remedial actions in areas affected by accidents and incidents can differ from similar operations in areas contaminated by ionising radiation sources of another origin associated with the nuclear industry and power engineering, the use of nuclear technologies in agriculture, medicine, industry, research activities, etc., which are described in this book. Amongst such particularities the following should be specifically mentioned.

First, in radiation accidents (and, in part, incidents) accompanied by radionuclide release into the environment the affected areas may be huge (as was the case with the Kyshtym accident at the Production Plant 'Mayak' in the South Urals, USSR, in 1957 in which the contaminated zone, called the East Urals Radioactive Trail (EURT), covered an area of 330 × 50 km). In some cases (the accident at the Chernobyl NPP (ChNPP) in 1986) radioactive contamination affected a number of countries (not only the former USSR-CIS but also some European states). Consequently, remediation of this contamination took place across nations due to transfer of radioactive materials and required international attention.

Second, these accidents and subsequent remediation actions may involve thousands and millions of people.

Third, severe radiation accidents with contamination of the environment (for instance, the Chernobyl accident in 1986) are reckoned amongst the largest anthropogenic catastrophes on the Earth.

Fourth, in some radiation accidents ionising radiation affects the health directly, and in this case densities of radioactive fallout reach the levels that cause radiation-induced death of, and injuries to, not only individual plant and animal species but also a population or a community of living organisms (including radiation-induced death of some ecosystems).

Fifth, radiation accidents and incidents, especially if accompanied by the release of radioactive substances into the environment, are always the focus of attention of the general public and the mass media. The latter has a two-way effect. On the one hand, an exaggerated concern of the society about the need for professional assessment and relevant measures to mitigate the influence on public health stimulates prompt allocation of financial resources for large-scale research and remedial programmes, and activation of national and international bodies responding to radiological emergencies. On the other hand, the huge interest of the public and biased media, especially in an anti-nuclear age, often provokes insufficiency and inconsistency of information on details of the accident or incident and its real danger, and can be detrimental to correct and timely decisions and can cause false (re)actions in the consequence management (particularly at the early stage of recovery).

And finally, in a societal perception, due to some subjective and objective reasons the discussion of which is beyond the scope of the present chapter (and maybe the whole book), the word 'radiation' hazard is categorically associated with the group of disastrous events in the anthropogenic activity of contemporary mankind. This may also be a cause of an unsound perception of realities.

1.2. Distinctive features of remediation in radiation accidents

Radiation accidents and incidents accompanied by radioactivity release into the environment are variable, thereby requiring distinct remediation actions. Moreover, every radiation accident or incident is specific and unique. Amongst the reasons for radiation accidents or incidents may be the human factor (sometimes simultaneous with structural defects and failures, as was the case with the RBMK reactor at the ChNPP), lack of process control, the worn-out state of equipment, violation of the technological regime (the 1957 Kyshtym accident), crash of the US Air Force aircraft carrying nuclear weapons in Spain, the reactor accident and fire in Sellafield (Windscale), UK and finally, illegal acts such as the stealing of a sealed ^{137}Cs source for medical purposes (Goiânia, Brazil) with its subsequent dismantling and thereby dispersion of the radioactive contents in the environment.

A particular case is the remediation of the Techa River, a cascade of water bodies and the adjacent territory contaminated by radioactive substances in the Russian Federation. Strictly speaking, this event may even not be named a radiation accident or incident (conventionally, it may be termed 'an authorised accident'). In the USSR in 1949–1953, the Production Plant 'Mayak' in the South Urals performed routine (partly regulated) discharges into the Techa River. The decision making on these discharges was dictated by political motives, namely the *speedy production of weapons-grade plutonium*. Certainly, to some extent at that time relatively poor knowledge of radiation medicine, biology and radioecology existed, as well as inadequate technologies of radioactive waste management.

Remediation is achievable through the implementation of re-cultivating and protective measures. The former means, for example, mechanical removal of radioactive materials or mass containing radionuclides. The latter are various actions aimed at dose reduction in the affected areas. In this case spatial distribution of radioactive materials in the affected region remains unchanged. It may be illustrated by measures in the agricultural production (e.g. addition of fertilisers and ameliorants to the soil) aimed at reducing or preventing radionuclide accumulation in agricultural products and, consequently, reducing doses from internal (and consequently total) exposures of the population.

Objectively, remediation often results in positive alterations of the re-cultivated landscape in radioactively affected regions; for example, remediation is accompanied by various land clearance operations (e.g. forest re-planting, area planning and regulation of the river system) which lead to improvement of general landscaping.

An essential factor, which dictates the length and other features of remediation, is the radionuclide composition of environmental contamination. In terms of remediation, the situation is classified as relatively simple when the radionuclide mixture lacks long- or very long-lived radionuclides (the practical aspect of greatest interest is the release to the environment of biologically important radionuclides such as ^{90}Sr ($T_{1/2} = 28.1$ years), ^{137}Cs ($T_{1/2} = 30.17$ years), ^{239}Pu ($T_{1/2} = 24{,}000$ years) and some transuranic elements). When the deposited radionuclide mixture after an accident is presented only by short-lived radionuclides (e.g. ^{131}I and other short-lived iodine radionuclides), the problem of area remediation as such does not arise, since rapid decay of short-lived radionuclides allows quick return of the radiological parameters in the environment to pre-accident levels. However, the presence of long- and very long-lived radionuclides in the mixture can be a great challenge to remediation. Hence, environmental remediation deals mainly with areas contaminated by long-lived radionuclides.

One important question in remediation is to define the start and the end of remedial activities. The detection of radionuclide release into the environment (which often relates to the end of release) or the moment when the contaminated zone is identified may be considered with certain reservations when remediation needs to start. However, the earlier the detection of any contamination (long-lived or short-lived radionuclides) the better it is as remedial actions during the fallout might even result in tremendous dose savings. Much more difficult and debatable is the question of what may be considered the end of remediation. Radical ecologists, representatives of different NGOs such as the Greenpeace movement, suggest that the contaminated area must be restored to the green lawn or at least brown lawn level (i.e. radioactive materials completely removed from the affected area).

Such an approach certainly seems very attractive from a humanitarian viewpoint. However, when facing the situation more realistically, first of all from the economic position, *restoration* of the affected areas to the original levels (or close to them) is probably possible and economically justified only for very small areas. But for remediation of large areas (the Chernobyl and Kyshtym accidents), the remediation pattern will be much more complex, costly and, of course, time-consuming. Undoubtedly, various social, economic, psychological, demographical, ethical and radioecological aspects of remediation will arise.

Accidents and incidents that release radioactive substances into the environment are always random and unpredictable. This practically excludes possible preventive measures that could make subsequent remedial actions easier. But in a nuclear explosion, military or industrial, there is an opportunity to locate the radioactive trace in some optimised direction in terms of possible radiation impacts, for instance by considering meteorological conditions during the explosion.

The first remediation stage of large contaminated regions is comprehensive radiation monitoring of the environment. Availability of detailed radiological information (deposition density, radionuclide composition of fallout, heterogeneity of spatial distribution of radionuclides and other factors) is essential for remediation. Usually, area remediation after radiation accidents and incidents has several stages, each being characterised by its own length and aims. The particularities of each such stage are much dependent on the economic resources available for the mitigation of its consequences. Therefore, the availability of technical devices and tools and qualified professionals is also of paramount importance. Crucial here is the state of scientific database and existence of sound scientific expertise.

At the same time, the stage-by-stage nature of the remediation depends on not only the economic resources but also the availability of professional staff and technical facilities. The experience gained in the mitigation of consequences of the two large radiation accidents in the USSR (Kyshtym and Chernobyl) has demonstrated that it is vital from the very first days to exclude production and consumption of agricultural produce that does not meet the radiological standards (i.e. products where radionuclide concentration is above the adopted limits). At one of the remote stages a question may arise of reducing the collective (exported) dose from consumption of foodstuffs produced in the contaminated regions. These stand-alone radiological tasks may be included in the objectives of some remediation stages.

Normally, the least-affected areas are subject to the remedial actions first. The advantage of this is the highest radiological and economic effects from remediation during the shortest period of time. Later remedial actions cover areas with higher levels of radionuclide contamination. At this stage, more sophisticated and costly options are applied, leading to an increase in the costs of protective measures.

A distinctive feature of remedial actions at the later stages of the radiation accidents, compared to the early ones, is a reduction in prohibitive measures (such as restrictions imposed on some kinds of farming activity and the use of certain products) and wider use of the recovery measures (e.g. amelioration of contaminated lands). Experience shows that imposition of prohibitive (restrictive) measures usually meets social resistance whereas constructive measures are approved and supported.

It is worth noting that the cost of remedial actions calculated, for example from a radiological parameter of saved or averted dose to the population per unit recovered re-cultivated area, usually grows with time after an accident or incident. This may be explained by some biogeochemical processes which cause stronger sorption of radioactive substances on different natural media (substrates) such as soil. As a result, the rate of reduction in radionuclide incorporation into the biological chains of migration leading to radioactivity accumulation by humans can drop significantly. All this emphasises the need for the earliest start of remediation in the affected areas.

The remediation procedure usually implies the introduction of some permissible levels of radionuclide concentration in products, environmental objects, etc. These standards may have different names: derived intervention levels (DILs), temporary permissible levels (TPLs), etc., and are considered as limits of radionuclide concentrations in various environmental objects adopted for a certain time. These permissible levels vary (usually become more severe) with time in the course of remediation. It should also be realised that the temporary permissible concentrations of radionuclides in environmental objects depend not only on the radiological parameters but also largely on the economic and social constraints.

Radiation accidents (especially large), which release radioactive substances, contaminate different ecosystems (agricultural land, forests, rivers, meadows, pastures, etc.). The particularities of remediation are closely dependent on the characteristics of the natural environments affected by radioactive contamination and subject to recovery. It is evident that remediation specifics for contaminated terrestrial ecosystems are significantly different from that for contaminated aquatic ecosystems. Within the terrestrial biogeocenose, equally large differences will be typical in remediation actions for the agricultural, natural and semi-natural ecosystems (meadows and haylands) and forests. When deciding on the priorities in remediation actions, one must first (bearing in mind the resource limitations) identify the main pathways and doses of human exposure. For instance, in the Kyshtym and Chernobyl accidents, the highest priority was given to foodstuff production on contaminated lands (i.e. introduction of protective measures to reduce radionuclide transfer to farm products).

Probably the greatest challenge is areas with the highest densities of radionuclide contamination. These are the Chernobyl Exclusion Zone, that is the 30-km zone, and the EURT in the Kyshtym accident (the location of the tank with radioactive waste from where radioactive material was released to the environment as a result of thermal explosion). Re-cultivation of such zones (1) requires large economic costs and (2) assumes considerable exposures of the personnel involved in the remediation actions.

As outlined earlier, remediation of territories affected by radiation accidents and incidents firstly addresses the safety of residents and areas where the activities of humans are affected. At the same time it is justified to use a wider, more holistic approach, that is radiation protection of not only humans but also non-human species (biodiversity, maintenance of the established ecological equilibrium, etc.). The thesis 'If Radiation Standards Protect Man Then the Environment is Also Protected' as formulated in ICRP 26 and 60 (ICRP, 1977, 1991) received some criticism in recent years and was taken further; it is reflected in ICRP Publication 91 'A Framework for Assessing the Impact of Ionizing Radiation on Non-Human Species' (ICRP, 2003). This aspect might result in revised approaches other than the egocentric approach adopted so far and require careful consideration further on.

2. Radioactive Contamination of the Techa River (USSR – Russia)

2.1. Description of the source of radioactive environmental contamination

Radioactive contamination of the Techa River was a result of the activities of the Production Plant 'Mayak', the first USSR (Russia) nuclear facility to produce weapons-grade plutonium. The absence of proper knowledge and experience in managing radioactive wastes, as well as complex social–political and economic conditions in the world and USSR in the 1950s, the initial period of nuclear industry development in the USSR (Russia), led to the accelerated use of improper technology for plutonium production and massive discharge of radioactive wastes into the Techa–Iset–Tobol–Irtysh–Ob river system (Figure 1). The volume of radioactive and chemical substances defined as radioactive waste accumulated in the Techa River network water bodies has no equivalents in the world (Akleev, 2000; Alexakhin et al., 2004).

The operations at Production Plant 'Mayak', comprising the reactor and associated radiochemical and metallurgical facilities to produce

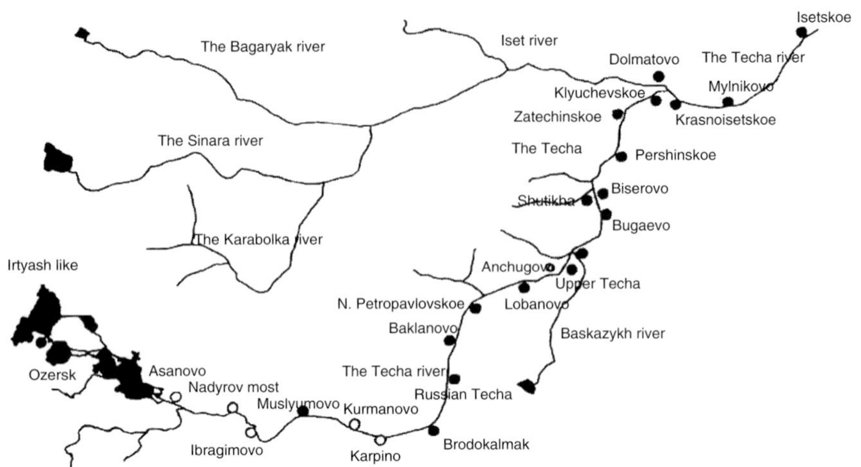

Figure 1 Location map of settlements along the Techa River (Akleev, 2000).

Table 1 Activity releases of liquid radioactive discharge into the Techa River (Alexakhin et al., 2004).

Period	α-Emitters	β-Emitters
Average daily discharge in 1949–1951 (GBq)		
January–November 1949	1.1	2.6×10^3
December 1949–February 1950	1.8	32×10^3
March 1950–November 1951	2.2	160×10^3
Annual discharge in 1949–1956 (PBq)		
1949	4.2×10^{-4}	1.8
1950	7.8×10^{-4}	52
1951	7.4×10^{-4}	52
1952	–	0.35
1953	–	0.074
1954	–	0.030
1955	–	0.018
1956	–	0.048
Total in 1949–1956 (approximately)	2×10^{-3} PBq (54 Ci)	110 PBq (3.0 MCi)

plutonium, led to the generation of huge amounts of radioactive wastes as given in Table 1. The reduction in radionuclide concentrations in these wastes below the permissible levels of 3.7 MBq l^{-1} (10^{-4} Ci l^{-1}) was then an unfeasible task and disposal of these wastes to the open hydrographic network was considered inevitable so that, over the period 1949–1956, some 110 PBq (3.0 MCi) of radioactive wastes were discharged into the Techa River. When summing up all the information available and taking into account the uncertainties of such a calculation partly due to data inconsistency, the estimated releases to the environment in 1949–1956 varied from 111 to 259 PBq (3–7 MCi) of radionuclides (State Duma, 2006). As a consequence, the Techa was contaminated all along its course (about 240 km); in addition, part of the Iset River from the point where the Techa joins the Iset proved to be contaminated, as well as large areas of the coastal regions.

Before the end of 1951, the radiochemical production was mainly responsible for much of the liquid radioactive waste discharges into the Techa (the contribution of the reactor and metallurgical productions was insignificant). In terms of total β-activity, the rate of liquid radioactive waste discharges into the Techa River was several tens of teraBecquerels per day and reached 160 TBq day^{-1} (Alexakhin et al., 2004). The disposal of liquid radioactive wastes into the Techa River ceased in October 1951. Radioactivity was discharged into the closed Lake Karachay, and only non-production wastes were disposed into the Techa, which was also stopped in 1956 (a total of 520 TBq were discharged during 1952–1956).

The most active discharges of liquid radioactive wastes into the Techa River (1949–1956) were caused by plutonium extraction using acetate-fluoride technology. Uranium fission products were responsible for almost all the β-activity of wastes, and consisted mainly of medium- and long-lived radionuclides due to ageing of the irradiated blocks. The estimated concentration ratio of α- to β-emitters in wastes was about 2.0×10^{-5} (Table 2).

Table 2 Radionuclide composition of LRW discharged into the Techa River in 1949–1956 (Alexakhin et al., 2004).

Period	$^{89}Sr+^{140}Ba$	^{90}Sr	$^{95}Zr+^{95}Nb$	$^{103,106}Ru$	^{137}Cs	REN[a]
Ratio to long-lived ^{90}Sr, which portion is equalled to 1						
January–November 1949	0.44	1	7.3	13.5	2.7	–
December 1949–February 1950	0.45	1	0.59	3.0	1.4	0.37
March 1950–November 1951	0.76	1	1.2	2.2	1.1	2.3
December 1951–December 1956	–	1	0.50–0.69	0.62–1.7[b]	0.28–0.41	–
Integral discharge (PBq or kCi)						
1949–1956	10 (270)	12 (330)	14 (390)	28 (760)	13 (350)	28 (760)

[a]REN – rare earth nuclides (Ce, Y and others).
[b]Together with rare earth nuclides.

During the period 1949–1956, ^{90}Sr and ^{137}Cs disposals were responsible for the long-term contamination of the Techa system and amounted to 12 and 13 PBq (0.33 and 0.35 MCi), respectively, the remaining radionuclides accounting for some 80 PBq (2.2 MCi).

2.2. Radiological situation

The exposure of the population in the coastal villages of the Techa River was prolonged and resulted in external and internal doses in the range of several milliSieverts to 4,000 mSv for red bone marrow.

In the 1950s–1960s, a cascade of reservoirs and dikes was constructed in the upper river for restriction (and future planned elimination) of radionuclide disposal into the Techa River. This cascade was named the Techa cascade and concentrated large amounts of radioactive wastes. These reservoirs and lakes are under permanent radioecological and hydrological control; regular remedial works are performed to maintain these hydraulic constructions.

Since the 1960s, radioactive contamination of the Techa River has been caused only by two long-lived radionuclides, ^{90}Sr and ^{137}Cs.

To date, concentrations of these radionuclides are still significant, 100 times and more above the levels typical for the rivers in the adjacent regions in Russia caused by global fallout. The highest contamination is found in bottom sediments and the river floodplain, with ^{137}Cs being the main contaminant. This can be explained by the fact that ^{90}Sr in the Techa system is largely in a dissolved (ionic) form, and ^{137}Cs is usually absorbed in bottom sediments and in the solid phase of the floodplain soils.

Overall, ^{137}Cs contamination densities in the floodplains of the Techa and the Iset are declining downstream. Thus, while in the bogs near the waste disposal site these values range between 5,550 and 20,350 kBq m^{-2} (150–550 Ci km^{-2}), in the middle Techa banks and at the confluence site they drop down to 740–1,100 kBq m^{-2} (20–30 Ci km^{-2}) and 185 kBq m^{-2} (5 Ci km^{-2}), respectively. No continuous decrease in ^{90}Sr contamination density of floodplain soils is observed downstream; even at a distance of 200–300 m from the riverbed in the upper Techa banks ^{90}Sr densities up to 259–370 kBq m^{-2} (7–10 Ci km^{-2}) were measured. The highest rates of exposure to γ-radiation were observed in the period of massive waste disposal in 1950–1951, 50,000 μR s^{-1} near the disposal site to 100 μR s^{-1} near a settlement of Techa–Brod (18 km downstream the disposal site). In the settlement of Muslyumovo, the rates in 1952 amounted to 0.1–0.25 μR s^{-1} (Shoigu, 2002).

In the early years following the waste disposal, dietary uptake of ^{90}Sr and other radionuclides by humans in settlements in the upper banks of the Techa River was very high (above 20 kBq in total β-activity in 1950–1952; in this case ^{90}Sr+^{90}Y contributed 20% to the total β-activity) (Table 3).

Table 3 $^{90}Sr + ^{90}Y$ uptake with some food products, % of total content in the diet in the Techa region (Akleev, 2000).

Settlement	Years	Fish	Poultry	Milk	Meat	Potato	Cabbage, vegetables	Eggs	Total β-activity (Bq)
Upstream Muslyumovo	1950–1952	45.5	13.7	20.0	0.1	16.0	4.1	0.6	21,830
(including Muslyumovo)	1953–1955	27.0	4.9	28.0	0.2	24.6	14.6	0.7	4,810
	1956–1958	–	14.3	36.0	0.5	17.2	32.0	–	407
Downstream Muslyumovo	1950–1952	33.0	6.1	7.3	0.1	42.5	11.0	–	4,070
	1953–1955	26.4	13.2	11.4	0.2	27.7	21.1	–	1,258
	1956–1958	–	9.0	26.5	0.2	33.3	31.0	–	555

Table 4 Estimated ^{90}Sr concentration in food products in settlements of the upper and lower reaches of the Techa River (Bq l^{-1} or kg^{-1}) (Akleev, 2000).

Settlement	Years	Milk	Meat	Potato	Cabbage, vegetables
Upstream Muslyumovo (including it)	1950–1952	1,456	436	2,910	3,056
	1953–1955	449	192	962	2,340
	1956–1958	49	40	58	434
Downstream Muslyumovo	1950–1952	100	81	144	1,492
	1953–1955	48	50	287	922
	1956–1958	49	22	116	573

The ^{90}Sr concentration in milk produced in villages situated upstream, for example Muslyumovo (population in 1950 was 1,958 people), which is the first settlement downstream from the disposal site (79 km), exceeded the permissible level by 66 times in the early years and 2–20 times in later years. The ^{90}Sr activity concentrations in potato were above the permissible level in settlements of the upper and lower Techa by 20–1,100 times and 45–55 times, respectively (Table 4). It is worth noting that the role of water as a means of radionuclide uptake by humans was significant: in 1950 the water fraction in radionuclide uptake was 92%, which declined in 1955 to 33%. Since 1956, the use of water from the Techa River for food production and drinking purposes has been prohibited; radioactive substances were transferred to the human diet only through foodstuffs.

2.3. Remedial actions and their efficiency

Remedial actions in the Techa region may be divided into three main groups (Shoigu, 2002):

1. Technical measures aimed at reducing radioactive contamination of the water network and exposure of the local population
2. Technical, organisational and other measures that involve reduction in radiation exposure or its exclusion from the contaminated water network of the coastal population
3. Measures to control the radiation situation in the coastal regions

The key measures to reduce dose in the affected region of the Techa River were of prohibitive nature, which was reflected in the establishment of a sanitary-protective regime. First of all, the use of the river water (for drinking purpose and for domestic usage), fish and farm products was restricted. Unfortunately, these measures were implemented too late.

In the early 1950s, large irrigated garden farms were operational with water supply directly from the Techa River. These farms had vegetable gardens where livestock grazed and hay was produced. In 1953 for example, only one such farm supplied the regional centre Chelyabinsk with some 250,000 l of milk, 1.4 tons of butter and cream, 28.2 tons of meat, and 300 tons of vegetables; 594 tons of hay was made on this farm (Akleev, 2000).

The most urgent measure, the ban on the use of river water for drinking, was realised 3–4 years after the river contamination, and water pipelines were constructed only 6–7 years after the ban. As far as restrictive measures are concerned, the water supply to the population was partially changed to artesian ones. The largest non-compliance with radiological protection rules was cattle pasturing in the coastal fenced zone, which resulted in high consumption of milk with increased ^{90}Sr levels.

Most effective in reducing the exposure of the population were the actions that reduced discharges of radioactive wastes from the Production Plant 'Mayak' into the Techa River system, for example construction of a cascade of dikes and reservoirs. One of the most radical (and costly) measures was the evacuation of the population from the coastal area. In 1954–1960, 7,500 people were resettled (some 30% of the total number of residents in the Techa region) from settlements located on the Techa River banks. The practical realisation of this countermeasure came, however, 5–7 years later, which was too late hence making it rather ineffective. By the resettlement time, the coastal residents had already received the major dose of both external and internal irradiation.

Of all the measures for public protection, the most difficult was the maintenance of a continuous sanitary-protective regime in the Techa zone over a prolonged time period. Its introduction in the 1970s reduced the exposure by 40-fold. Before 1970, the effectiveness of the sanitary regime was even higher, which was achieved in particular by restricting the use of river water for drinking and the use of the contaminated floodplain for milk production. At present, restrictions and in some cases bans are still in force for free cattle pasturing and fodder and hay making in the affected floodplain of the Techa, as well as fishing (Akleev, 2000).

The main factor responsible for the persisting rigid sanitary-protective regime of the Techa River and adjacent coastal regions is the presence of water bodies near the 'Mayak' production plant that still contain huge amounts of radionuclides. Radionuclides percolate into the Techa cascade and Lake Karachay, and are remobilised from deposits on the boggy floodplain and bottom sediments. The remedial works in the Techa floodplain are still in progress and will not be stopped as long as the radionuclides accumulated in the river present a high risk for the population and the environment.

To clean up the Techa riverbed, an attempt was made to 'wash' it with non-contaminated water. During two successive years the riverbed was flushed with volumes of clean running water (for 2–3 months), much higher than the low stream flow; however, no positive results were achieved. As a countermeasure, creation of an artificial new riverbed was considered within the most affected regions while burying the old one. This idea was, however, turned down due to technical difficulties and extraordinary financial costs (Shoigu, 2002).

The improvement of the radiation situation in the Techa region may be considered as a very specific remediation case. Remedial measures such as decontamination of the soil cover or reduction of radionuclide transfer into the food chains, that is the soil–plant–animal chain, and other similar measures were applied on a very limited scale because of their low technical feasibility. Besides, in the 'acute period' of disposal (1949–1953), the picture of the total radiation hazard remained relatively vague as the radioactive contamination of the river was dictated by *force-majeure* of a political nature.

One important issue – although often forgotten – is the self-healing capacity of nature. This proved to be a crucial remediation element in this area (same as in the 30-km zone of the ChNPP accident, see Section 6). Autorehabilitative processes in the Techa region include radioactive decay and biogeochemical processes resulting in the redistribution of radionuclides in natural and agricultural ecosystems (e.g. deeper penetration of radionuclides into soil and bottom sediments) and reduction in radionuclide chemical mobility and bioavailability, for example 'ageing' of radionuclides. Over 50 years after termination of active discharges of liquid radioactive wastes, the Techa River system has undergone rather large self-clearance due to both physical and biological/ecological decay processes.

To date, the radioecological situation in the Techa region has substantially improved. Thus, in Muslyumovo, exposure rates are currently around 0.1 mSv year^{-1} and only for the critical groups these may reach 1 mSv year^{-1}. In recent years, ^{90}Sr and ^{137}Cs concentrations in milk in the Techa floodplain area have noticeably reduced. Thus, in Muslyumovo, ^{90}Sr concentrations in milk in 1968 and 1995 were 18.5 and 3.7 Bq l^{-1} while ^{137}Cs concentrations were 65.5 and 14.0 Bq l^{-1}, respectively. In 1979 and 1988, annual ^{90}Sr uptake with milk reached the annual limit for this radionuclide only in 2% of the Muslyumovo population. However, it has been concluded that '… neither at present nor in the foreseeable future is it realistic to overcome totally the whole complex of radioecological problems of the Techa cascade (for both financial and technical–ecological reasons)' (State Duma, 2006).

3. The Radiation Accident at the 'Mayak' Production Plant (the Kyshtym Accident) in the USSR in 1957

3.1. Description of the accident

The radiation accident at a military plant producing weapons-grade plutonium in Chelyabinsk-40 (now Ozersk), Eastern Urals, which occurred on 29 September 1957, was one of the most serious catastrophes that resulted in radioactive releases into the environment (Nikipelov et al., 1987; Burnazyan, 1990; Sokolov and Krivolutsky, 1993). Scored on the 7-point scale (INES scale of events), this accident has been classified as category 6.

As a result of the thermal explosion of a high-activity liquid waste tank, radioactive substances were released into the environment affecting a vast territory (Nikipelov et al., 1987). The explosion cloud contained about 740 PBq (20 MCi) of radioactive fission products. Part of the radioactive substances was injected to a height of 1,000 m, with 90% of the radionuclides depositing on-site and the remaining 74 PBq (2 MCi) in the adjacent areas of the Chelyabinsk, Sverdlovsk and Tyumen regions. The radioactive trail formation was completed within 10–11 h. That affected area is known as the East Urals Radioactive Trail (EURT) and the accident itself is often termed in the literature as the *Kyshtym* accident (after the name of the nearest town).

The EURT area is part of the Trans-Ural mountain platform adjacent to the Eastern Ural Mountains, a poorly drained wave-shaped plain with less-developed river network and forest steppe vegetation. The climate is continental and the soil coverage very patchwork-like; most of the territory is dominated by chernozems, besides which there are soddy-podzolic soils, solod and meadow soils. The affected area was an extensively used agricultural region for grain and meat/milk production.

In the initial radioactive contamination of the environment ^{90}Sr was the radiologically most important radionuclide in the medium and long terms; therefore, it was considered as a reference radionuclide in relation to other radionuclides. The EURT area was restricted by a ^{90}Sr contamination density of 3.7 kBq m^{-2} at the time of the accident, which is equivalent to twice the background from the global fallout of this radionuclide after nuclear tests. In the EURT area (23,000 km^2) there were 217 settlements with a population density of 270,000 inhabitants. The criterion for identification of the territories that needed application of measures was a contamination density of 74 kBq m^{-2} and above for ^{90}Sr. The territories with a ^{90}Sr contamination density exceeding 74 kBq m^{-2} formed a narrow band, 4.5–6 km in length and 105 km in width (as shown in Figure 2) and covering an area of 1,000 km^2 (some 5% of the EURT area). Any economic activity within the contamination isolines of 74–150 kBq m^{-2} of ^{90}Sr was stopped (the restriction zone amounted to 700 km^2) (Alexakhin et al., 2004).

Most of the environmental contamination in the early period was caused by relatively short-lived radionuclides, and therefore, as early as the first year after the accident, there was a significant decline in the contamination of the environment ((% of initial radionuclide composition and release in PBq in brackets): ^{89}Sr – traces, ^{90}Sr + ^{90}Y – 5.4 (2.0),

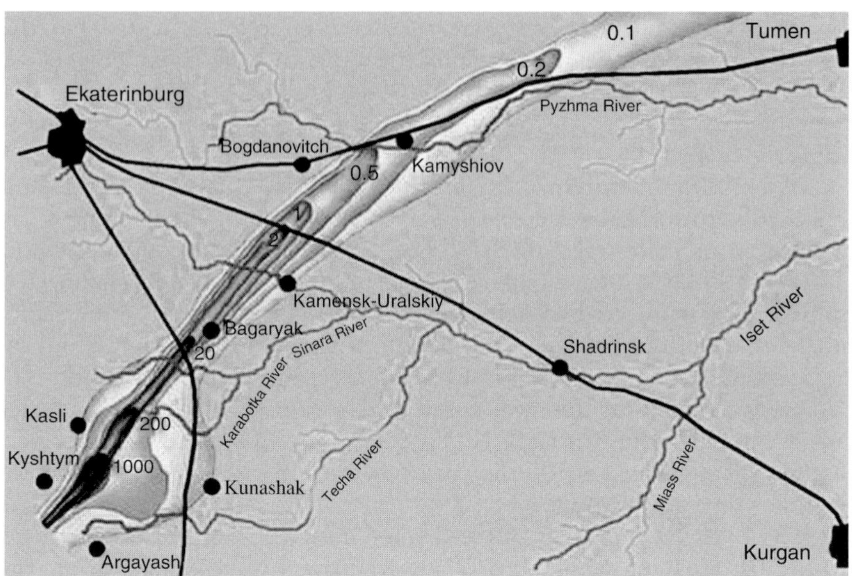

Figure 2 Map of the East Urals Radioactive Trail. Figures denote isolines of initial ^{90}Sr contamination density in $Ci\,km^{-2}$ ($1\,Ci\,km^{-2} = 37\,kBq\,m^{-2}$) (Alexakhin et al., 2004).

$^{95}Zr + ^{95}Nb$ – 24.8 (18.4), $^{106}Ru + ^{106}Rh$ – 3.7 (2.7), ^{137}Cs – 0.35 (0.26), $^{144}Ce + ^{144}Pr$ – 65.8 (48.7), ^{147}Pm – traces, ^{155}Eu – traces, Pu – 0.002 (0.0014) (Alexakhin et al., 2004) (Figure 3)).

In the first phase, the major contributor to exposure was ^{144}Ce uptake by humans through foodstuffs (Akleev and Kiselev, 2001). The maximum concentration of radionuclides in agricultural products on land closest to source areas (up to 20 km) reached 10–10,000 $kBq\,kg^{-1}$, ^{144}Ce and ^{95}Zr being the main contributors to these values except for milk where ^{90}Sr presented 70% of the activity concentration. The radionuclide concentration in all environmental media, including farm products, declined with time reflecting the ecological half-life of the affected territory. In 5–8 years (after ^{95}Zr, ^{106}Ru and ^{144}Ce decay), only ^{90}Sr and, to a very small extent, ^{137}Cs were responsible for contamination of the environment.

The major radiation dose contributor during the first month following the accident was external γ-radiation which dropped in the second to third months by 10-fold. In the most affected settlements, the exposure dose rate reached 200–400 $\mu R\,s^{-1}$, which corresponded to 0.015 $\mu R\,s^{-1}$ (1.3 mR day^{-1}) per 1 $kBq\,m^{-2}$ of ^{90}Sr initial contamination. The dose contribution of γ-radiation in the first year was 86% of the overall dose calculated over the period until 1990; 97% of the total dose was accumulated by the fifth year (Alexakhin et al., 2004).

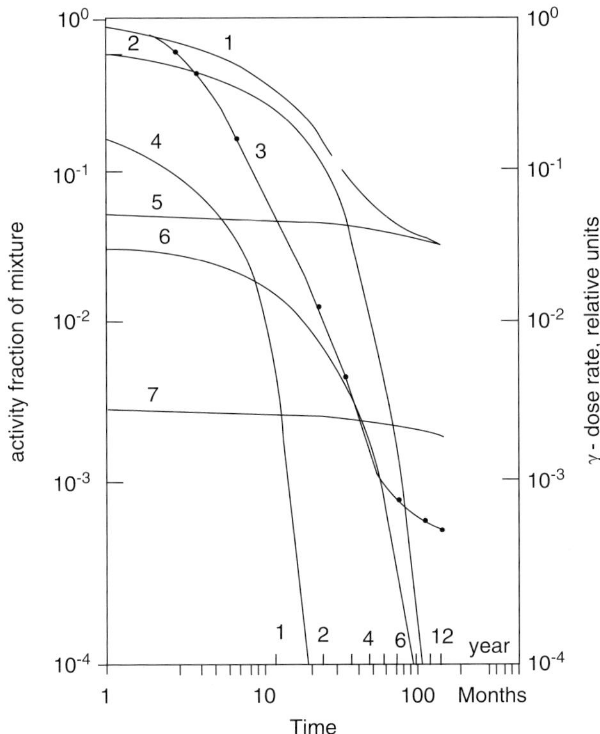

Figure 3 Change of activity ratio in fallout radionuclides resulting from the Kyshtym accident and the dynamics of γ-dose rate within the first 12 years. 1 — total β-activity; 2 — activity of $^{144}Ce+^{144}Pr$; 3 — dose rate; 4 — activity of $^{95}Zr+^{95}Nb$; 5 — activity of $^{90}Sr+^{90}Y$; 6 — activity of $^{106}Ru+^{106}Rh$; 7 — activity of ^{137}Cs (Alexakhin et al., 2004).

More than 40 years after the accident, radioactive contamination within the EURT area fell by more than 50-fold, and dose rates decreased 4,000 times due to radionuclide decay and vertical migration within the soil profile. In 2000, 80–85% of ^{90}Sr was still found in the upper 0–20 cm soil layer. The predicted concentration of ^{90}Sr in the soil root layer will, by 2045, be reduced by a factor of 18–20 from its initial values, covering by then three physical half-life periods. Analyses show that its biological availability for uptake from soil to plant dropped 7–10 times over 40 years. In the first 5 years after the accident, a twofold decrease in ^{90}Sr concentration in milk was observed in one year; 42 years later the milk half-life had increased to 15 years (Akleev and Kiselev, 2001). Over the whole post-accident period, radionuclide transfer to humans via foodstuffs continuously declined, and over 30–40 years the annual ^{90}Sr and ^{137}Cs uptake dropped on average by 200 and 2,000 times, respectively (Alexakhin et al., 2004).

3.2. Theoretical background of remediation in agriculture, forestry and water management

In terms of remediation, the Kyshtym accident may be considered as a challenge to rural management because consumption of food products contaminated by radionuclides has become one of the key sources of exposure of the people living in the affected area. In this context, reduction in dose from internal irradiation was the focus of remediation actions and changes of agricultural practices (as well as forestry and freshwater management) were the main remediation elements. In fact, remediation in the Kyshtym affected region meant a decrease in ^{90}Sr doses to the population from according to the adopted limits, and reduction in ^{90}Sr concentration in environmental media to the existing radiation standards.

The main factor responsible for the presence of any radionuclide – and here the ^{90}Sr content – in the agricultural chain 'soil–plant–animal' is the deposition density and condition of the particular radionuclide. At the same time, ^{90}Sr accumulation in plant products is influenced by the soil agrochemical characteristics, for instance exchangeable calcium as a non-isotopic chemical analogue. Since plants lack selectivity in root uptake for Sr or Ca, ^{90}Sr uptake by plants is inversely proportional to the content of exchangeable Ca in the soil, this being almost 40 times higher than the combined influence of all other soil factors. The resulting ^{90}Sr content in plant products depends on the demand for Ca, and consequently ^{90}Sr, by plants or their productive organs.

V.M. Klechkovsky (Arkhipov et al., 1969) suggested to predict ^{90}Sr accumulation by a numerical universal complex parameter, P_c, which is the ratio of ^{90}Sr concentration in plants (estimated per unit Ca quantity) to the so-called effective contamination density defined by the ratio of ^{90}Sr concentration in soil and the content of exchangeable Ca:

$$P_c = \frac{^{90}\mathrm{Sr}_{\mathrm{plant}}\ (\mathrm{Bq/g})/\mathrm{Ca}_{\mathrm{plant}}\ (\mathrm{g/g})}{^{90}\mathrm{Sr}_{\mathrm{soil}}\ (\mathrm{Bq/g})/\mathrm{Ca}_{\mathrm{exchangeable,soil}}\ (\mathrm{mg-eq/100\ g})}$$

Estimation of P_c for a variety of crops on identical soils has demonstrated that, at the same density of radioactive contamination of lands, the maximum ^{90}Sr concentration of all plant products is in grass or hay from natural lands (haylands, pastures), whereas the minimum ^{90}Sr content is in potato and root vegetables. Intermediate are cereal and legume crops, with differences in ^{90}Sr accumulation of various crops varying by a factor of 300.

In view of the peculiar features in the Kyshtym accident's affected region, the basic approach for farming management on the contaminated agricultural lands was based on the selection of sites with high ^{90}Sr deposition density for potatoes, root vegetables and cereal production and sites with lower ^{90}Sr activity concentrations in soil for fodder production and grazing animals.

The ^{90}Sr activity concentrations in farm animals and their derived products were directly proportional to the ^{90}Sr content in the fodder, generally following regularities of mineral metabolism. Since ^{90}Sr is an osteotropic radionuclide, it is mainly deposited in animal bones and the rate of its removal from the bone tissues is very slow (Alexakhin et al., 2004). This fact explains the long-term residence of ^{90}Sr in the body as well as long-term excretion of this radionuclide with milk (this factor is enhanced upon chronic lifetime ^{90}Sr intake by animals).

As opposed to plants which do not show selectivity differences in root uptake for Ca and Sr, the animal body distinguishes these elements by their chemical nature, giving preference to Ca. As a result, for per unit Ca taken up by the animal, meat was 2.5-fold and milk was 10-fold 'less-contaminated' compared to fodder. However, due to large differences in Ca concentrations in animal tissues and organs (for instance, in cattle Ca concentrations amount to 150, 1.0 and 0.1 $g\,kg^{-1}$ in bones, milk and muscular tissues, respectively), the least ^{90}Sr concentrations were found in animal muscles (soft tissues) while milk was about 40-fold more contaminated compared to meat. Therefore, meat production is recognised as the best option amongst animal farming practices. Overall, for ^{90}Sr-contaminated agricultural lands, animal products exhibited significantly lower concentrations of this radionuclide than plant products (Romanov et al., 1993).

3.3. Remedial actions and their effectiveness

3.3.1. Agriculture

The EURT area was subject to application of a complex remediation regime in agriculture from the very first growing season in spring of 1958 onwards. In crop farming, the most effective measure was primary ploughing of soil with burial of its top contaminated layer into subarable horizons (deep ploughing) immediately after radioactive fallout (Table 5). To this end, special-purpose equipment were developed – modernised ploughs and plough-shifters of soil horizons (Alexakhin et al., 2004). The modernised plough buried the contaminated soil layer to a depth of 30–40 cm (thereby reducing ^{90}Sr concentration in the arable layer by 80%) and the plough-shifter of soil horizons to a depth of 30–70 cm (reducing by 10–50-fold the ^{90}Sr content in the arable layer). With this measure, the ^{90}Sr accumulation in products dropped, for instance, by 75% in wheat and 99% in potato. Ploughing was especially effective on chernozem soils with a thick humus horizon (up to 0.5 m) whereas its effect was noticeably lower on low-fertility soils.

On small plots, direct decontamination of land, that is when the top contaminated soil layer was removed by special machines, such as bulldozers, graders and scrapers, with subsequent burial in special burial facilities, proved to be quite effective. The decontamination effect from

Table 5 Effectiveness of countermeasures for different crop types (Alexakhin et al., 2004).

Treatment	Soil type	Reduction of ^{90}Sr in root uptake
Ploughing		
Ploughing, depth of cultivation 20–25 cm	Mineral	1.0
Mouldboard ploughing, depth of cultivation 50 cm	Mineral	1.3–2.3
Ploughing with turnover of upper layer	Mineral	1.6
Liming		
0.5 H_g[a]	Mineral	1.1–2.1
1 H_g		1.4–2.7
2 H_g		1.7–3.1
1 H_g	Organic	1.0–1.1
Mineral fertilizers		
$N_{90}P_{180}K_{90}$	Mineral	1.1–1.4
$N_{60}P_{90}K_{120}$		1.3–1.9

[a]H_g hydrolytic acidity.

deep ploughing was enhanced by regular addition of mineral fertilisers to the arable layer. This technique, which is widely used in practice, reduced the ^{90}Sr content in the yield by a factor of 10 compared to conventional ploughing. A successful but costly method such as removal and disposal of the upper contaminated 5–10 cm layer of soil achieved a 5–15-fold decrease in ^{90}Sr transfer to crops. It was not, however, widely applied due to the high costs. In the various agrochemical measures taken to reduce ^{90}Sr transfer to plants, soil liming and mineral fertilising were widely used. These improved the physico-chemical properties of soil which resulted in increased soil fertility, increased crop yield and decreased radionuclide uptake to the crop yield. Mineral fertilising at balanced rates (according to plant demand for mineral nutrients) reduced ^{90}Sr uptake by plants on grey forest soils and leached chernozem by factors of 1.5–2. Liming of acid soils resulted in a reduction factor of 3. Further, reduction in ^{90}Sr uptake by humans via plant products (and via fodder by animals) was achieved through cultivation of plants with minimum ^{90}Sr accumulation (crop selection). The reduction factor in ^{90}Sr accumulation by the main farm crops amounted to 10, species difference accounting for a reduction of up to 60 times.

In addition to reducing ^{90}Sr uptake to plants, ploughing immediately after radioactive fallout reduced the dose rate of γ-radiation. Normal ploughing (20–25 cm) applied first resulted in a 1.1–2.4-fold decrease in dose rate of γ-radiation while ploughing with burial of the top layer decreased the external dose rate by up to three times.

There is great potential to reduce ^{90}Sr concentration in farm products via processing of raw products. Most common is the processing of milk to fermented milk products and butter, which reduces by 20-fold the ^{90}Sr concentration in butter compared to raw milk. Conventional processing of grain to flour and grouts reduces the ^{90}Sr concentration in the final products by two to three times. Starch and alcohol as well as vegetable oil can be produced practically 'clean' from the original contaminated agricultural produce.

Animal-based measures to reduce ^{90}Sr accumulation in animal products mainly consisted of selection of fodder rations with minimum ^{90}Sr contents. The most effective way was to feed animals with potatoes, root vegetables and grain which contained lower ^{90}Sr concentrations than fodders from natural lands, thereby changing the pre-accident practice of intensive animal husbandry based on the use of feeds from natural lands. The decline in ^{90}Sr uptake by farm animals was achieved through the introduction of Ca additives or the use of fodder rich in Ca (e.g. legume crops). This method resulted in a 10-fold decline in ^{90}Sr concentrations in milk. If cattle were planned for slaughter, ^{90}Sr reduction in meat was achieved by pre-slaughter feeding with 'clean' or less-contaminated fodders. Implementation of such ameliorative measures resulted in a decrease of ^{90}Sr content in products of specialised agricultural farms; in meat and milk the concentrations were two to seven and three to four times lower compared to non-ameliorated farms. These countermeasures proved to be less effective in private holdings, where feeding ratios were based on natural grass (hay) characterised by higher ^{90}Sr compared to feeds from arable lands.

3.3.2. Forestry

The following countermeasures were recommended for forestry:

(1) Establishment and introduction of radiation limits for forest products including berries and mushrooms and game. The adopted maximum permissible level for the ^{90}Sr contamination density of a forested area was $92\,kBq\,m^{-2}$ when used for pasturing and haymaking and $3.7\,MBq\,m^{-2}$ for production of industrial wood.

(2) Reduction of areas under pasture and for haymaking and restriction of their use by the population. This problem was solved either by forest planting in regions with ^{90}Sr density above $370\,kBq\,m^{-2}$ or by transfer of priority right to use these lands to specialised farms, while controlling access of the population.

(3) Restriction on the use of wood as a fuel by the population from the areas with ^{90}Sr contamination density above 74 kBq m^{-2}. For heating working places it was permitted to use wood produced on territories with ^{90}Sr density up to 1.11 MBq m^{-2} combined with an obligatory burial of ash outside the agricultural area.
(4) The use of industrial wood for economic needs but not for civil buildings. In this case debarking was recommended just at the harvest site to remove radioactive substances contained in the bark.
(5) Establishment of specialised forestry firms in 1960, whose activity has provided total compliance of the usual forestry practice with the above requirements.

3.3.3. Aquatic systems

The initial restrictions on the economic use of water bodies (fishing, use of aquatic vegetation as fodder for farm animals) were imposed on lakes in the area with ^{90}Sr density above 74 kBq m^{-2}. Due to water self-clearance (^{90}Sr half-life is 5–6 years), lakes located in the peripheral EURT area could be re-utilised for fish culture by 1970.

3.4. Remediation strategy and its implementation

Because of the Kyshtym accident, large areas were excluded from economic use. By 1959, this area amounted to 106,000 ha, the agricultural land accounting for 55% of the total area (29% arable land). The remaining 45% of the territory was covered by forests (36%) and lakes (9%). Hence, the agriculture, forestry and water economy suffered great direct economic damage from losses of relevant products due to the abandonment of the areas and the restrictions on some materials produced outside the abandoned (sanitary) zone. As to the economic use of the contaminated area, the remediation strategy presumed that lands must not be vacant (with the observance of occupational standards and dose limits to individuals) and their exploitation must compensate (if only partially) damage from radioactive contamination. Even the use of the relocated areas was assumed.

The long-term remediation strategy focused on offering two solutions: land-use optimisation and development of new economic structures to be applied to every study region or, if possible, every farm. Land-use differentiation in agriculture, forestry and other branches of the economy using soil, plant and water resources was based on estimations of permissible levels of ^{90}Sr-contaminated produce in the territory.

The structure of the economic use of the affected area was developed as follows. Should farming be impossible in the affected region (due to exceeding intervention levels), the economic activity in it was refocused on the production of industrial (non-food) goods, development of forest and local industry or exploitation of peat, sand, gravel and other mineral resources. Some examples include the production of crops as raw materials for non-foods, production of crops as raw materials for industrial applications (e.g. production of potatoes for starch and alcohol, grain for alcohol) and production of seeds of cereals, potato, vegetables and grasses.

Remediation of agricultural lands within the EURT area followed a step-by-step approach:

- Less-contaminated areas (74–185 kBq m^{-2} for ^{90}Sr) were allocated for food crops.
- Animal husbandry was intensified by controlling animal diet and excluding fodder from natural lands and rough feed, and also by enriching fodder rations with concentrates, potatoes and root vegetables.
- Fodders for dairy cattle were produced in areas with ^{90}Sr levels three to four times lower than that for meat cattle.
- Pig and poultry production (these having lowest ^{90}Sr transfer to muscle) were given preference.
- Grain and potatoes with ^{90}Sr concentration above the permissible limit were used only for industrial applications (see the preceding text).
- Special attention was paid to monitoring of ^{90}Sr activity concentrations in products in the private sector where the radionuclide concentration was noticeably higher than that in the collective sector for a number of reasons (cattle ranching on more contaminated lands, lesser scales of countermeasures, etc.).

The proposed specialisation was dictated by the need for radiation protection of the population, which potentially could consume any products obtained from the contaminated lands (Table 6) (Romanov et al., 1993). Simultaneously, the ^{90}Sr TPLs were introduced in agricultural foodstuffs for the first time (Table 7).

Table 6 Densities of ^{90}Sr contamination of lands determining the possibilities of various farming specializations (kBq m^{-2} or Ci km^{-2})[a] (Romanov et al., 1993).

Animal production	Without specialized systems of farming, using natural lands	With special systems of farming
Meat production	370 (10)	740 (20)
Milk production	93 (2.5)	185 (5)
Pig production	3,700 (100)	3,700 (100)

[a]In 1985, the ^{90}Sr limits in products were as follows: for milk – 55 Bq kg^{-1}; grain, meat, vegetables – 185 Bq kg^{-1}; fodder and forage – 3,700 Bq kg^{-1}; seed grain – 1,850 Bq kg^{-1}. In 1968, these were reduced: for milk – 12.6 Bq kg^{-1}; for meat – 11.8 Bq kg^{-1}. In 1979, for food grain – 7.4 Bq kg^{-1}; for milk – 5.5 Bq kg^{-1}; meat and vegetables – 3.7 Bq kg^{-1}; for potatoes – 1.85 Bq kg^{-1}.

Table 7 Temporary permissible levels (TPSs) for ^{90}Sr in agricultural food products within the EURT area[a] (Bq kg^{-1}) (Alexakhin et al., 2004).

Time of the TPL implementation	Grain	Potatoes	Vegetables	Milk	Meat
January 1958	92	92	92	28	28
Mid 1958	185	185	185	56	185
January 1959	74	74	74	37	92
1976 (NRB-76/87, 1988)	7.4	1.8	3.7	5.6	3.7
1977 (NRB-96, 1996)	140	240	240	25	50

[a]TPLs introduced in January and mid 1958 were approved by the USSR Ministry of Health. These were established based on the maximum annual ^{90}Sr uptake, 52 kBq (1.4 μCi). The 1976 TPLs were introduced by resolution of the Chelyabinsk authorities in accordance with NRB-69 (NRB-69, 1970) and NRB-76 (NRB 76/87, 1988), maximum uptake is 12 kBq (0.32 μCi), with limits of annual uptake being cut threefold. DILs in 1977 correspond to SanPiN 2.3.2.590-96 (SanPiN-96, 1997).

As a result of the remediation, a substantial decrease in effective annual doses to the local population was achieved and it was observed that these doses were less compared to the radiation safety standards existing in the USSR at that time.

Taking into consideration the strong migration relationship of ^{90}Sr and its biological carrier, Ca, an initial limit of 200 strontium units was introduced (1 strontium unit = 1 pCi $^{90}Sr\ g^{-1}$ Ca); later, this standard was made three times more strict, that is 66 strontium units. Also for the safe residence of the population, the maximum permissible ^{90}Sr contamination density in the environment (soil) was determined to be 74 kBq m^{-2} (or 300 μR h^{-1} of the initial radiation dose rate) (Alexakhin et al., 1996). During the first year after the accident, the maximum ^{90}Sr uptake by humans was 52 kBq $year^{-1}$.

3.5. Return of abandoned lands to economic use

Since 1958–1959, remediation has covered an area of 20,000 ha, where ploughing, partial destruction of dwellings and their burial, and forest planting were applied. By the radiation standards of the 1960s, radiation safety of residence was guaranteed to the population in areas with ^{90}Sr deposition below 150 kBq m^{-2}. The remediation began in 1961, when all lands in the Sverdlovsk region (^{90}Sr deposition density below 300 kBq m^{-2}) and 2,000 ha in the Chelyabinsk region were returned to economic use and given to specialised farms. By 1982, half of the contaminated 32,000 ha of agricultural land with ^{90}Sr deposition of 74–3,700 kBq m^{-2} was successfully returned.

The remediation consisted of several stages in terms of involvement of the agricultural land into economic use: the first stage – lands with ^{90}Sr

contamination density up to 300 kBq m^{-2}; the second stage – lands with ^{90}Sr contamination density in the range of 300–920 kBq m^{-2} (the 1960s); the third stage –lands with ^{90}Sr contamination density in the range of 1,850–3,700 kBq m^{-2}. By 1993, within the 74 kBq m^{-2} isoline, more than 80% of lands were returned to use.

An exception was the most affected EURT region where the ^{90}Sr content ranged between 3.7 and 150 MBq m^{-2}. In 1966, the East Urals State Reserve was established for long-term field observations in this area. The total reserve area is 16,616 ha, with a perimeter length of 90 km. It was the first worldwide experience of establishing a reserve associated with radioactive contamination of the environment. Later on, similar sites were organised in the Chernobyl-affected area.

4. The Radiation Accident at Goiânia, Brazil

4.1. Description of the accident

A serious radiation accident occurred in September 1987 in Goiânia, the capital of the state of the same name in Brazil with a population of 800,000. The city is situated 1,348 km from Rio de Janeiro and 919 km from São Paulo. One of the hospitals of this city possessed a 50.9 TBq (1,375 Ci) ^{137}Cs (^{137}CsCl) sealed source for therapeutic applications. A detailed description of the accident and its radiological consequences is given in Eisenbud and Gesell (1997) and IAEA (1998).

In 1985, the Institute of Radiotherapy in Goiânia moved to a new building leaving the source in the old abandoned building. After reconstruction of the incident, it turned out that several unemployed young people entered the unguarded building and stole the device. They thought that the 50 kg metallic product was a material of value. They separated it from the biological shield, opened the source container, and sold the source at a junkyard, the owner of which considered the material to be precious because of its blue luminescence.

Those who came to look at the object 'and even took a fragment of it' prompted dispersion of the radioactive substance. Some people took parts of that source as souvenirs, while others painted their faces with the blue powder for carnival. A girl aged 6 years even ate a sandwich playing with a fragment of the source. As a result she received a dose of 4 Gy and the estimated ^{137}Cs uptake by her was 10^9 Bq.

4.2. Radiological consequences

On 2 October 1987, dose rates near the unsealed source were measured to be 1.1 Gy h^{-1}. As a result of a two-week period of free circulation, the source was broken into small fragments and widely distributed. Radioactivity was also dispersed by the wind, which was high at that time, over a variety of vegetable gardens, houses and roofs of buildings. Consequently, radioactivity from that source was detected at distances as far as 100 km from the junkyard. Radioactive material even reached the main river in the region, the Meia-Ponte River. ^{137}Cs was found in the bottom sediments, 12 km from the source site. Extensive monitoring of the environment began three weeks after the incident.

In this incident, some 250 persons were exposed to external and internal irradiation, 50 of whom developed symptoms of whole-body and acute local irradiation, as well as external and internal contamination with radioactive substances. Fourteen of these 50 persons showed marrow failure. Doses of internal exposure in the affected people ranged from 0.046 to 0.97 Gy.

4.3. Remediation

The remediation work after the Goiânia accident was carried out on an international scale. Some 755 professionals took part in remediation of the contaminated sites and were exposed to different extents. However, doses to around 70% of the participants were below 1 mSv with a maximum dose of 16 mSv (IAEA, 1998).

The initial idea of which sites needed urgent remediation was based on examination of the residents and the contamination levels. Very high contamination was found in 85 houses and the residents from 41 houses were evacuated. The total area subject to radiation monitoring within the urban territory covered 67 km^2. In the city, a helicopter γ-survey was performed followed by vehicular γ-survey.

Once the contaminated sites had been identified, the most laborious work began. For mechanical decontamination, various chemical solvents and abrasives were used and physical removal of radionuclide-containing materials was performed. Standards were elaborated for decontamination of the affected objects (Table 8).

Table 8 Dose limits for decontamination of the environment during the Goiânia incident (Da Silva et al., 1991).

Objects and physical parameters	Limits
Different surfaces	0.1 nCi cm^{-2} (3.7 Bq cm^{-2})
Dose rate of γ-radiation in dwellings	50 μR h^{-1} (~0.5 μGy h^{-1})
Dose rate of γ-radiation outside dwellings	100 μR h^{-1} (~1 μGy h^{-1})

The limits of contamination density chosen for the remediation actions in Goiânia (3.7 Bq cm^{-2} or 37 kBq m^{-2}) were the same as for the classification of the Chernobyl-affected area (1986) (areas with ^{137}Cs levels of 37 kBq m^{-2} or 1 Ci km^{-2}).

The identification of the potential location to store the collected radioactive materials faced a highly negative public response. Ultimately, the site for temporary storage was chosen in the thinly populated town of Abadia de Goids (20 km from the city of Goiânia), and for the future, the building of a permanent waste storage facility was considered and approved. For waste storage, 1.2 m^3 tanks with a maximum capacity of 5 tons were used.

The storage facility was ready by November 1987, when remediation actions began. In the most affected places houses were destroyed, soil was removed and the decontaminated site was covered with concrete. In less-affected buildings, walls, floors and roofs were decontaminated. Personal items that contained radioactive substances were considered as wastes, since people refused to use these, albeit decontaminated, any longer. In addition to houses, 45 public places including parts of streets, squares, parks and shops were decontaminated. Low-level wastes were placed in 2001 industrial canisters or metallic barrels, as well as in special wheeled 32 m^3 containers. Medium-level wastes were stored in 2001 containers which were placed in special cylindrical concrete constructions with 200-mm thick reinforced concrete walls. By the end of December 1987, the main affected sites were decontaminated. A total of 3,800 tanks, 1,400 barrels, 10 containers and 6 concrete wells were constructed. Tanks and barrels were covered with plastic coatings.

In the subsequent years the wastes from the incident were repacked and two permanent repositories with a total area of 1,600,000 m^2 were constructed near the temporary storage site. The first repository contained some 40% of the total volume of low-level wastes. According to Brazilian laws, such types of wastes can be located in urban repositories. The second storage repository contained high-level wastes. Both repositories were covered by herbaceous vegetation; these are in forested areas and are under environmental control. The radiological condition is constantly monitored; radiological laboratories and an information centre are in full operation. Quantitatively, remediation after the Goiânia incident involving wide-scale ^{137}Cs dispersion resulted in the bulk of the ^{137}Cs source (50.9 TBq) to be localised in the 1.6 km^2 waste disposal area.

The remediation in Goiânia boiled down to the mechanical clean-up of the contaminated territories, collection of the various contaminated materials which were later considered as radioactive waste and their disposal for storage at specially allocated sites.

5. The Palomares Accident (Spain)

5.1. Description of the accident

In January 1966, during air refuelling over Palomares near the southeastern coast of Spain, two American Air Force planes collided. One of the four nuclear bombs that were onboard the aircraft was found three months later in the Mediterranean Sea. The second bomb, also dud, fell with parachute on agricultural fields. The remaining two bombs were physically destroyed due to explosion of their chemical detonators. Pyrophoric metallic plutonium combusted and the resulting radioactive cloud, which contained plutonium oxide, moved westwards covering an area of 262 ha in a region of intensive farming occupied by rural dwellings, agricultural fields and forest plants. Figure 4 illustrates the contamination of the affected area.

5.2. Remediation

The main remedial actions were applied to soils. At Pu levels greater than 1.2 MBq m^{-2}, the soil was removed, placed into containers and shipped to

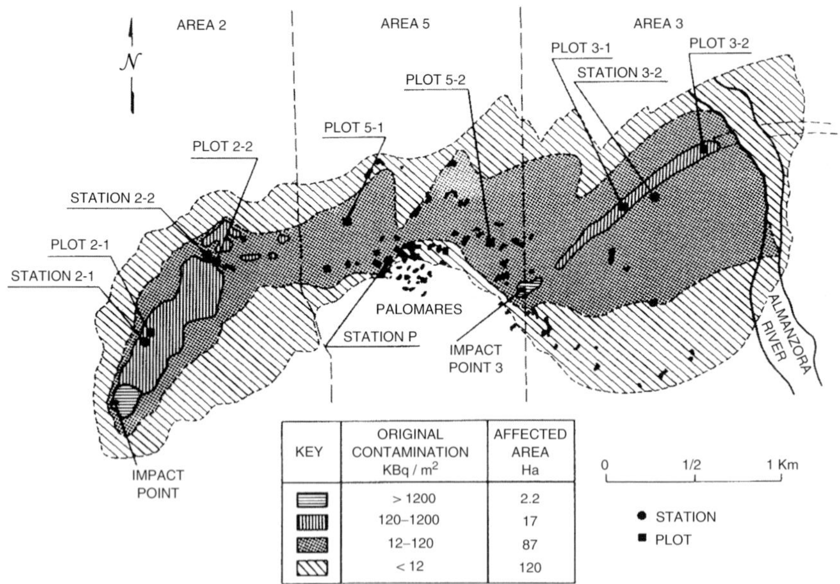

Figure 4 The post-accident levels of plutonium contamination in the vicinity of Palomares, Spain, showing the locations of sampling and experimental field plot stations (Iranzo and Richmond, 1987).

the United States for disposal at the Savannah River plant. The total area of the decontaminated soil was 2.2 ha, the number of exported containers being 6,000 (250 l each). The arable soil with Pu levels below 1.2 MBq m^{-2} was re-ploughed to a depth of 30 cm. Altogether around 17 ha was treated by this method. In rocky areas, where ploughing was impossible and Pu levels exceeded 120 kBq m^2, soil was removed by hand and shipped to the United States (Iranzo and Richmond, 1987).

Upon completion of mechanical decontamination, radiation monitoring of the air, soil, plants, farm animals and the population was organised.

The major farm crops produced in the affected area were tomatoes, barley, alfalfa, maize, pepper and fruits such as melon. The time of the fallout was within the period of the last tomato harvest but the contamination levels of all these products were too insignificant to cause serious harm to the public. Twenty-three years after the accident, approximately 99% of the Pu still remains in the 0–5 cm surface soil layer.

The accident produced no serious health impacts, since doses were insignificant and the exposed cohort was rather small. Only in 124 persons (1,815 assays), Pu concentrations in urine were found to be above 0.37 mBq day^{-1} (detection level). The Pu content in the whole human body accordingly proved to be below the detectable level (814 Bq). The 50-year effective dose of the exposed population exceeded 0.05 Sv only in 33 cases, and only 5 of these received doses of 0.15–0.2 Sv.

6. The Chernobyl Nuclear Power Plant Accident

6.1. Description of the accident

The accident at the ChNPP on 26 April 1986, in terms of both the amount of radioactivity released into the environment and the area affected by the radioactive contamination, was the largest in the history of nuclear power engineering. The cause of the accident was the human factor, relating to a failure of the technical staff during planned reactor discontinuance and engineering tests of a generator. As a consequence, an explosion occurred followed by a fire. Prior to 6 May, the destroyed reactor continued to release radioactive materials into the environment, the discharge being of multi-stage nature (IAEA, 2006a) leading to widespread transboundary contamination, mainly of Europe. The total release of fission products in the ChNPP accident (ignoring inert radioactive gases) amounted to 5.3×10^{18} Bq, including 8.5×10^{16} Bq ^{137}Cs and ^{134}Cs, 1×10^{16} Bq ^{90}Sr and 3×10^{15} Bq Pu (IAEA, 2006a).

The radiation situation in the affected area in the early period after the accident was dependent on short-lived products of fission and neutron activation, including ^{131}I, and in the later periods on ^{137}Cs (as well as ^{134}Cs in the first years after the accident) and in some areas in the vicinity of the ChNPP also on ^{90}Sr. The main dose-forming radionuclide for most of the affected territory in the medium and long terms was ^{137}Cs. The amount of ^{137}Cs deposited within the former USSR amounted to 4.0×10^{16} Bq (in Belarus, 41%; Russia, 35%; Ukraine, 24%; in other republics, under 1%). The Chernobyl ^{137}Cs release was approximately 6% of that released from all other sources, including global fallout due to atmospheric nuclear weapons tests in the 1950s–1960s (total 1.0×10^{18} Bq) (NCRP, 2007). The radiological importance of Pu was low.

^{137}Cs contamination levels exceeding 37 kBq m^{-2} (1 Ci km^{-2}) were taken as the limit confining the affected zone. The areas with these levels amounted to 150,000 km^2 (3.2% of the former USSR territory) and was approximately seven times greater than the area affected by the 1957 Kyshtym accident in the USSR (Izrael et al., 1994).

More than 15,000 settlements were affected, with a population of about 6,000,000. In the region with ^{137}Cs contamination density above 555 kBq m^{-2} there were 640 settlements (about 270,000 residents). In addition to the former USSR territory, high levels of radioactive fallout were reported outside its borders (especially in the Nordic countries, the UK, Germany, Poland, etc.).

Because of the influence of a range of physico-chemical and meteorological factors responsible for the fallout pattern, contamination of the environment was of a complex nature. In the near-ChNPP zone (10–30 km), the radionuclide composition of the fallout was close to that in the nuclear fuel; outside this zone significant radionuclide fractioning was observed, in particular considerable enrichment with more fugitive ^{131}I, ^{134}Cs and ^{137}Cs (Figures 5 and 6). Most of the 'hot particles' and Pu were deposited in the near zone.

The population dose in the affected region comprised external and internal (due to consumption of foodstuffs containing radionuclides) exposures. In the early days and weeks after the accident, the radionuclide concentrations in different environmental samples reached hundreds of thousands of Becquerel per kilogram or litre (IAEA, 1991, 2006a; Prister et al., 2007). Depending on the transfer rate of radionuclides from soils (their granulometric composition), the contribution of external irradiation to the dose was dominant (up to 93% of the overall dose) in soils with heavy mechanical composition, while in light and medium loamy soils it was 80–85%; in sandy and sandy loam soils it was much less, that is 53–57%, and in peat-boggy soils it did not exceed 20% (Panov et al., 2006) of the total dose (Figure 7).

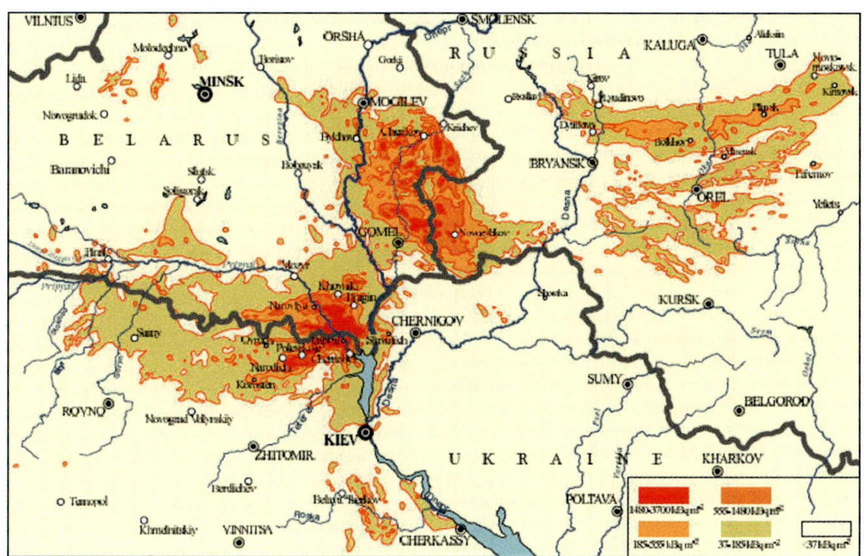

Figure 5 Deposition of ^{137}Cs in areas of Ukraine, Belarus, and Russia affected by the Chernobyl accident (IAEA, 1991).

Figure 6 Deposition of ^{90}Sr in areas of Ukraine, Belarus, and Russia affected by the Chernobyl accident (IAEA, 1991).

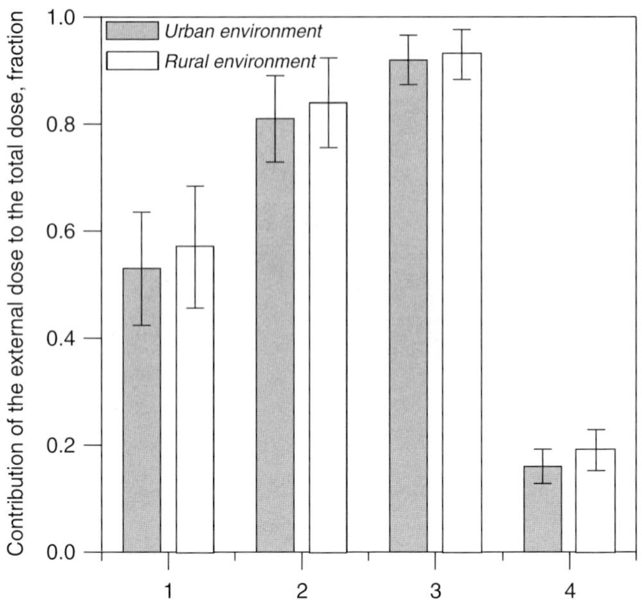

Figure 7 Contribution of external exposure to the total dose when farming on sandy, sandy loam (1), light- and medium-loam (2), clay (3), and peat-boggy (4) soils with similar contamination of urban and rural settlement areas (Panov et al., 2006).

6.2. Theoretical remediation background in the agriculture, forestry and freshwater ecosystems

Considering the economic, ecological, radiological and social importance, primary attention in the remediation strategy was paid to agriculture and, to a certain extent, to the forestry and water economy. The regulation of dose burdens to the population by application of agricultural countermeasures provided the largest reserve for radiological improvement of the situation.

Much of the soil cover affected was represented by low-fertility soddy-podzolic soil, sandy and sandy loam, as well as peaty soils. Typical for these soils is an increased mobility of man-made radionuclides (including ^{90}Sr and particularly ^{137}Cs). A small part of the exposed territory is occupied by more fertile soils (grey forest clay and soddy-podzolic loamy soils) where radionuclide availability is lower.

Forests in the affected region occupy about 30% of the area (60–70% – coniferous forests, 30–40% – deciduous woods). The region is permeated with a system of rivers and lakes which belong to the basin of one of the largest rivers in Europe, the Dnieper. The accident affected the cascade of large water reservoirs of the Dnieper basin (the Dnieper cascade) (IAEA, 2006b).

The ChNPP accident and the Kyshtym event in 1957 in the South Urals (see Sections 3.2 and 3.3) may be classified as *rural* accidents. The total area of the most contaminated lands in the three republics of the former USSR (Belarus, Russia and Ukraine) was agricultural land amounting to 45,000 km^2. First, consumption of agricultural products that contained radionuclides was one of the major (sometimes dominant) sources of public internal exposure. Second, the rural population (with 'rural' diet type) was the main cohort of people living in the affected region who consumed locally produced agricultural products. Third, the dose originating from internal exposure was economically and technologically more effectively handled than the external.

The key dose-forming radionuclide in the affected region was ^{137}Cs, and consequently reduction in the migration rate of this radionuclide via the trophic chains in the soil–plant–farm animal–man system was the main challenge.

The behaviour of ^{137}Cs in the environment as an alkaline element is strongly dependent on the presence of potassium, which is its stable non-isotopic carrier. Many countermeasures to reduce ^{137}Cs uptake by plants (and in some cases animals) are therefore based on the antagonism in ^{137}Cs and K transport via the trophic chains, and restriction in ^{137}Cs transfer is achieved through saturation (e.g. of soil) with potassium.

In the soil, ^{137}Cs is strongly absorbed by its solid phase. In this case, of special importance is its sorption by a number of soil clay minerals such as clinoptilolite, where ^{137}Cs is fixed in the interlayer spaces of its lattice. Typical of ^{137}Cs uptake by soil is the so-called 'ageing' factor which represents a gradual increase in strength of its sorption by the soil solid phase accompanied by a decrease in its bioavailability for root uptake. Thus, the ^{137}Cs uptake from soil and its fixation within soil are closely related to the soil properties. In organic (peaty) soils, as well as in low-fertility soils of light mechanical composition (sandy and sandy loam), ^{137}Cs availability for plant uptake is 5–10 times higher than that in soils of heavy mechanical composition. ^{137}Cs is accumulated in huge amounts in plants, with the ^{137}Cs plant/soil concentration ratio varying between 0.02 and 1.1. In soil-to-plant transfer, ^{137}Cs discrimination against K is reported; overall there is a similarity to the behaviour of ^{90}Sr-Ca, although less stringent.

In farm animals ^{137}Cs absorption with fodder amounts to 80–100% in the gastrointestinal tract if in the ionic form. In comparison to ^{90}Sr, which is accumulated in the bone tissue, Cs is evenly distributed in the body and is fairly readily removed from both the muscular tissue of animals and milk with half-lives of about 12–14 days.

6.3. Protective and remedial actions in the agriculture, forestry and freshwater ecosystems

6.3.1. Agriculture

6.3.1.1. Early period (1986–1987). Since consumption of radioactively contaminated products contributed an essential dose to the exposed population in the Chernobyl-affected regions, from the early days after the event onwards, agriculture as well as forestry and water resources (to a lesser extent) was heavily treated with countermeasures. The main goal of these actions was to reduce individual and collective doses to the population through reduction in radionuclide concentrations in the consumed products.

The radiation accident at the ChNPP occurred in an extremely unfavourable season for agriculture from the radiological point of view: in the late spring to early summer. It was the start of the grazing period (low pasture productivity in that period caused high radionuclide concentration in vegetation and, consequently, milk); sowing was completed, as well as planting of the major crops, which was one of the causes of high contamination of products in the autumn of 1986. In addition, the reserves of clean winter feeds for the livestock stored from previous years became exhausted soon.

In the early period (the first two months) after the accident the ^{131}I posed the main radiological hazard. A somehow delayed introduction of the key agricultural countermeasure at that time, namely the exclusion of ^{131}I-containing feeds (first of all termination of grazing), resulted in a reported increase of thyroid cancer incidence, the main radiation-induced health impact in the affected regions. To reduce ^{131}I effects on the population, the following countermeasures were applied:

- Radiation monitoring and subsequent rejection of milk at processing plants in which ^{131}I content was above the adopted limit (3,700 Bq l^{-1} at that time).
- Processing of rejected milk (mainly converting it into storable products such as condensed or dried milk, cheese or butter).
- Exclusion of contaminated pasture grasses from animal diet (changing from pasture to indoor feeding with uncontaminated fodder).

During the first year after the accident, the major agricultural countermeasures were mainly of a restrictive nature. In the first few months, severely contaminated land was taken out of use and suitable countermeasures were developed that would allow continued production on less heavily contaminated land. In the most affected regions, a ban was imposed on keeping dairy cattle. To reduce contamination levels in crops, an effective method was to delay harvesting of forage and food crops. Radiation monitoring of products was introduced at each stage of food production, storage and processing (Alexakhin, 1993; Fesenko et al., 2007).

Based on a radiological survey performed from May to July 1986, approximately 130,000, 17,000 and 57,000 ha of the agricultural land were initially excluded from economic use in Belarus, Russia and Ukraine, respectively (Fesenko et al., 2007). The criterion used to define such land was that ^{137}Cs deposition density exceeded 1,480 kBq m^{-2}.

From June 1986, other countermeasures aimed at reducing ^{137}Cs uptake to farm products were implemented:

- Ban on cattle slaughter in regions where ^{137}Cs levels exceeded 555 kBq m^{-2} (animals had to be given clean feed for 1.5 months before slaughter)
- Minimisation of external exposure and formation of contaminated dust by omitting some procedures normally used in crop production
- Restriction on the use of contaminated manure for fertilisation
- Preparation of silage from maize instead of hay
- Restriction of private milk consumption
- Obligatory radiological monitoring of agricultural products
- Obligatory milk processing

During several months after the event, in addition to the evacuated population, 50,000 head of cattle, 13,000 pigs, 3,300 sheep and 700 horses were evacuated from the 30 km ChNPP zone; 20,000 domestic and farm animals were killed. Because of feed shortage and other problems, 95,000 head of cattle and 23,000 pigs had to be slaughtered later (IAEA, 2006a).

6.3.1.2. Late period (1988–2007). The major long-term exposure pathways in the Chernobyl-affected areas were external irradiation and ingestion of contaminated foods. For long-term remediation, intervention in the agricultural systems was a more practical measure to reduce doses of internal exposure than decontamination of settlements aimed at reducing external exposure. Therefore, wide-scale application of countermeasures in agriculture was, and continues to be, a priority (Figure 8).

One of the paradigms of agricultural radionuclide fluxes is the thesis that the higher the soil fertility, crop yield and animal productivity, the lower is the radionuclide content in products. Hence, most of the remedial actions were aimed at increasing soil fertility, crop yield and animal productivity.

When introducing countermeasures in agriculture in the later period, based on the results from large-scale radiation monitoring, the territory was divided into zones according to the ^{137}Cs contamination levels. Four zones were identified: 37–185, 185–555, 555–1,480 and above 1,480 kBq m^{-2}. In the least contaminated zone (37–185 kBq m^{-2}), the agricultural practice was practically unchanged, only the peaty soils with enhanced ^{137}Cs transfer to plants receiving countermeasures (application of increased rates of mineral fertilisers). In the 185–555 kBq m^{-2} zone, countermeasures were used on a larger scale, while in the 555–1,480 kBq m^{-2} zone, these were at

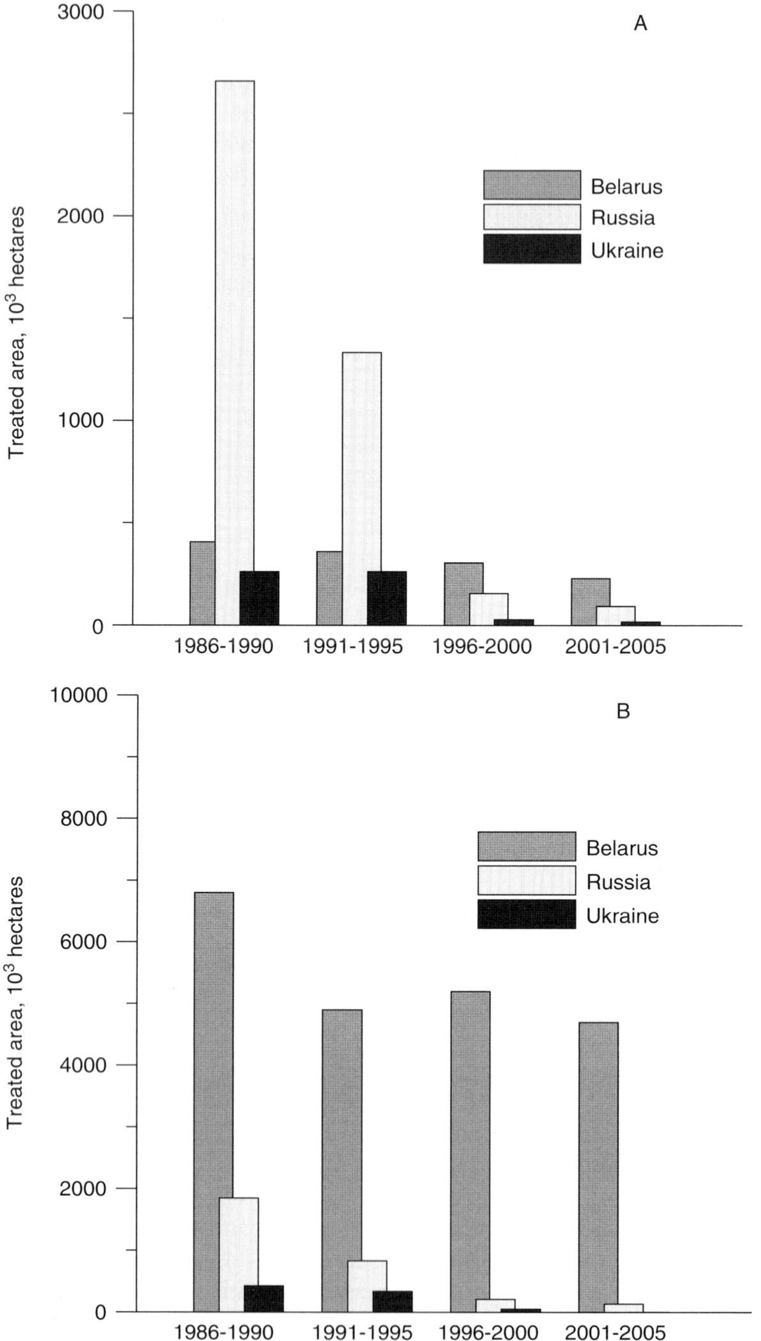

Figure 8 Changes with time in the extent of agricultural areas treated with liming (A) and mineral fertilisers (B) as a countermeasure in the USSR/FSU countries contaminated by the Chernobyl accident (Fesenko et al., 2007).

maximum (Fesenko et al., 2006). In the area with ^{137}Cs levels above 1,480 kBq m^{-2}, in Belarus, Russia and Ukraine, 1,300, 173 and 570 km^{2} of land, respectively, were excluded from agricultural use.

A regional soil classification scheme for estimating ^{137}Cs contamination was developed, which allowed estimation of the radiation safety of products (Table 9). The permissible contamination of agricultural lands for one food product changed 10-fold or more when taking into account the variability in the physico-chemical and agrochemical properties of soils.

In crop production, the simplest and most effective way to reduce radionuclide concentration in plants was by soil ploughing (particularly ploughing with deepening of the contaminated layer to 25–30 cm and below), which reduced the ^{137}Cs content in plants by factors of 1.5–2.0. However, soil ploughing as a countermeasure is effective only when used for the first time after contamination; repeated ploughing can be considered as of minor efficiency. The first ploughing after the event and, to a lesser extent, subsequent ploughings were also applied to reduce dose due to external exposure. Especially effective in reducing ^{137}Cs contents in plants (1.5–3.0 times) was mineral fertilising and liming of acid soils which decreased ^{137}Cs concentrations in plants by about half (Table 10).

In fodder production, amelioration of meadows and pastures proved to be highly efficient. Re-ploughing of low-fertility meadows and pastures with mineral fertilising and liming and their change to cultivated meadow lands reduced the ^{137}Cs content in vegetation by 3.6–16.0 times (Table 11).

In animal production, pre-slaughter feeding of animals with 'clean' or relatively 'clean' fodder (4–10 weeks before slaughter) was widespread

Table 9 Maximum permissible ^{137}Cs levels of agricultural lands and forests as a function of properties of soils where products can be obtained in compliance with the SanPiN 2.3.2.1078-01[a] (Ministry of Health, 2002) as ^{137}Cs inventory (kBq m^{-2}) (Panov et al., 2006).

Product	Soil			
	Sandy, sandy loam	Light- and medium-loam	Clay	Peat-boggy
Milk	410	1,000	2,000	120
Beef	240	420	720	65
Pork	330	660	1,600	90
Potato	1,500	2,000	7,500	500
Grain (bread)	230	460	1,750	115
Mushrooms	50	365	795	30

[a]Maximum permissible ^{137}Cs concentrations were: milk – 100 Bq l^{-1}, beef, pork – 160 Bq kg^{-1}, potato – 120 Bq kg^{-1}, grain – 60 Bq kg^{-1}, mushrooms – 500 Bq kg^{-1}.

Table 10 Summary of reduction factors of soil-based countermeasures used in the FSU countries (Alexakhin, 1993; Bogdevich et al., 2002; Fesenko et al., 2006).

Countermeasure	^{137}Cs	^{90}Sr
Normal ploughing (first year)	2.5–4.0	–
Skim and burial ploughing	8–16	–
Liming	1.5–3.0	1.5–2.6
Mineral fertilizers	1.5–3.0	0.8–2.0
Organic fertilizers	1.5–2.0	1.2–1.5
Radical improvement (first application)	2.0–9.0 (2.0–3.0)[a]	1.5–3.5 (1.5–2.0)
Removal of soil (external dose)	1.5	Not relevant

[a]Multiple applications.

Table 11 Efficiency of countermeasures on meadow and pasture lands (Prister et al., 1996; Alexakhin et al., 2004; Fesenko et al., 2007).

Remedial action	^{137}Cs reduction factors in plants	
	Mineral soils (sandy loam)	Organic soils (peaty)
Drainage	–	2–4
Disking or rototilling	1.2–1.5	1.8–3.5
Re-ploughing	1.8–2.5	2.0–3.2
Re-ploughing with layer turnover and its relocation at a depth of 35–40 cm	8–12	10–16
Liming	1.3–1.8	1.5–2.0
Application of mineral fertilizers		
Nitrogen	Coefficient increase up to 1.1–3.0	
Potassium (60–240 kg ha^{-1})	1.5–3.0	1.5–3.0
Nitrogen, phosphorus, potassium (1:1.5:2)	1.2–2.0	1.5–2.0
Surface improvement	1.6–2.9	1.8–14
Radical improvement	3.0–12	4.0–16

(Table 12). The use of ferrocyn compounds that bind ^{137}Cs in the gastrointestinal tracts of animals and thus prevent absorption in the gastrointestinal gut turned out to be very effective. Thus, ferrocyn treatment reduced ^{137}Cs in cows' milk by 1.5–14 times and in mutton by 3–16 times (Ratnikov et al., 1998). Processing of milk to milk products such as cheese and butter reduced the ^{137}Cs concentration in the final consumable product significantly (at a ^{137}Cs activity concentration of

Table 12 Summary of reduction factors of animal countermeasures used in the FSU countries (Fesenko et al., 2006).

Remedial action	^{137}Cs	^{90}Sr
Change in fodder crops	3–9	2–8
Clean feeding	2–5 (time dependent)	2–5
Administration of Cs binders	2–5	–
Processing milk to butter	4–6	5–10

1 Bq l^{-1} in milk, its concentration in butter and cheese is 0.02 and 0.4 Bq kg^{-1}, respectively).

6.3.2. Forestry

In the affected region, the forested area with ^{137}Cs levels above 37 kBq m^{-2} exceeded 40,000 km^2, while about 500 km^2 of the forested area had a density of ^{137}Cs contamination above 1,480 kBq m^{-2}. In Russia alone, about 50,000 forestry workers and members of their families (Alexakhin et al., 2004) lived in the affected areas. In terms of remediation, forested areas have two aspects:

i. Forests play a prime role in the sustainability of landscapes, limiting or even preventing radionuclide dispersion.
ii. Forest products are an important source of the population's external and internal exposures.

The role of forest products as sources of human diet can be significant (consumption of mushrooms, berries, sometimes game). Unlike agricultural products, natural clearance of 'forest products' from ^{137}Cs is much slower and associated with peculiar features. The effective half-lives for ^{137}Cs in mushrooms and berries are much higher than 20 and 10 years, respectively (Shutov et al., 1996); ^{137}Cs in wild deer grazing such areas, specifically when feeding on mushrooms, is even season/climate dependent with an ecological half-life which can exceed the physical half-life (IAEA, 2009). As a consequence, in the long-term after the accident, the contribution of 'forest products' (especially mushrooms) as a source of ^{137}Cs in the human diet can exceed that of critical foodstuffs such as milk (particularly in settlements near forests) (Fesenko et al., 2001b). Thus, the contribution of mushrooms to the internal dose in one of the most affected settlements, Smyalch in the Bryansk region of Russia, increased in 1995 relative to 1991–1992 from 20% to 43%, whereas the milk contribution dropped from 66% to 39% (Fesenko et al., 2001a; Jacob et al., 2001).

Unfortunately, countermeasures in forests were mainly of a restrictive or prohibitive character, the role of active remediation methods being extremely small. In the affected forested areas, bans were imposed on collection of mushrooms and berries, hunting, private cattle-grazing on forest pastures and clearings; or fire-control was implemented. Active countermeasures in contaminated woods (collection and removal of forest litter, leaves, reforestation, etc.) are either technologically ineffective or economically expensive and ecologically unjustified.

6.3.3. Aquatic systems

The Chernobyl accident has affected a wide system of water bodies in the river and lake networks as well as in the Dnieper catchment area, amongst which there is the most contaminated water reservoir, the ChNPP cooling pond. Some 8 million people were exposed to additional radiation due to the consumption of drinking water and 32 million due to the consumption of fish and the use of water for the irrigation purposes (Alexakhin et al., 2004; IAEA, 2006b).

There were two major aspects of the remediation strategies in the water bodies: first, introduction of countermeasures to restrict the radionuclide distribution in water and, second, reduction in the dose burdens to the population from the consumption and use of aquatic foodstuffs and drinking water.

In the contaminated catchment area different hydro-technical constructions were implemented (dikes, dams, etc., as well as special channels) to control and limit radionuclide migration and various sorbents were employed to incur radionuclide sedimentation from water masses. As a rule, these measures were rather costly and their effectiveness was low (Alexakhin et al., 2004; IAEA, 2006a); in addition their contribution to the dose is minor when compared to other pathways.

6.4. Effectiveness of remedial actions

In regions with maximum ^{137}Cs deposition activity, concentrations of this radionuclide above the permissible level affected 80% of grain and milk production in 1986 (Figure 9). As milk and meat are considered as critical foodstuffs, they have become the focus of attention for countermeasure application. The introduction of a large-scale agricultural countermeasures regime has abruptly reduced the volumes of products that exceeded the action levels (Fesenko et al., 2007). As early as 1991, only 10% of agricultural products did not meet the radiation standards (Fesenko et al., 2006).

The effectiveness of remedial actions and the choice of remediation strategy depend on a large number of factors. Amongst these in particular are biogeochemical conditions of the environment in the affected region, primarily soil cover peculiarities and soil properties which are highly influential.

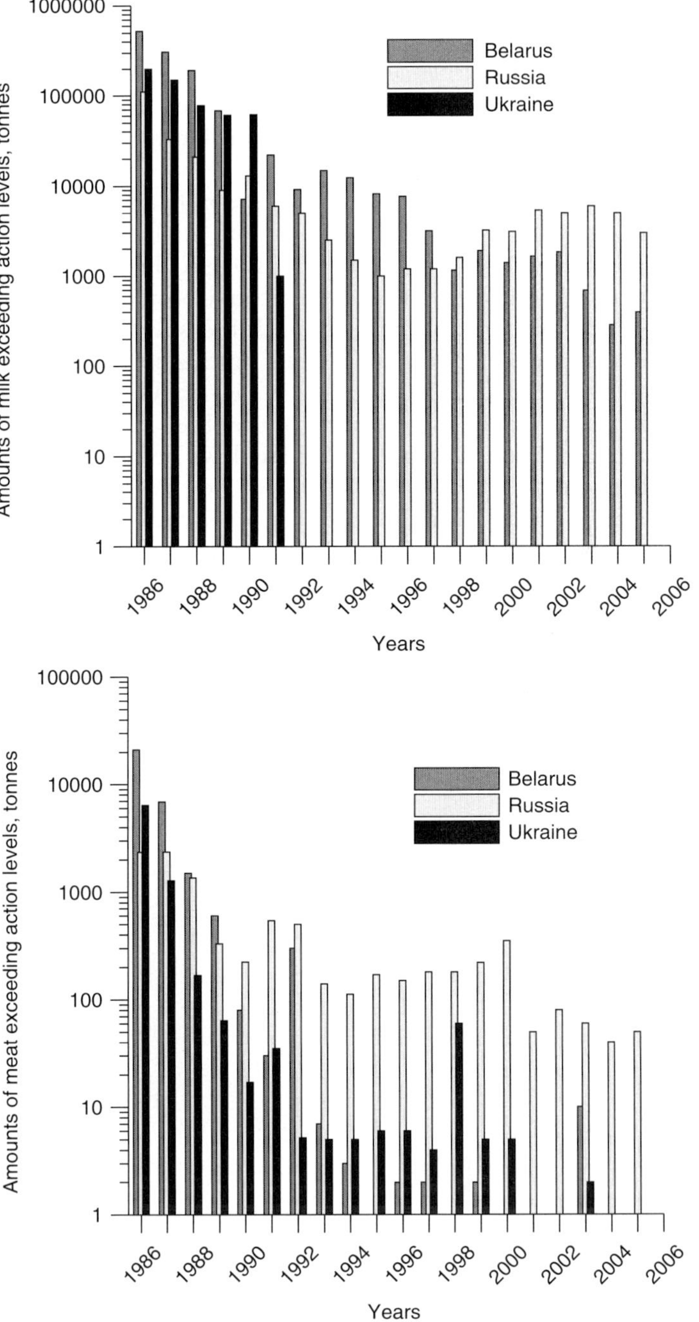

Figure 9 Amounts of milk and meat exceeding action levels in Russia (all milk and meat – collective and private), Ukraine and Belarus (only milk and meat entering processing plants) after the Chernobyl accident in tons (Fesenko et al., 2007).

In regions where agricultural countermeasures were started in due time and on sufficient scales, ^{137}Cs effective half-lives in the critical foodstuff milk were 1.0–2.8 years. In areas where countermeasures were delayed or of a limited scale, ^{137}Cs decrease in milk was mainly related with biogeochemical processes that reduce radionuclide mobility, its half-lives being 2.3–4.8 years (Fesenko et al., 1995).

The agricultural countermeasures were implemented in both the collective and private sectors. In the latter case, collective and individual doses to the population consuming local agricultural products were reduced, while in the former case, doses could be reduced for products exported outside the contaminated region. If the effectiveness of countermeasures is estimated in terms of averted doses, the collective sector is the main subject for dose reduction of the population. Overall, the implementation of agricultural measures in the three former Soviet Union (FSU) republics, Belarus, Russia and Ukraine, have provided a reduction of the collective dose by 12,000–19,000 man-Sv (Fesenko et al., 2006). Thus, the introduction of agricultural remedial actions was estimated to reduce collective doses derived from internal irradiation (except for the thyroid dose) by 30–40% and the total dose approximately by 20–25% for the population living in the affected region compared to a scenario where no countermeasures would have been applied (Fesenko et al., 2007).

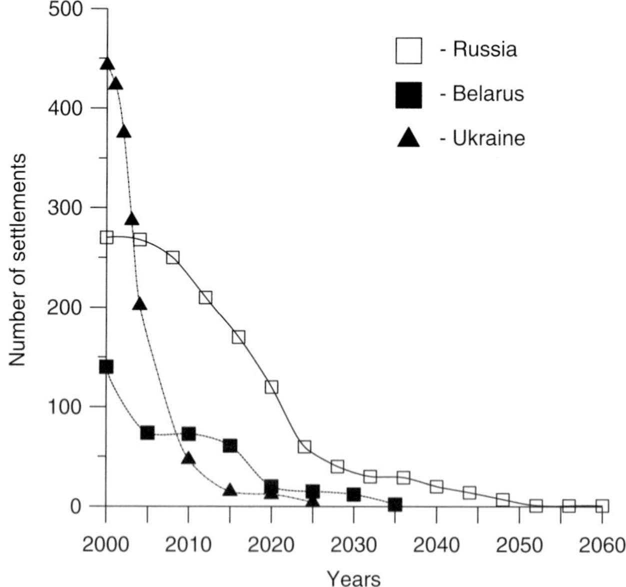

Figure 10 Predicted changes with time in a number of rural settlements with dose exceeding 1 mSv year^{-1} (Fesenko et al., 2006).

Despite the evident progress in the improvement of the radiological situation in the Chernobyl-affected area, there were, in 2007, still several hundreds of settlements in the contaminated districts with a population of 100,000 persons whose annual effective dose is above 1 mSv, internal exposure being an important contributor (Figure 10). It is predicted that by 2035 in Belarus and Ukraine and by 2050 in Russia, when using an exposure half-life of 15.0 (internal) and 18.8 (external) years, respectively (Jacob et al., 2001), there will not be any settlements where the effective dose will exceed 1 mSv year^{-1} (Fesenko et al., 2007).

6.5. The abandoned 30-km ChNPP zone

In the post-accidental management and remediation process of the Chernobyl-affected region, one of the most complex and difficult issues is the management of the 30-km zone. This is the area with the highest contamination densities (including the presence of plutonium and transuranic radionuclides), as well as the disposal site of the Chernobyl radioactive materials of different origins. The distribution of radioactive substances over the abandoned zone is very uneven, and the radiological situation in the zone needs long-term (tens of years) sanitary-hygienic restrictions. There are fundamentally different approaches to assessing ways of management of this area. In particular, proposals were made to use it as an industrial zone for long-term storage of radioactive wastes, for their geological burial for the Ukrainian NPP, or as a natural reserve and study site for quantifying long-term radiation effects in the environment. Finally, it cannot be excluded that residents will return to this place after implementation of remediation countermeasures. Hence, the final decision on the fate of the 30-km ChNPP zone depends on a range of closely interconnected radiological, economic, ecological, social and political factors (IAEA, 2006a).

REFERENCES

Akleev, A. V. (2000). In: *Medical-Biological and Ecological Impacts of Radioactive Contamination of the Techa River* (Eds A. V. Akleev and M. F. Kiselev). RF Health Ministry, Moscow, 532pp. (in Russian).

Akleev A. V., and M. F. Kiselev. (2001). In: *Ecological and Medical Impacts of the 1957 Accident at the PP "Mayak"* (Eds A. V. Akleev and M. F. Kiselev). RF Ministry of Health, Moscow, 294pp. ISBN 5-8099-0006-2 (in Russian).

Alexakhin, R. M. (1993). Countermeasures in agricultural production as an effective means of mitigating the radiological consequences of the Chernobyl accident. *Science of the Total Environment*, 137, 9–20.

Alexakhin, R. M., L. A. Buldakov, V. A. Gubanov, Ye. G. Drozhko, L. A. Ilyin, I. I. Kryshev, I. I. Linge, G. N. Romanov, M. N. Savkin, M. M. Saurov, F. A. Tikhomirov, and Yu. B. Kholina. (2004). In: *Large Radiation Accidents: Consequences and Protective Countermeasures* (Eds L. A. Ilyin and V. A. Gubanov). IzdAT Publisher, Moscow, 555pp.

Alexakhin, R. M., V. Fesenko, and N. I. Sanzharova. (1996). Serious radiation accidents and the radiological impact on agriculture. *Radiation Protection Dosimetry*, 64(1/2), 37–42.

Arkhipov, N. P., A. V. Egorov, and V. M. Klechkovsky. (1969). On the estimation of strontium-90 transfer from soil to plants and its accumulation in crop yield. *Doklagy VASHNIL*, 1, 2–4.

Bogdevich, I., N. Sanzharova, B. Prister, and S. Tarasiuk. (2002). Countermeasures in natural and agricultural areas after the Chernobyl accident. In: *Role of GIS in Lifting the Cloud off Chernobyl* (Ed. J. Kolejka). Kluwer-Academic, Amsterdam, pp. 60–73.

Burnazyan, A. I. (1990). In: *Results of the Studies and Experience of Mitigation of Consequences of the Accidental Contamination of the Territory by Uranium Fission Products* (Ed. A. I. Burnazyan). Energoatomizdat, Moscow (in Russian).

Da Silva, C. J., J. U. Delgado, M. T. B. Luiz, P. G. Cunha, and P. D. de Barros. (1991). Considerations related to the decontamination of houses in Goiânia: Limitation and implications. *Health Physics*, 60, 87–90.

Eisenbud, M., and T. Gesell. (1997). *Environmental Radioactivity from Natural, Industrial and Military Sources*. Academic Press, San Diego, 4th edition, ISBN 0-12-235154-1, 656pp.

Fesenko, S. V., R. M. Alexakhin, M. I. Balonov, I. M. Bogdevich, B. J. Howard, V. A. Kashparov, N. I. Sanzharova, A. V. Panov, G. Voigt, and Yu. M. Zhuchenko. (2006). Twenty years' application of agricultural countermeasures following the Chernobyl accident: Lessons learned. *Journal of Radiological Protection*, 26(4), 1–9.

Fesenko, S. V., R. M. Alexakhin, M. I. Balonov, I. M. Bogdevich, B. J. Howard, V. A. Kashparov, N. I. Sanzharova, A. V. Panov, G. Voigt, and Y. M. Zhuchenko. (2007). An extended review of twenty years of countermeasures used in agriculture after the Chernobyl accident. *Science of the Total Environment*, 383, 1–24.

Fesenko, S. V., R. M. Alexakhin, S. I. Spiridonov, and N. I. Sanzharova. (1995). Dynamics of ^{137}Cs concentration in agricultural products in areas of Russia contaminated as a result of the accident at the Chernobyl nuclear power plant. *Radiation Protection Dosimetry*, 60, 155–166.

Fesenko, S., P. Jacob, R. Alexakhin, N. Sanzharova, A. Panov, G. Fesenko, and L. Cecile. (2001a). Important factors governing exposure of the population and countermeasure application in rural settlements of the Russian Federation in the long term after the Chernobyl accident. *Journal of Environmental Radioactivity*, 56, 77–98.

Fesenko, S. V., A. V. Panov, and R. M. Alexakhin. (2001b). A methodological approach to justifying countermeasures in rural settlements in the long term after the Chernobyl NPP accident. *Radiation Biology. Radioecology*, 41, 415–426. (in Russian).

IAEA. (1991). *The International Chernobyl Project. Assessment of Radiological Consequences and Evaluation of Protective Measures*. Report of the International Advisory Committee, Vienna, 640pp.

IAEA. (1998). *Dosimetric and Medical Aspects of the Radiological Accident in Goiânia in 1987*. IAEA-TECDOC–1009. IAEA, Vienna, ISSN 1011-4289.

IAEA. (2006a). *Environmental Consequences of the Chernobyl Accident and Their Remediation: Twenty Years of Experience*. Report of the Chernobyl Forum Expert Group Environment, IAEA, Vienna, ISBN 92-0-114705.

IAEA. (2006b). *Radiological Conditions in the Dnieper River Basin: Assessment by an International Expert Team and Recommendations for an Action Plan*. Radioecological Assessment report Series, 185 pp., IAEA, Vienna, ISBN 92-0-104905-6. STI/PUB/1230.

IAEA. (2009). *Quantification of Radionuclide Transfer in Terrestrial and Freshwater Environments for Radiological Assessments*. IAEA-TECDOC. IAEA, Vienna, in press.

ICRP. (1977). *Recommendations of the International Commission on Radiological Protection*. Publication 26. *Annals of the ICRP*, 1, Pergamon Press, Oxford.

ICRP. (1991). *1990 Recommendations of the International Commission on Radiological Protection.* ICRP Publication 60. *Annals of the ICRP*, 21. Elsevier, Amsterdam.

ICRP. (2003). *A Framework for Assessing the Impact of Ionizing Radiation on Non-Human Species.* ICRP Publication 91. *Annals of the ICRP*, 33. Elsevier, Amsterdam.

Iranzo, E., and C. R. Richmond. (1987). *Plutonium contamination twenty years after the nuclear accident in Spain.* Oak Ridge National Laboratory, Tennessee.

Izrael, Yu. A., E. V. Kvasnikova, I. M. Nazarov, and Sh. D. Fridman. (1994). Global and regional contamination of the territory of the European part of the former USSR by ^{137}Cs. *Meteorology and Hydrology*, 5, 5–9. (in Russian).

Jacob, P., S. Fesenko, S. K. Firsakova, I. A. Likhtarev, C. Schotola, R. M. Alexakhin, Y. M. Zhuchenko, L. Kovgan, N. I. Sanzharova, and V. Ageyets. (2001). Remediation strategies for rural territories contaminated by the Chernobyl accident. *Journal of Environmental Radioactivity*, 56, 51–76.

Ministry of Health of the Russian Federation. (2002). *Hygienic Requirements to Safety and Nutritive Value of Food Products.* Sanitary-epidemiologic code. SanPiN 2.3.2.1078-01. Moscow, 166pp. (in Russian).

NCRP. (2007). *Cesium-137 in the Environment: Radioecology and Approaches to Assessment and Management. Recommendations of the National Council on Radiation Protection and Measurements.* NCRP Report 154, pp. 1–382.

Nikipelov, B. V., G. N. Romanov, L. A. Buldakov, N. S. Babaev, Yu. B. Kholina, and E. I. Mikerin. (1987). The radiation accident in the South Urals in 1957. *Atomnaya Energia*, 67, 74–80. (in Russian).

NRB 76/87. (1988). *Radiation Safety Standards (NRB-76/87) and the Basic Sanitary Code for Work with Radioactive Substances and Other Sources of Ionizing Radiation*, OSP-72/87. 3rd revised and enlarged edition. Moscow, Energoatomizdat, 1988, 160pp. (in Russian).

NRB-69. (1970). *Radiation Safety Standards (NRB-69).* Atomizdat, Moscow, 2nd edition, 112pp. (in Russian).

NRB-96. (1996). *Radiation Safety Standards (NRB-96).* Hygienic standards HS 2.6.1.054-96, Moscow, Goskomsanepidnadzor of Russia, 127pp. (in Russian).

Panov, A. V., S. V. Fesenko, N. I. Sanzharova, and R. M. Alexakhin. (2006). Remediation of local areas of radioactive contamination. *Atomnaya Energia*, 100(2), 125–134. (in Russian).

Prister, B. S., R. M. Alexakhin, V. G. Bebeshko, I. M. Bogdevich, P. I. Zamostyan, Ya. E. Kenigsberg, I. A. Likhtarev, V. A. Poyarkov, V. M. Shestopalov, and A. F. Tsyb. (2007). In: *The Chernobyl Disaster: Effectiveness of Measures to Protect the Public, Experience of International Cooperation* (Ed. B. S. Prister). Kiev, Centre of technical information "Power Engineering and Electrification" (in Russian).

Prister, B. S., Yu. A. Ivanov, L. V. Perepelyatnikova, and V. A. Pronevich. (1996). Post-accidental problems in the Ukraine. *Agrarnaya Nauka*, 3, 8–11. (in Russian).

Ratnikov, A. N., A. V. Vasiliev, R. M. Alexakhin, E. G. Krasnova, A. D. Pasternak, B. J. Howard, K. Hove, and P. Strand. (1998). The use of hexacyanoferrates in different forms to reduce radiocesium contamination of animal products in Russia. *Science of the Total Environment*, 223, 167–176.

Romanov, G. N., E. G. Drozhko, B. V. Nikipelov, I. G. Teplyakov, and V. P. Shilov. (1993). Summing up: Restoration of the economic activity. In: *Environmental Impacts of Radioactive Contamination in the South Urals* (Eds V. E. Sokolov and D. A. Krivolutsky). Nauka, Moscow. ISBN 5-02-005719-3 (in Russian), pp. 324–332.

SanPiN-96. (1997). *Hygienic Requirements to Quality and Safety of Raw Food and Foodstuffs.* Sanitary code, SanPiN 2.3.2.560-96. Moscow, 270pp. (in Russian).

Shoigu, S. K. (2002). In: *Impacts of Technogenic Radiation Effects and Problems of Remediation of the Urals Region* (Ed. S. K. Shoigu). Ministry of Civil Defence, Emergencies and Elimination of Consequences of Natural Disasters, Moscow, 287pp. (in Russian).

Shutov, V. N., G. Ya. Bruk, L. N. Basalaeva, V. N. Vasiletskiy, N. P. Ivanova, and I. S. Karlin. (1996). The role of mushrooms and berries in the formation of internal exposure doses to the population of Russia after the Chernobyl accident. *Radiation Protection Dosimetry*, 67, 55–64.

Sokolov, V. E., and D. A. Krivolutsky. (1993). *Environmental Impacts of Radioactive Contamination in the South Urals* (Eds V. E. Sokolov and D. A. Krivolutsky). Nauka, Moscow, 336 pp. ISBN 5-02-005719-3 (in Russian).

State Duma of the Russian Federation. (2006). Findings by the established in accordance with the resolution of the Ecological Committee of the State Duma of the Federal Assembly of the Russian Federation commission of independent expert examination of the ecological situation within the Techa cascade and Techa river area, resulting from the PP "Mayak" activity. Moscow (in Russian).

CHAPTER 5

Remediation of Sites Contaminated by Nuclear Weapon Tests

Piero R. Danesi*

Contents

1. Introduction

The testing of nuclear weapons in the atmosphere started in 1945 and continued until 1980. The total number of tests carried out amounted to 541, corresponding to a total yield of about 440 megatonnes (Mt). It has been estimated that about 29 Mt of fission yield was associated with debris locally deposited at the test sites (Bennet, 2000). This debris was the major cause of local contamination. An additional cause of local contamination was the conduction of the so-called safety tests. In this case, assembled nuclear weapons were exploded by conventional explosives to simulate a possible accident. Although in safety tests no or very little fission took place, a considerable amount of fissile material was scattered at the site of the test.

*Corresponding author. Tel./Fax: +43-1-7968936
E-mail address: piero@danesi.at

Arsenal, Objekt 3/30, A-1030 Vienna, Austria

Radioactivity in the Environment, Volume 14
ISSN 1569-4860, DOI 10.1016/S1569-4860(08)00205-2

This led, in several cases, to long-lasting contamination by long-lived actinide isotopes.

The atmospheric nuclear tests were conducted by the United States, the Soviet Union, the United Kingdom, France and China at a total of 16 sites, namely the Nevada Test Site (USA); Bikini and Enewetak (Marshall Islands); Johnston Island (USA); Christmas and Maiden Islands (Kiribati); Emu, Maralinga and Montebello Islands (Australia); Mururoa and Fangataufa (French Polynesia); Reggane (Algeria); Semipalatinsk (Kazakhstan); Novaya Zemlya and Kapustin Yar (Russia) and Lop Nor (China).

However, proper remediation was only conducted at the atoll of Enewetak, in the Republic of the Marshall Islands (test conducted by the United States), and at Maralinga, in the Victoria desert of Australia (tests conducted by the United Kingdom). At the Bikini Atoll, radioecological and remediation studies were conducted and optimal remediation technologies were identified. Nevertheless, complete remediation is still pending and resettlement of the local population has not yet taken place. Therefore, only the three cases of Bikini, Enewetak and Maralinga will be discussed in this chapter.

Remediation at all other sites was either considered not necessary (e.g. at the atolls of Mururoa and Fangataufa in French Polynesia and in the desert location of Reggane in Algeria) due to the low residual contamination and the remoteness of the locations from populated areas or postponed to other times because of financial reasons (e.g. Semipalatinsk in Kazakhstan) or because the sites are still used for military purposes (e.g. Lop Nor in China). However, it must be mentioned that, even at these sites, some cleanup operations were conducted, particularly where considerable plutonium residues, resulting from safety tests, were present.

For the three cases of Bikini, Enewetak and Maralinga, we will provide (a) a short historical overview of what happened at the test location, (b) a description of the radiological condition existing some time after the tests and (c) the remediation actions that were planned and/or implemented and the improvements they led to.

2. The Bikini and Enewetak Atolls in the Republic of the Marshall Islands: Tests Conducted by USA

The Republic of the Marshall Islands (see Figure 1) consists of two archipelagic island chains of 29 atolls comprising 1,152 islands and islets in total.

In general, an atoll is an island consisting of several small islands (islets) of different dimensions arranged along the rim of the volcano from which the atoll originated. The largest of these islets, often called islands as well,

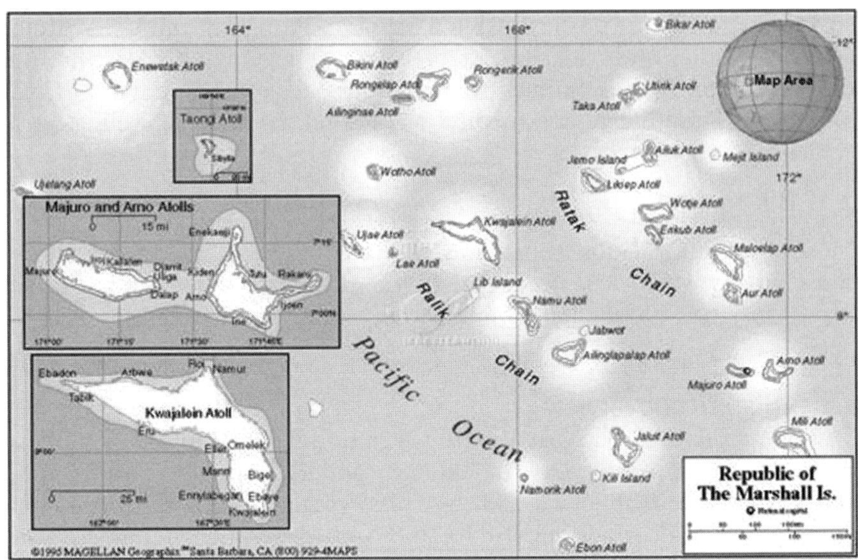

Figure 1 Map of the Republic of the Marshall Islands.

are the parts of the atoll where generally (but not always) the population resides.

The Marshall Islands are situated about 4,000 km southwest of Honolulu, about halfway between Hawaii and Papua New Guinea, in the tropical waters of the northern Pacific Ocean, north of the equator and west of the International Date Line. The land area amounts to only 181 km^2, but the total sea territory is vast, stretching over 3,000 km between its north-western and south-western extremes. The capital is Majuro, in the Majuro Atoll.

Immediately after the Second World War, the United States created a Joint Task Force to develop a nuclear weapon testing programme. Planners examined a number of possible locations in the Atlantic Ocean, the Caribbean and the Central Pacific. Eventually it was decided that the coral atolls in the northern Marshall Islands offered the best advantages of stable weather conditions, fewest inhabitants to relocate and isolation, with hundreds of miles of open ocean to the west where trade winds were likely to disperse radioactive fallout. During the period between 1945 and 1958, a total of 67 nuclear tests were conducted at Bikini and Enewetak Atolls and adjacent regions within the Republic of the Marshall Islands (UNSCEAR, 2000; USDOE, 2000; Simon and Robison, 1997). The most significant contaminating event was the Castle Bravo test conducted at the Bikini Atoll on 1 March 1954. The Bravo was an experimental thermonuclear device with an estimated explosive yield of 15 Mt that led to widespread fallout contamination also over the inhabited islands of Rongelap and Utrik Atolls, as well as other atolls to the east of Bikini. The United States through the Office of Health Studies still continues to provide environmental monitoring, health care and medical services on the affected atolls.

2.1. Bikini Atoll

Most of the information concerning this atoll has been taken from Simon and Graham (1995), IAEA (1998a, 1998b), McEwan (2000) and Marshall Islands Program (2006) and additional references as indicated.

2.1.1. Historical overview

Testing at Bikini Atoll started with 'Operation Crossroads' in 1946. This experiment was organised by the US Navy and included the so-called Able and Baker shots. It involved 242 ships, 156 aircraft and more than 42,000 military and civilian personnel. Five thousand experimental animals were also involved. From July 1946 until February 1954, Bikini Atoll remained inactive as a test site. In 1948, 1951 and 1952, nuclear tests were conducted on the Enewetak Atoll. Then, in February 1954, Bikini Atoll was reactivated as a test site with the 'Castle' series of tests. In 1956 they continued with the 'Redwing' series and terminated the tests in 1958 with the 'Hardtack I' series. The tests with highest yield were those in the 'Castle' series, which included the 'Bravo' shot, a thermonuclear device with 15 Mt equivalent yield of TNT. A considerable fraction of the debris from near-surface denotations and all the debris from high-altitude air bursts entered the global environment, producing a worldwide pattern of global-fallout deposition. The regional fallout from the Bravo test also caused widespread contamination of the Bikini Atoll and forced the evacuation of Marshallese people living on Rongelap and Utrik Atolls. It is estimated that about 50% of the fission yield associated with near-surface nuclear detonations was deposited on a local or regional scale (Hamilton, 2004). Figure 2 shows a map of the Bikini Atoll and the site of the crater created by the 'Bravo' explosion.

Prior to the Able test in 1946 – the first nuclear test in the Bikini Atoll – the 167 Bikinians then living on Bikini Island were evacuated to the Rongerik Atoll, about 200 km to the east, where they were supposed to reside until the completion of the testing. The Bikinians remained on Rongerik Atoll for a period of two years. In 1948, they were moved briefly to Kwajalein Atoll and later in the same year to Kili, a small (0.8 km^2) isolated island. In August 1968, following a number of radiological surveys that had been carried out since 1958 to assess the impact of the United States' programme of nuclear weapon testing, the United States announced that the Bikini Atoll was safe for habitation, and the resettlement of the

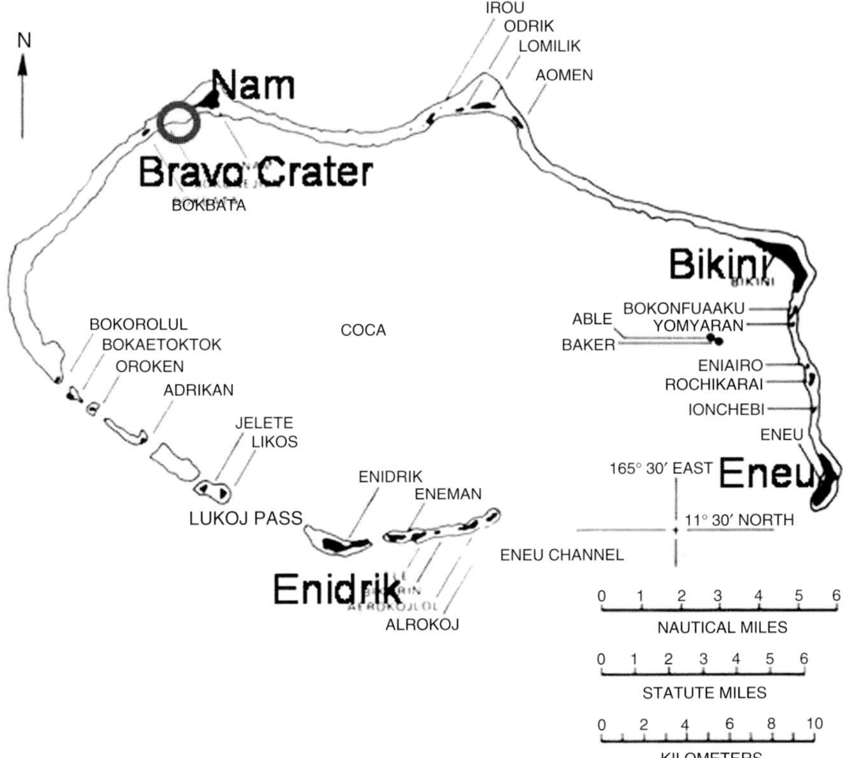

Figure 2 Map of the Bikini Atoll.

Bikinian people on the atoll was approved. From February to October 1969, the atoll was cleared of debris. Fruit trees, including coconut, breadfruit, pandanus, papaya and banana, were replanted. A further radiological survey of the Bikini Atoll was carried out in 1970. The same year about 100 people returned to the atoll. However, the Bikinian people remained unconvinced of the safety of the atoll, and in 1975 they initiated a lawsuit against the US government to stop the resettlement effort until a satisfactory and comprehensive radiological survey had been carried out.

Therefore, in 1975 a further radiological assessment of Bikini Atoll was conducted (Robison et al., 1977). However, at that time the trees planted in 1969 had not yet grown to maturity and few samples were available for making a reliable estimate of radionuclide concentrations in food crops. In 1976, an external radiation survey programme was conducted for five northern atolls, which included some radiological measurements at Bikini. A continuing sampling and analytical programme was begun at Bikini Atoll in 1978 to gather additional data as a basis for more precise radiation dose estimates for the residents of Bikini and Eneu Islands. Whole body radiation measurements for the purpose of estimating the intake of radioactive materials by Bikinian residents started in April 1977. In 1978, it was determined that for the inhabitants of Bikini Atoll, a 10-fold increase in the body content of the radionuclide ^{137}Cs had occurred (Miltenberger et al., 1980). This increase was the result of a combination of the coconut trees starting to bear fruit and a drought that led to increased consumption of coconut fluid due to the limited availability of drinking water. In August and September 1978, in response to the high uptake of caesium by the population, the Bikinians who had returned to Bikini Atoll were relocated to Kili Island and to Ejit Island at Majuro Atoll.

2.1.2. Radiological conditions before remediation

At the time of the second relocation, a new radiological survey was started in 11 northern atolls of the Marshall Islands, sponsored by the US Department of Energy. Samples of vegetation, marine foods, animals and soil were collected and analysed, and revised radiation dose evaluations were published in 1980 and 1982 (Robison et al., 1980, 1981a, 1981b, 1982). These indicated that the terrestrial food chain would be the most significant exposure pathway to future inhabitants. This dose assessment was then updated on the basis of ongoing monitoring programmes at the atoll (Kehl et al., 1995; Robison et al., 1997a, 1997b).

A separate radiological assessment was also commissioned by the government of the Republic of the Marshall Islands. In general, the study confirmed the findings of earlier measurement programmes (Simon and Graham, 1997). Nevertheless, the government of the Marshall Islands subsequently requested the IAEA to conduct an independent international review of the radiological conditions at Bikini Atoll. This information was published in 1998 (IAEA, 1998a, 1998b). The three studies reached essentially the same conclusions. The major findings of these studies are summarised in the following sections.

2.1.2.1. Residual radionuclides. The significant residual radionuclides from the nuclear tests that were found to remain in the soil and surroundings of Bikini Atoll were ^{90}Sr, ^{137}Cs, $^{239+240}Pu$ and ^{241}Am. They were found in varying degrees in both the terrestrial and marine environments.

2.1.2.2. Soil characteristics. The coral soil of Bikini Atoll was found to be composed mostly of calcium carbonate ($CaCO_3$), with some magnesium carbonate ($MgCO_3$) and essentially no clay. The soil was highly alkaline, with a pH ranging from 7.7 to 9.0. The surface horizons had high amount of organic matter, but this dropped significantly with depth in the soil column. As a result, most of the natural nutrients and the water retention capacity of the soil were found to be confined to the top 25–40 cm of the soil column. The soil was low in exchangeable potassium (generally less than 50 ppm) and marginal in phosphorus and trace mineral content. Some native plant species and most introduced species showed definite signs of potassium deficiency. The soil had a pattern of ^{137}Cs and ^{90}Sr availability to plants very different from that for which most data were reported in the literature, corresponding to the aluminium silicate clay soils generally present in Europe, South and North America.

2.1.2.3. Radionuclide concentrations. The Bikini Island, the primary island for habitation at Bikini Atoll, was found to have the highest activity concentrations of ^{137}Cs per unit mass of soil and vegetation. The average ^{137}Cs concentration varied over a considerable range in the various atoll islands. Examples of activities of radionuclides in soil in the Bikini Island are shown in Table 1 (Robison et al., 1997a).

The average ^{137}Cs concentration in soil and vegetation on Eneu Island, the other main island of residence, was found to be about 10%–13% of that of Bikini Island. Nam Island, one of the two other islands large enough for possible residence, had a ^{137}Cs concentration in soil of about 70% of that of Bikini Island. The ^{137}Cs concentration in soil on Enidrik Island, the other large island, was about 15% of that of Bikini Island. The concentrations of transuranic radionuclides ($^{239+240}Pu$ and ^{241}Am) and their ratios to

Table 1 Median activity concentrations of ^{90}Sr, ^{137}Cs, $^{239+240}Pu$ and ^{241}Am in soil on Bikini Island (Bq g^{-1}, dry weight).

Soil depth (cm)	^{90}Sr	^{137}Cs	$^{239+240}Pu$	^{241}Am
Interior of the island				
0–5	2.3	1.7	0.32	0.26
5–10	1.2	2.0	0.29	0.19
10–15	0.58	1.5	0.15	0.081
15–25	0.19	0.73	0.053	0.026
25–40	0.071	0.47	0.0081	0.012
40–60	0.018	0.32	0.011	0.017
0–40	0.70	1.1	0.17	0.11
Village area				
0–5	1.2	1.0	0.20	0.1
5–10	1.0	1.2	0.30	0.13
10–15	0.81	1.5	0.22	0.12
15–25	0.53	0.9	0.14	0.064
25–40	0.18	0.62	0.064	0.059
40–60	0.028	0.32	0.0058	0.012
0–40	0.67	1.6	0.24	0.13

Source: Robison et al. (1997a, p. 104).

concentrations of ^{137}Cs and ^{90}Sr were found to vary around the atoll, reflecting the difference in the design of the nuclear devices detonated in the various parts of the atoll and different release scenarios. The concentrations of transuranic radionuclides in the soil on Nam Island exceeded those on Bikini Island, while those on Enidrik Island were somewhat lower than those on Bikini Island. In general, the radionuclide concentrations decreased rapidly with depth in the soil column, although there were exceptions in parts of some islands.

Of the residual radionuclides present in soil, those found to be of greatest potential significance for the inhalation exposure pathway were $^{239+240}$Pu and ^{241}Am incorporated into surface soil particles which could be resuspended by the wind action. The average resuspension of the surface soil was considered very low, with the resuspension factors ranging from 10^{-10} to $10^{-11}\,m^{-1}$. On the basis of the measured activities of these radionuclides in the soil and the resuspension factors mentioned above, the air concentrations of these radionuclides were found to be very low, and consequently, the expected contribution to doses from radiation exposure via inhalation pathways was judged to be negligible. Samples of various locally available foods and water were also collected and analysed for their content of residual radionuclides. The highest ^{137}Cs concentrations were found in coconut (mean value $= 12\,Bq\,g^{-1}$ of fresh product) and some other fruits such as pandanus (mean value $= 39\,Bq\,g^{-1}$ of fresh weight) and breadfruit (mean value $= 47\,Bq\,g^{-1}$ of fresh weight).

The ^{90}Sr activities were found to be less than 10% of the respective ^{137}Cs activities in the relevant foodstuffs, and the $^{239+240}$Pu and ^{241}Am activities were even lower than those of ^{90}Sr. The food stuff containing the highest activities per unit mass of ^{90}Sr, ^{137}Cs, $^{239+240}$Pu and ^{241}Am are given in Table 2 (Robison et al., 1997a). Sr-90, ^{137}Cs, $^{239+240}$Pu and ^{241}Am were also found in the lagoon. Cs-137 was always present at very low concentrations both in the lagoon sediment as well as in the lagoon water and fish. Sr-90 remained in the lagoon environment, primarily in the

Table 2 Mean activities of ^{90}Sr, $^{239+240}$Pu and ^{241}Am per unit wet weight in selected Bikinian food (Bq/g).

Type of food	^{90}Sr	Type of food	$^{239+240}$Pu	^{241}Am
Papaya	4.9×10^{-2}	Clam	8.3×10^{-4}	4.6×10^{-4}
Squash	6.8×10^{-2}	Trochus	8.3×10^{-4}	4.6×10^{-4}
Pumpkin	6.8×10^{-2}	Tridacna muscle	8.3×10^{-4}	4.6×10^{-4}
Banana	4.9×10^{-2}	Jedrul	8.3×10^{-4}	4.6×10^{-4}
Arrowroot (cooked)	6.8×10^{-2}	Pork liver	1.2×10^{-4}	5.2×10^{-5}
Citrus	4.9×10^{-2}			

Source: Robison et al. (1997a, p. 103).

carbonate matrix, being chemically bound in the growing coral and in the coral sediment. Consequently, ^{90}Sr was relatively unavailable to marine biota species. Pu-239+240 and ^{241}Am were also found in the coral sediments. The best estimates for the total inventories of $^{239+240}$Pu and ^{241}Am in Bikini Atoll sediments were 103 ± 25 TBq and 93 ± 10 TBq, respectively (Noshkin et al., 1975). The studies indicated that plutonium was transferred into the aquatic ecosystem in small but measurable concentrations through the action of biogeochemical processes acting on contaminated components of the sedimentary reservoir at the atoll. However, the observed transfer of these radionuclides through the marine food chain to human foodstuffs was very low, to make negligible any associated radiological impact.

*2.1.2.4. **Radiation doses.*** Measurements of absorbed dose rates in air due to gamma radiation were also made (Simon and Graham, 1995). The annual absorbed dose in air measured in 1978 at 1 m above the ground varied from about 0.01 to 5 mGy. The average effective dose contribution due to external gamma radiation was estimated by means of direct measurements inside and outside houses, measurements in the village area and aerial survey measurements. The dose estimate was based on occupancy assumptions made on the basis of discussions with the Marshallese people and direct observations. They spent 10 h day^{-1} inside the houses, 9 h day^{-1} outside in the village area, 3 h day^{-1} in the interior region of the island and 2 h day^{-1} on the beach or lagoon. The average annual effective dose based on this occupancy model, and decay corrected to 1999, was 0.4 mSv. For internal doses the assessment used conversion factors from activity intake into effective dose, which were compatible with those established in the Basic Safety Standards (IAEA, 1996).

The overall dose, that is the sum of the annual effective dose due to external radiation and the committed effective dose due to intakes – assuming that the diet consisted of both imported and locally derived foods – was estimated to be 4.0 mSv (plus the dose due to natural background radiation). For a diet consisting of only locally derived foodstuffs, the annual overall dose amounted to 15 mSv (plus the dose due to natural background radiation).

The uptake of ^{137}Cs into terrestrial foodstuffs was found to account for the largest fraction of the total estimated dose, and the external gamma exposure pathway accounted for most of the remainder. The contribution of ^{90}Sr to the total dose was minor and the contributions from $^{239+240}$Pu and ^{241}Am insignificant. The intake of marine foods, stored rainwater and groundwater, and inhalation of resuspended soil together accounted for less than 1% of the dose. As the dominant contribution to dose derives from ^{137}Cs, the annual doses from living on Bikini Atoll will decline with time.

2.1.3. Remedial measures

As mentioned earlier, although remedial measures and resettlement of the local population have not yet taken place, several remedial options for Bikini were studied in detail. The planned interventions were based on the potential exposure of a hypothetical critical group living on the island consuming a high-calorie diet entirely derived from foods produced locally. This assumption was considered conservative, as it is more likely that future diets will include imported foodstuffs. At present, the diets generally adopted at the Marshall Islands have a large component of imported foodstuffs, and it seems unlikely that the trend in this direction will be reversed in the near future. However, the possibility was not excluded that the consumption of locally derived foodstuffs could increase again if the financial conditions that presently permit the import of many foodstuffs were going to change.

Based on this assumption, the annual effective dose to the critical group was assessed to be about 15 mSv. This is higher than the generic action level of an annual effective dose of up to about 10 mSv. Therefore, the possibility of implementing remediation strategies was considered to be highly justified. Several remedial techniques to reduce the dose from ^{137}Cs were proposed, such as:

- irrigation of the soil with large quantities of salt water for the purpose of leaching the ^{137}Cs from the soil;
- addition of binding agents such as zeolites to soil to trap the ^{137}Cs and make it unavailable for plant uptake;
- cropping and disposal of vegetation to remove ^{137}Cs accumulated in plants;
- the removal and disposal of the top 40 cm of soil where most of the activity was present; and
- treatment of the soil with fertilisers with high potassium concentration.

Of these techniques, the last two, soil removal and treatment with potassium fertilisers, were found to be the most effective.

Specifically, the removal of the top 40 cm of soil from Bikini Island was effective in reducing radiation exposure due to ^{137}Cs and other residual radionuclides in the terrestrial environment. However, as the natural activity levels of the local topsoil were very low in comparison with those of most continental soil, if the topsoil was replaced by soil brought in from outside the Marshall Islands, the annual effective dose due to natural background radiation would consequently increase. Moreover, the removal of hundreds of thousands of tonnes of soil could have adverse environmental and social consequences, especially because the tree crops that constitute the natural food supply require the fertile topsoil. The removal of the topsoil would necessitate removing about 30,000 mature coconut, pandanus and papaya trees and breadfruit. Therefore, the soil-removal option was considered an effective but very complex and extremely expensive operation.

Experiments on the treatment of the local soil with potassium fertiliser indicated that this approach could be the most practical and effective one in reducing the uptake of ^{137}Cs into foods (Robison and Stone, 1992). The results from one of several field trials at Bikini Island are shown in Figure 3.

A reduction in ^{137}Cs concentrations in coconut milk (and many other food items) to 5% of their original values was achieved in several experiments. The reduced concentration of ^{137}Cs was found to persist for nearly five years after the application of potassium fertiliser was terminated, and the subsequent increase in uptake of ^{137}Cs was very slow. However, the additional application of potassium fertiliser was considered necessary every four or five years to maintain the low ^{137}Cs activity levels in local foods, and radioactive decay would have reduced the activity to insignificant levels. Consideration was also given to the potential ecological effects of the proposed potassium fertiliser treatment. At the planned levels of potassium treatment, no possibility of altering significantly the soil chemistry was foreseen, and the transfer of potassium to groundwater was found to be very low (IAEA, 1998a, 1998b).

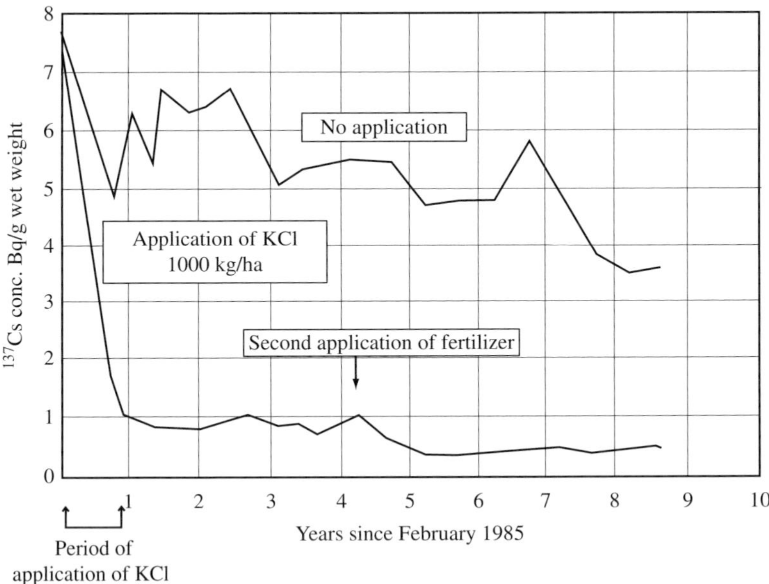

Figure 3 Reduction in ^{137}Cs levels in drinking coconuts at Bikini Island (lower curve) after an initial and a second application of potassium chloride fertilizer. Each application: 1,000 kg ha^{-1}. Very similar results were obtained with two applications of 2,000 kg ha^{-1} of KCl or by replacing the second application with NP fertilizer (redrawn from Robison et al., 1997a).

In dwelling areas, two potentially significant exposure pathways were considered, namely the external exposure to the gamma radiation emitted by ^{137}Cs and the possible ingestion or inhalation of plutonium and americium from soil. Thus, the remediation strategy considered was the removal of surface soil up to a depth of 30 cm in the location where the village would be established. The best option was to remove soil from around each housing site and replace it with a layer of crushed coral to minimise external exposure and possible ingestion of the remaining soil. This combination of remedial actions was estimated to reduce the doses by a factor of about 10 from pre-treatment levels (Robison et al., 1997a).

An alternative to the removal of soil from the living areas was a process later employed successfully in the aftermath of the Chernobyl accident. This consisted of ploughing the uppermost 30–50 cm of soil into deeper soil layers. In this way, the most contaminated topsoil would become buried and replaced by the much less contaminated deeper soil. This would also allow avoidance of the problem of soil disposal. Nevertheless, the soil would still require the same potassium fertiliser treatment as for the rest of the land where crops were to be grown.

After the potassium treatment of the soil in areas where food crops could be grown and the replacement of soil from around the living areas, it was estimated that the dose rate from the consumption of an entirely locally derived diet would be significantly reduced. The annual effective dose estimate of 15 mSv (plus natural background dose) would change after remediation to 1.2 mSv (plus natural background doses). The mixed diet of imported plus local foods, which resulted in an annual effective dose estimate of 4 mSv without remediation (plus the doses due to the natural background), would, after the same remediation, result in a dose of 0.4 mSv (plus natural background dose) as indicated in Figure 4.

Finally, it must be mentioned that, by using empirical data from annual and semi-annual monitoring surveys of selected trees on Bikini and Eneu Islands, it was demonstrated that the environmental half-life of ^{137}Cs was more important than the radiological decay in controlling its fate and distribution in coral soils (Robison et al., 2003). For example, the estimated effective half-life of ^{137}Cs on Bikini, Enewetak and Rongelap Atolls was found to be from 8 to 9.8 years. These findings suggested that dose predictions based on simple radiological decay-corrected measurement data for calculating prospective integral doses were too pessimistic. It was also considered likely that labile ^{137}Cs in soil could become slowly fixed to the mineral phases present in the soil and with time would become less available for soil-to-plant uptake.

Applying a mean effective ^{137}Cs half-life of 8.5 years (Robison and Sun, 1997), the predicted population average effective dose for resettlement of Bikini in 2010, assuming the consumption of some imported foods, was conservatively estimated to decrease to 0.17 mSv year^{-1}. This means

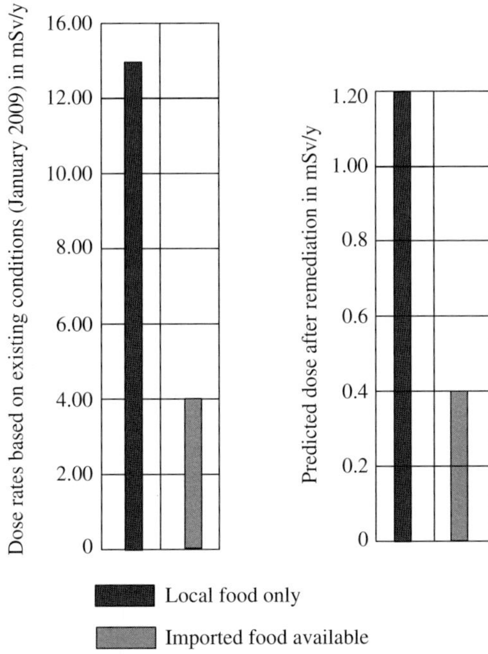

Figure 4 Annual radiation doses on Bikini before and after remediation in case of availability and non-availability of imported food (Robison et al., 1997a, 2003).

that the exposure conditions on Bikini could improve at an accelerated rate, making return of the population to the Bikini Atoll feasible in the near future.

2.2. Enewetak Atoll

Most of the information concerning this atoll has been taken from Hamilton et al. (2001), Marshall Islands Program (2006), Johannes et al. (2002) and additional references as indicated.

2.2.1. Historical overview

The Enewetak Atoll is located in the Equatorial Pacific Ocean in the north-western portion of the Republic of the Marshall Islands. This atoll was chosen as the site for nuclear testing from 1948 to 1951 because of its remote location and its geological features. The map of the atoll is shown in Figure 5.

After the first series of nuclear tests on the Bikini Atoll in 1946, the 136 local inhabitants of the Enewetak Atoll were relocated to the Ujelang Atoll in December 1947, in the course of the preparation for the tests on Enewetak. Operation Sandstone started in April 1948 and included three tests conducted on 60-m-high steel towers located on the islands of

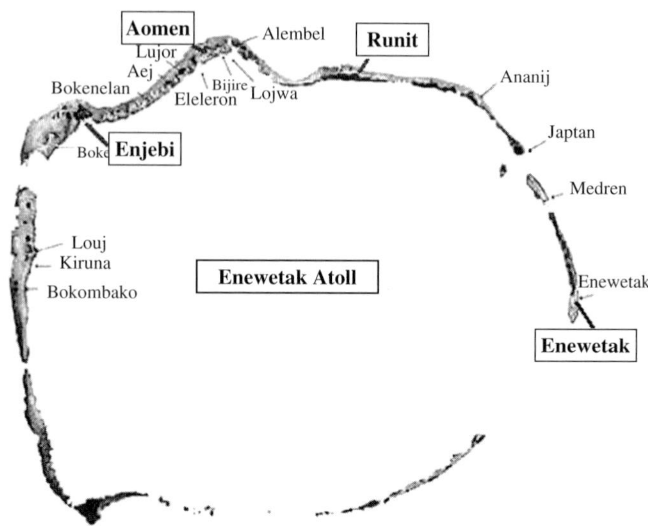

Figure 5 Map of the Enewetak Atoll.

Enjebi, Aomen and Runit. Four additional near-surface tests were conducted on towers as part of Operation Greenhouse during 1951. Operation Ivy, in 1952, initiated the tests of large thermonuclear devices. The Mike thermonuclear blast of 31 October 1952, having an explosive yield of 10.4 Mt, vaporised the island of Elugelab and left behind a deep crater about 1 km in diameter. Early analysis of Mike fallout debris showed the presence of two new isotopes of plutonium, ^{244}Pu and ^{246}Pu, and led to the discovery of the new heavy elements, einsteinium and fermium. Operation Castle involved a single test on Enewetak in 1954. A total of 11 nuclear tests were also conducted on Enewetak in 1956 as part of Operation Redwing, including an air burst from a balloon located over water. In 1958, the United States in anticipation of suspending atmospheric nuclear testing assembled a large number of devices to be tested before the moratorium became effective. From April through August 1958, 22 near-surface nuclear denotations on platforms, barges or underwater were conducted on the Enewetak Atoll. The majority of nuclear tests were conducted in the northern parts of the atoll. They produced highly localised fallout contamination of neighbouring islands and the atoll lagoon. By the time the test moratorium came into effect on 31 October 1958, the United States had conducted 42 nuclear tests on the Enewetak Atoll.

2.2.2. Radiological conditions before remediation

Enewetak Atoll continued to be used for military operations until a cleanup and rehabilitation programme started in 1972. Forty-two nuclear tests had produced local fallout that contaminated the islands and lagoon of the atoll with radioactive fission products, activation products and fissioned nuclear material. Large amounts of concrete, metal, cables, bunkers, buildings and other miscellaneous materials, some contaminated and some not, were also left on the atoll after the testing programme came to an end. As a result of the radiological contamination of the atoll, plutonium activities in the top 2 cm of the soil ranged from 0.4 to 7 mBq g^{-1} with a median of 4.4 for Enewetak Island.

The Runit Island and the adjoining reef were used for several nuclear tests. These took place on the surface, on towers and in the atmosphere, and many occurred on barges located in the lagoon. As a result, this location was the most severely radiologically contaminated of the Enewetak Atolls.

In 1958 the safety test called Quince was also conducted on Runit. In this test the nuclear weapon was detonated only by conventional high explosives. Therefore, no fission/fusion reactions took place. This resulted in scattering the plutonium nuclear fuel over a large area of the island. In preparation for the next explosion, scheduled only 12 days later in the same location, 3–5 in. of the plutonium-contaminated soil was removed from the site by bulldozers and disposed of in the lagoon immediately offshore from the centre of the island (Noshkin and Robison, 1997; Hamilton et al., 1982). During a radiological survey conducted in 1972 (Nelson and Noshkin, 1973), the transuranic elements (TRUs) resulting from the barge events and the bulldozing operations were well identified in the near-shore sediments and quantified. The survey showed that the mean quantity of TRUs distributed over the surface sediments (to a depth only of 2.5 cm) in a 0.7 km^2 region extending 0.8 km lagoon-ward from the island was about 64 GBq. The mean concentration and inventory of the transuranics in the surface sediment offshore Runit and in the entire lagoon (Nelson and Noshkin, 1973; Noshkin, 1980) are summarised in Table 3 (Noshkin and Robison, 1997).

The lagoon sediments were found to be the largest reservoir of plutonium at the atoll. As these sediments are exposed to the bottom waters of the lagoon,

Table 3 Surface concentration of transuranic elements in the lagoon sediments (GBq km^{-2}).

	$^{239+240}Pu$	^{238}Pu	^{241}Am	Total TRU
0–2.5 cm depth	9.9	1.4	3.0	14.3
0–16 cm depth	44,000	6,200	17,600	67,800
0–2.5 cm depth offshore Runit (0.86 km^2)	56.5	14.1	5.7	76.3
0–16 cm depth offshore Runit (0.86 km^2)	225	57	30	312

Source: Noshkin and Robison (1997, p. 237).

the radionuclides are continuously remobilised to the hydrosphere from the sedimentary source term. Mean water concentrations of plutonium measured in the lagoon over time are shown in Table 4 (Noshkin and Robison, 1997). The data demonstrate that remobilisation is continuously occurring from the sediments as the inventory of plutonium in the lagoon water stays approximately constant even decades after testing. At Runit, the plutonium in the near-shore sediments was also found to mobilise, becoming available for uptake by near-shore organisms for many years (Nelson and Noshkin, 1973; Noshkin, 1980; Noshkin et al., 1974, 1976; Noshkin et al., 1981).

In May 1958, on an artificially constructed extension of Runit Island on the lagoon side of the reef, an 18 kt weapon was detonated (Cactus test). The test produced a crater about 112 m in diameter and 10 m deep. When the device exploded, some of the pulverised material fell back into the crater so that the original hole was deeper than 10 m. The crater had the shape of a spherical segment with a flat base and had a volume of approximately 7,000 m^3. Much of the surrounding rock was heavily fissured from events detonated nearby. The majority of the crater rim was on land, but about a quarter of the eastern circumference was open, permitting exchange of water between the crater and the ocean during periods of high tide. Quantities of different radionuclides were distributed non-uniformly throughout the sediment, as indicated by the analysis of the samples collected in the 10–15-m-thick fall-back zone of altered carbonate materials (Ristvet et al., 1978; Robison and Sun, 1997).

Additional studies (Noshkin, 1980; Marsh et al., 1978) demonstrated that most of the water in the Cactus crater was lost by overflow during periods of high tide and that water eventually flowed into the lagoon through a break in the land extension some 400 m northwest of the crater. The residence time of the water in the crater was found to be a function of the tidal range and could be predicted using available tide data. The studies also showed that only small amounts of crater water entered the island's groundwater or flowed subterraneanly into the lagoon. It was established

Table 4 Variation with time of the average $^{239+240}$Pu lagoon concentrations and its inventory.

	Soluble $^{239+240}$Pu ($Bq\,m^{-3}$)	Particulate $^{239+240}$Pu ($Bq\,m^{-3}$)	Total inventory (GBq)
October–December 1972	0.81	0.3	53.9
July 1974	0.93	0.70	74.5
May 1976	0.59	0.48	48.9
May 1982	0.63	–	–

Source: Noshkin and Robison (1997, p. 237).

that plutonium and other radionuclides were supplied to the crater water by three processes:

(a) transportation associated with surface ocean water advecting over the reef;
(b) release to the bottom interstitial water from the contaminated bottom sediments; and
(c) interactions involving resuspended bottom sediments with the crater water.

The latter two mechanisms contributed most of the plutonium radionuclides to the crater water column. Seawater samples from different depths in the crater and filtered interstitial water and short sediment cores were also analysed for their radionuclide content. Table 5 summarises selected concentrations of $^{239+240}$Pu, ^{241}Am and ^{137}Cs (Noshkin et al., 1976). Using an estimated exchange rate and size of the plutonium sediment reservoir, it was calculated that 11.5 MBq of $^{239+240}$Pu and approximately half this amount of ^{238}Pu were annually released from the crater bottom sediments.

This plutonium mixed with the seawater along the reef and subsequently merged with the inventory contained in the lagoon water mass. The crater source contributed about 0.03% to the annual average lagoon water inventory of soluble $^{239+240}$Pu. By filling the crater with solid debris and closing the access of ocean water on the eastern perimeter during cleanup (see next section), this small contribution of plutonium and other radionuclides to the lagoon water was eliminated.

2.2.3. Remedial actions

In 1972, the US government announced that it would conduct a cleanup and restoration operation to return the atoll to the Enewetak people. Planning for the cleanup extended from 1972 to 1977. The final project was implemented between May 1977 and April 1980. The radiological cleanup concentrated on reducing the surface soil levels of the TRUs (^{238}Pu, $^{239+240}$Pu and ^{241}Am) on some of the islands that might eventually be used for residence or growing of subsistence agricultural products. Only the quantities of transuranics were measured during the field operations. However, soil relocation also involved moving unknown quantities of long-lived fission and activation products present in the carbonate soil. From 1972 onwards, major efforts were made to remove the top 30 cm of soil from the island. The soil was eventually buried on Runit Island, located in the north of Enewetak Atoll. The cleanup guidelines were as follows:

(a) soil should be removed if the plutonium concentration exceeded 15 Bq g^{-1};
(b) soil could be left in place if the concentration was less than 1.5 Bq g^{-1}; and
(c) for concentrations ranging between 1.5 and 15 Bq g^{-1}, a decision had to be made on a case-by-case basis.

Table 5 Selected data on the concentrations of radionuclides in Cactus crater water, sediments and interstitial water before the remediation.

	$^{239+240}Pu$ solution	$^{239+240}Pu$ particulate	^{241}Am particulate	^{137}Cs solution
Mean surface water	2.6 ± 0.7 Bq m^{-3}	4.0 ± 2.1 Bq m^{-3}	<1.5 Bq m^{-3}	8.9 ± 2.6 Bq m^{-3}
Mean bottom water	3.6 ± 1.7 Bq m^{-3}	16.1 ± 8.6 Bq m^{-3}	<1.5 Bq m^{-3}	8.9 ± 2.5 Bq m^{-3}
Interstitial sediment pore water (May 1977)	–	8.8 ± 0.4 Bq m^{-3}	–	10.1 ± 1.3 Bq m^{-3}
Bottom sediments, 0–5.7 cm	–	3.04 ± 0.07 Bq g^{-1} dry weight	0.34 ± 0.07 Bq g^{-1} dry weight	0.46 ± 0.04 Bq g^{-1} dry weight
Bottom sediments, 11.4–17.2 cm	–	3.74 ± 0.26 Bq g^{-1} dry weight	0.39 ± 0.07 Bq g^{-1} dry weight	0.56 ± 0.04 Bq g^{-1} dry weight

Source: Noshkin and Robison (1997, p. 238).

Surveys were conducted with *in situ* monitoring equipment to measure surface levels (to a depth of 3–5 cm) of ^{241}Am (as a plutonium indicator) over established grids on each island. Soil samples were collected for laboratory analysis of plutonium radionuclides and americium, to identify ratios between the TRUs and ^{241}Am. These analyses provided the data to develop radiological contour maps that were used for the cleanup activities. Once sectors of the islands requiring cleanup were identified, the solid contaminants were removed. This operation was followed also by the removal of several centimetres of topsoil. The material was loaded onto barges and transported to Runit where it was off-loaded in a stockpile near the Cactus crater. The crater was surrounded by a concrete key wall to reduce scouring and undermining by wave action. The soil was filtered through a 3.8 cm screen to remove larger particles and was mixed with cement and attapulgite (a hydrous Mg–Al silicate mineral) to form a mixture designed for use in the 'tremic' method of placing a concrete mixture under water. A concrete pump transferred the slurry through a pipe to the bottom of the crater, displacing the overlying water. In the end, about 41,600 m^3 of soil filled the crater to the low-tide water level. Above the water level, the soil was blended with cement and each layer was compacted. Following this procedure, a dome-shaped mound of soil was formed over the Cactus crater. This procedure prevented communication with subterranean groundwaters. A central 'donut' hole was left in the soil dome. This space was reserved for introducing contaminated debris and soil coming from other parts of Runit Island. After the hole was filled, the 46-cm-thick concrete cap (the dome) was completed.

The amount of soil and TRUs placed inside the crater and above the ground under the dome (NRC, 1982) is shown in Table 6 (Noshkin and Robison, 1997). This table also provides the quantity of soil and TRUs removed from the different islands. Using a dry weight soil density of 1.29 g cm^{-3} (USDOE, 1982), the average TRU concentration in the soil filling the crater was calculated to be 2.4 Bq g^{-1}.

Table 6 TRU activity and volume of soil removed and placed in Cactus crater and under the dome on Runit Island.

Island	TRU activity (GBq)	Soil removed to crater (m^3)	Soil removed to dome (m^3)
Aomen	81.4	8,442	7,130
Enjebi	96.2	32,890	7,633
Runit	267.4	0	8,210
Others	100	322	14,865
Total	545	41,645	37,838

Source: Noshkin and Robison (1997, p. 240).

Approximately one-half of the total inventory of TRU originating from five northern islands was put under the concrete dome. The remaining material was surface material removed from one or more areas on Runit and dumped above the ground in the crater donut hole. The inventory of the TRUs in this waste disposal site is equivalent to only 0.8% of the total TRU inventory in the lagoon sediment to a depth of 16 cm. Therefore, if the content was to find its way into the lagoon because of a catastrophic event, the inventory of the TRUs in the lagoon would not change by any significant amount and no unacceptable hazard would result (NRC, 1982).

A report on the radiological cleanup of Enewetak Atoll provides a detailed account of the construction and filling of the dome during cleanup. The structure is referred to as a waste disposal site since it covers material contaminated with quantities of long-lived radionuclides.

A groundwater programme was also conducted to study the hydrology and groundwater geochemistry on selected islands of the atoll including Runit. Average concentrations of $^{239+240}$Pu and ^{137}Cs were measured in groundwater during pre- and post-cleanup periods. The results showed no change with time in the concentration of these radionuclides. Moreover, the ^{137}Cs concentration in the lagoon surface water off the Cactus crater was many times lower than in the groundwater wells and was comparable to the surface concentrations found in California surface waters during the late 1970s.

In order to study radionuclide activity concentrations in the edible flesh of fish, different species of fish (mullet, surgeonfish and goatfish, all used in the local marine diet by Marshallese people) were collected and analysed (Noshkin and Robison, 1997). Mullet are herbivorous and detritus feeders. Surgeonfish are herbivorous browsers, feeding on algae fronds and filamentous algae. Goatfish also consume small clams, crustaceans and small benthonic fish. Differences encountered in the concentrations of radionuclides in the flesh were found to depend on fish species and size, the location where the fish were caught, the feeding habits, the concentrations in the material ingested and the trophic level. The results (Noshkin and Robison, 1997) indicated that the surgeonfish contained higher levels of ^{137}Cs in the flesh than mullet or goatfish, with concentrations decreasing from about 10 Bq kg^{-1} wet weight in 1980 to 2 Bq kg^{-1} wet weight in 1985. The concentrations of ^{137}Cs were similar for mullet and goatfish, and stayed approximately the same over the years and were equal or below the detection limit (1–20 Bq kg^{-1} of wet weight). Before the cleanup, the median level of $^{239+240}$Pu in the flesh of all fish was only a few milliBecquerel per kilogram wet weight. When fish were again collected from Runit and from islands to the north during the early 1980s and 1990s, the median concentration of $^{239+240}$Pu in the flesh of all fish was practically the same. This led to the conclusion that no TRUs from material in the disposal site had impacted on the marine resources.

Due to concerns expressed over the possible aquatic impacts from the radionuclides buried in this disposal site, in 1980 a committee of experts was requested to evaluate the effectiveness of the Cactus crater structure in preventing harmful amounts of radioactivity from becoming available for internal or external human exposure. The committee report, published in 1982 (Noshkin and Robison, 1997), concluded that the Cactus crater containment structure and its contents did not represent any present or future health hazard for the people of Enewetak. Nevertheless, the report also recognised that part of the radioactivity contained in the structure could become available for transport to the groundwater and subsequently to the lagoon. Consequently, an environmental surveillance programme was conducted (Noshkin and Robison, 1997). Comparison of the pre- and post-cleanup data indicated that there was no adverse radiological impact on the environment from the radionuclides contained in the Cactus crater structure (Noshkin and Robison, 1997).

Repatriation of the Enewetak population started immediately after the end of the cleanup programmes. At present, the largest inventory of plutonium on Enewetak Atoll remains in the sediments of the lagoon. Since 1957, members of the Marshall Islands Program at Brookhaven National Laboratory have routinely visited Enewetak to assess any radiation-induced hazard to the population living on the island by determining their internal radionuclide body content using whole body counting and urine analysis methods (Greenhouse et al., 1980; Miltenberger et al., 1981; Lessard et al., 1984; Sun et al., 1997). So far, no significant radiological health problem has been reported.

3. Maralinga in Australia: Tests Conducted by the United Kingdom

After the Second World War also the United Kingdom took the decision to acquire nuclear weapons and in 1947 started their development. In 1949 Britain made its first approaches to the Australian government regarding the possibility of testing nuclear weapons in Australia. The Australian government agreed, and on 3 October 1952, the United Kingdom tested its first atomic weapon, named 'Hurricane', at the Montebello Islands, off the coast of Western Australia (Figure 6).

As conducting nuclear tests in naval operations proved difficult, the land-based location of Emu Fields was preliminarily chosen. One year later, the first two atomic tests on the Australian mainland, namely Totem 1 (15 October 1953 – 10 kt) and Totem 2 (27 October – 8 kt), were conducted. Subsequently, due to concerns about fallout from the previous tests and the remoteness of Emu Field, the British government formally requested a more suitable permanent test facility. The Maralinga site, about

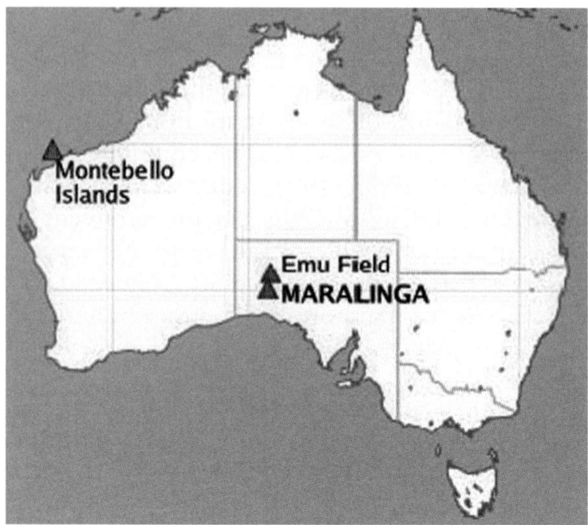

Figure 6 Map of Australia showing the locations where UK nuclear weapon tests were conducted.

200 km to the south, was then selected for this purpose in May 1955. The nuclear test ground was developed as a joint, co-funded facility between the British and Australian governments. Prior to the selection, the Maralinga site was inhabited by the Pitjantjatjara and Yankunytjatjara Aboriginal population, a large part of which was relocated to a new settlement at Yulata, and attempts were made to prevent access to the Maralinga site. Maralinga was supposed to become a permanent atomic weapons test range, but with the advent of the Nuclear Non-Proliferation Treaty and the banning of atmospheric nuclear tests, the site was abandoned in 1967. Moreover, the geology of Maralinga was not suitable for underground tests.

3.1. Maralinga

Maralinga is located on the Nullarbor Plain in South Australia. A total of seven nuclear tests were performed, with approximate yields ranging from 1 to 27 kt. The site was also used for hundreds of minor trials, many of which were safety tests intended to investigate the effects of fire or non-nuclear explosions on atomic weapons.

Most of the information concerning the events that took place at Maralinga has been taken from Symonds (1985), IAEA (1998a, 1998b), Lokan (2000), Parkinson (2000), Maralinga Cleanup (2002) and MARTAC (2003). These publications extensively describe the history of the tests, the residual contamination and the remediation operations that were later implemented.

3.1.1. Historical overview

Figure 7 (MARTAC, 2003) shows a layout of the Maralinga Range. The nuclear tests at Maralinga can be divided broadly into two categories referred to as 'major' trials, which were the explosions of nuclear weapons, and 'minor' trials, which were tests of components of nuclear weapons or tests for checking the safety of nuclear weapons.

The four series of minor trials were codenamed Kittens, Tims, Rats and Vixen. In all, these trials comprised up to 700 tests, with tests involving plutonium, uranium and beryllium. Operation Kittens consisted of 99 trials that were used for the development of neutron initiators. They involved the use of polonium-210 and uranium, and generated relatively large amounts of radioactive contamination. Operation Tims took place in 1955–1963 and involved 321 trials of uranium and beryllium tampers, as well as studies of plutonium compression. Operation Rats, consisting of 125 trials that took place between 1956 and 1960, investigated explosive dispersal of uranium and about 1 kg of plutonium. Operation Vixen consisted of several safety tests aimed at understanding what would happen to a nuclear device which burnt or was subject to a non-nuclear explosion. The 31 Vixen A trials, performed between 1959 and 1961, investigated the effects of an accidental fire on a nuclear weapon. The nuclear bomb was assembled on a large steel structure which was erected on a concrete firing pad. The device was then detonated in a manner which prevented a nuclear

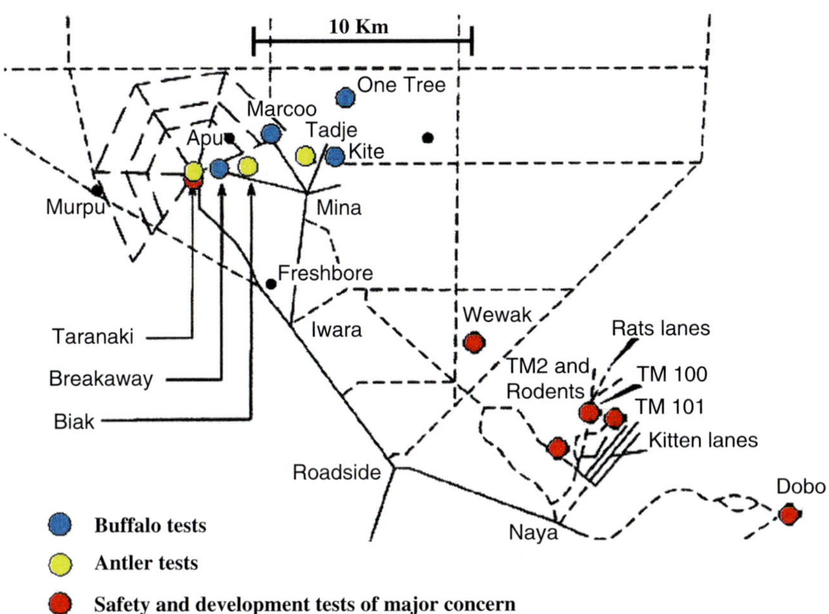

Figure 7 Major tests conducted at Maralinga (MARTAC, 2003).

explosion. However, the heat generated by the chemical explosion caused molten plutonium to be shot about 800–1,000 m into the air, where it was caught by the wind and carried several kilometres downwind. The plumes of radioactive contamination could be detected in three directions to the northeast, north and northwest of Taranaki. The force of the explosion was sufficient to damage the steel structure and the concrete firing pad so that it could not be used a second time. Moreover, these became contaminated with plutonium. The 12 Vixen B trials, conducted between 1960 and 1963, attempted to discover the effects of high explosives detonating a nuclear weapon and involved 22 kg of plutonium. They also produced 'jets' of molten plutonium extending hundreds of feet into the air. It was the residual plutonium from these minor trials – Vixen B especially – which created the major radiation problems at the site. The three main plutonium contamination plumes at Taranaki are shown in Figure 8 (Burns et al., 1994).

Seven nuclear bombs were exploded at Maralinga. Two major nuclear test series were conducted at the site: (a) Operation Buffalo and (b) Operation Antler. Operation Buffalo started on 27 September 1956. The operation consisted of the testing of four nuclear devices, codenamed One Tree, Marcoo, Kite and Breakaway. One Tree (12.9 kt) and Breakaway (10.8 kt) were exploded from towers, Marcoo (1.4 kt) was exploded at ground level and Kite (2.9 kt) was released by a RAF Vickers Valiant bomber from a height of 35,000 ft. The fallout from these tests was measured using sticky paper, air-sampling devices and water sampled from rainfall and reservoirs. The radioactive cloud from One Tree reached a height of 37,500 ft, exceeding the predicted 27,900 ft, and radioactivity was also detected in other locations in Australia. Operation Antler followed in 1957. Antler was designed to test components for thermonuclear weapons, with particular emphasis on triggering mechanisms.

Three tests began in September, codenamed Tadje, Biak and Taranaki. The first two tests were conducted from towers; the last was suspended from balloons. Yields from the weapons were 0.93, 5.67 and 26.6 kt, respectively. The Tadje test used cobalt pellets as a tracer for determining yield. Although the Antler series was better planned and organised than earlier series, intermediate fallout from the Taranaki test exceeded predictions. Nevertheless, the Taranaki site was left relatively uncontaminated from that test because of the specialised site preparation and because the device was exploded at a height of 300 m. Although at the sites of these nuclear explosions, some radioactivity can still be measured, the level of radioactivity is so low that it does not present any radiological hazard.

In addition to the major tests, a large number of minor trials were also carried out from June 1955 through to May 1963. They were to leave the most dangerous legacy at Maralinga. Figure 7 indicates where the various tests and trials were conducted. There were four main types of

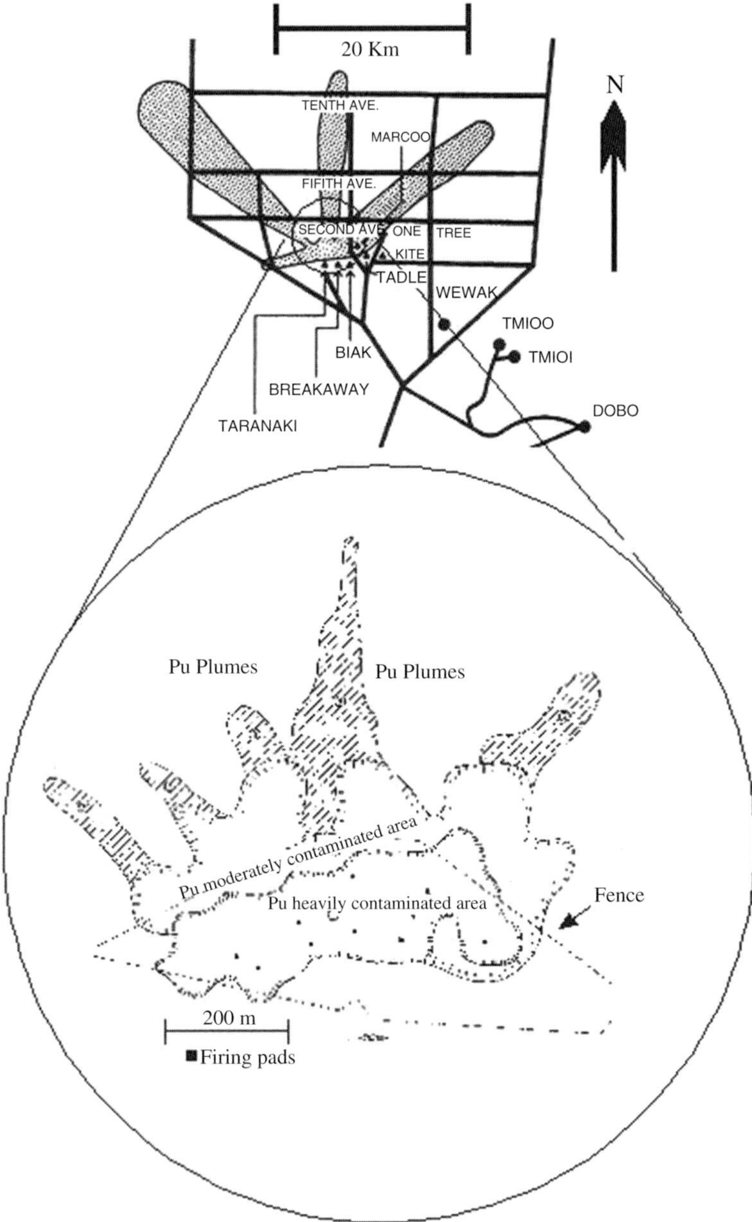

Figure 8 The Taranaki Pu-contaminated area (Burns et al., 1994).

minor trial. Three were aimed at the development of the bomb. The fourth was a safety trial.

3.1.2. Radiological conditions before remediation

Since the closure of the site in 1967, numerous surveys were carried out to map and characterise the contamination. Detailed studies were conducted in 1984–1985 by staff of the Australian Radiation Laboratory (this became ARPANSA in February 1999). These studies revealed that contamination levels at the site were greater than that recognised earlier (Lokan, 1985). The levels of plutonium in the soil exceeded by approximately one order of magnitude the values expected from the surveys made at the time of the experiments and during the initial cleanups. Significant radiation hazards still existed at many test sites, particularly at Taranaki where the Vixen B trials had been carried out. In contrast, the residual fallout and neutron activation radioactivity from the major trials were of little significance.

Following these discoveries, it became necessary to re-evaluate the risk posed to potential occupants of the area, and a Technical Assessment Group (TAG) was formed by the Australian government in 1986. TAG was charged with supervision of the gathering of the scientific information on the nature and extent of the contamination and advising on future remediation options (Church et al., 1990).

One of the problems faced by the radiation protection experts during the surveys was that none of the plutonium isotopes present in the contamination of Maralinga emitted gamma rays with sufficient intensity to permit practical measurements in the field. Measurement of alpha particles was possible only for freshly deposited contamination lying on smooth, clean surfaces; even then, the reliability of such measurements was questionable. The isotopes of plutonium emitted low-energy X-rays with considerable intensity, but these were heavily attenuated in only a few millimetres of soil and did not provide reliable information about weathered contamination which had migrated downwards into the soil. Therefore, a minor constituent of the plutonium, the short-lived ^{241}Pu (*half-life* = 14 years) that decays to ^{241}Am that emits a 59.5 KeV gamma ray with a probability of 36%, was used to measure Pu indirectly. The ^{241}Am gamma ray could travel through several centimetres of soil, or many metres of air, and permitted reliable and practical measurement in the surface layers of soil. Am-241 could serve as a useful indicator of plutonium in soil, providing the ratio of plutonium to americium was determined by separate laboratory experiments. This ratio and the isotopic composition of the plutonium were found to vary from site to site and even from one trial to another at the same site (Burns et al., 1994). Other radionuclides were used in various trials but, apart from ^{235}U, these were largely short lived and have since decayed to insignificant values.

The contamination was initially surveyed by hand-held scintillation detectors, ground-based high-resolution gamma-ray spectroscopy and helicopter-based aerial survey making use of a large array of sodium iodide detectors (Stover and Jee, 1972). However, this latter type of survey suffered from poor spatial resolution and, in areas near major trial sites, had difficulty in distinguishing ^{241}Am from other radionuclides. To obtain the fine details necessary to guide remediation, ground-based, *in situ* measurements using a high-resolution germanium detector were then made. These measurements were sufficiently accurate in areas where the soil had not been disturbed. However, in areas where the soil was mixed, the attenuation of the 59.5 keV gamma ray in soil caused the level of contamination to be underestimated. Nevertheless, the measured values provided a good indication of the inhalation doses, as inhaled dust primarily came from the same surface layer as did the observed gamma rays. The depth profile of the plutonium was measured by layered soil sampling and laboratory analysis.

The radiological survey, as well as the ongoing monitoring of the remediation operation that was conducted later, was essentially based on two types of measurements, namely *in situ* high-resolution gamma-ray spectrometry to measure the dispersed contamination by means of a germanium detector and low-resolution sodium iodide detectors for scanning areas of ground for radioactive particles and for determining their activities.

For the measurement of the large-scale average level of americium, and hence plutonium, in the surface layer of soil, a closed-end coaxial intrinsic gamma-ray detector suspended at a height of 4 m was used. For these measurements, a light truck was modified to incorporate an adjustable height boom to which the gamma-ray detector was fitted. The custom-built boom was hydraulically operated and allowed the detector to be positioned at varying heights in front of the truck. For most routine monitoring, the detector was held at 4 m above the ground, and for calibrations and testing, at about 1 m.

For the determination of discrete particulate contamination or contaminated fragments, an array of four sodium iodide detectors with 12.5 cm diameter and 1.6 mm thickness was mounted on the bull bar of another truck. The detectors were each connected to a single-channel analyser set to count gamma rays of 59.5 keV. The thin-crystal detectors provided significant rejection of the high-energy background but were still fully efficient at 59.5 keV. The detectors were mounted at a height of 25–30 cm above the ground, with their centres 0.5 m apart. This allowed the system to scan effectively a 2 m wide track, and, by driving at a speed of 5–6 km h^{-1}, all individual particles and fragments of 20 kBq could be detected with approximately 50% efficiency. By this procedure, the vehicle was able to thoroughly scan a hectare in 1–2 h. The vehicle was fitted with a differential GPS system that was interfaced with a computer so that the entire area covered by the searching process was accurately recorded. The position of all positive signals from any of the four detectors was recorded with an accuracy of 1 m.

At Taranaki, in areas where no previous attempt at rehabilitation were made, the plutonium was present at the surface of the predominantly sandy soil, with typically 85% of the activity being found in the top 10 mm. It lay along plumes starting near the site of each trial and extending for tens or even hundreds of kilometres in the direction of the wind at the time. The plumes, shown in Figure 9 (MARTAC, 2003), were well defined but often contained plutonium of different compositions from different trials. Discrete particles containing plutonium were found along the plumes more than 100 km from the test site. In other areas, where the preliminary initial cleanup operation had involved mixing of the surface layers with deeper layers of soil, the plutonium was distributed throughout the top 100–200 mm of soil.

In total, several square kilometres of land were found contaminated to levels exceeding 300 kBq m^2 of ^{239}Pu. Localised areas were 10 or 100 times more contaminated. The plutonium contamination was found to consist of the following isotopes: ^{238}Pu, ^{239}Pu, ^{240}Pu and ^{241}Pu. The minor trials involved negligible fission yield, so that the isotopic composition of the

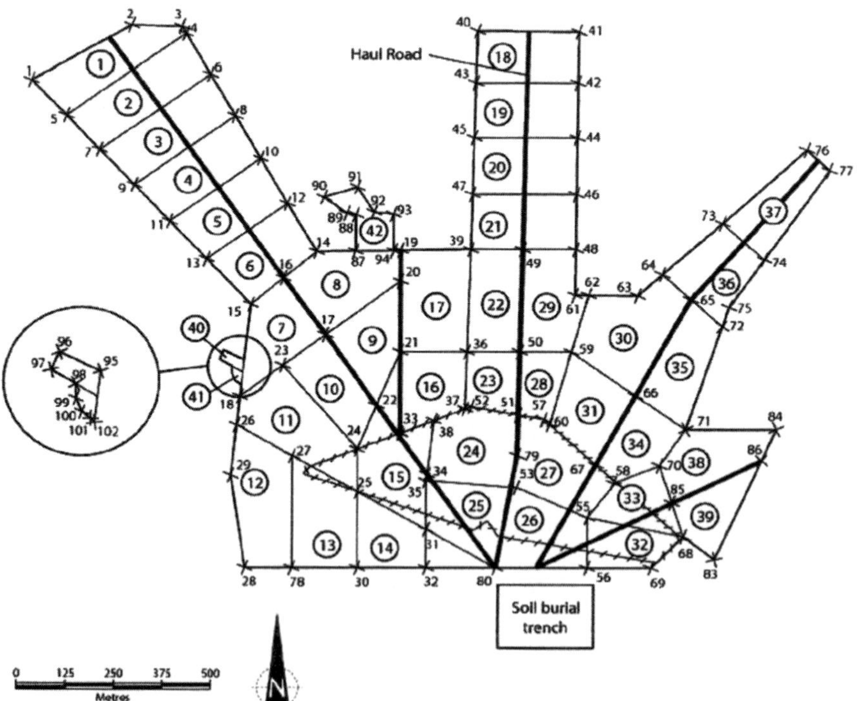

Figure 9 Taranaki cleanup Lots (MARTAC, 2003).

source material was unaffected by the explosion. However, radioactive decay had substantially removed the ^{241}Pu, replacing it with ^{241}Am.

Plutonium contamination was found to be present in three forms:

(i) as fragments, that is plutonium-contaminated debris and pieces of steel, plastic, wires, lead and so on, visually identifiable when lying on the surface;
(ii) as particles, that is sub-millimetre pieces of soil and other material incorporating plutonium oxide, indistinguishable from soil on casual inspection;
(iii) as dust, that is very finely divided, and potentially inhalable grains of plutonium oxide or contaminated soil. The fragments and particles had activities of plutonium ranging from a few kiloBecquerels to many megaBecquerels.

Field measurements that were part of this survey also led to important estimates of the doses that potential inhabitants of Maralinga could have received. Plutonium, being an alpha emitter, presents a health risk only if it enters into the body. Of the three pathways for entry into the body (inhalation, ingestion or through cuts and wounds), inhalation and subsequent retention in the lungs give rise to a risk of lung cancer. If plutonium enters the body through one of the other pathways, the greater risk is of bone cancer (osteosarcoma) or liver cancer (Stover and Jee, 1972). The degree to which each of these exposure pathways contributes to the potential dose, in general, depends on the type of lifestyle practised by inhabitants of the land (TAG Report, 1990).

The plutonium at Maralinga was found to be largely in the form of plutonium oxides that showed very low solubility in simulated lung fluid and practically no solubility in rainwater, as expected by the fact that the plutonium had remained very near the surface for 30 years after deposition (Williams, 1990; Stradling et al., 1992). Due to this insolubility, the ingestion pathway was considered much less important to potential doses than inhalation. Wound contamination was less likely to occur but had the potential to deliver large, single doses (Harrison et al., 1993) if contaminated dust or particles were to enter existing wounds or in the event of a simultaneous injury and wound contamination caused by a contaminated fragment (Lokan and Williams, 1995).

For nomadic Aborigines such as the Maralinga Tjarutja, living an outstation lifestyle, the inhalation dose pathway was considered by far the most significant for both adults and children (Haywood and Smith, 1992) in assessing the requirements for land rehabilitation. Consequently, specific experiments and measurements were conducted to best quantify the plutonium inhalation risk (Johnston et al., 1992; Williams, 1990). The fine dust was considered as the main hazard through the inhalation pathway.

Moreover, as some of the larger particles were found to be very friable if disturbed, they were considered a potential additional source of fine dust.

The potential doses due to inhalation of plutonium-containing dust were assessed by measuring (a) the aerodynamic diameter of the inhalable fraction, which determines the behaviour of the dust inside the respiratory tract, and (b) the activity of the various radionuclides within that fraction. Artificial resuspension experiments were undertaken to determine the activity median aerodynamic diameter (AMAD) as well as the relationship between the radionuclide concentration in inhalable dust and that in the soil from which it came. An AMAD of 5 μm and an average enhancement factor (defined as the ratio of activity concentration in the inhalable fraction to that in the bulk parent soil) of 6 for the outlying 'plume' areas (higher values were considered for the inner, more contaminated, areas) were used for the dose assessment (Johnston et al., 1992). Total activities for naturally resuspended dust collected with high- and low-volume air samplers were measured, and soils from several sites at Taranaki were sieved to determine mass and activity distributions. The 0–45 μm fraction was found to have a much larger proportion of activity than its mass would have suggested. These measurements permitted calculation of annual doses, which could considerably exceed 100 mSv $year^{-1}$ if continuous occupancy was to occur in some localised regions (IAEA, 1998a, 1998b). While very few areas represented a significant hazard to the casual visitor, doses in excess of 1 mSv $year^{-1}$ could be received over a very large area where the ^{241}Am concentration exceeded about 0.6 kBq m^{-3}. However, because of the narrow plume structure of the contamination, 100% occupancy of the contaminated areas by the very mobile Aboriginal people was considered extremely unlikely. On the other hand, in case of very specific behaviours, doses near the threshold for deterministic effects (about 500 mSv) could, in principle, also be received. The presence of plutonium in visually identifiable pieces of debris in megaBecquerel quantities indicated that the deliberate collection of highly contaminated objects had to be taken into account as well.

As a result of this thorough survey, a much more extensive cleanup, namely a rehabilitation project, was initiated at the site.

3.1.3. Remediation measures

The first cleanup operation at Maralinga was carried out in 1964. The remains from the firings, including numerous contaminated fragments and most of the remaining infrastructure, were buried in a series of pits, each approximately 2–3 m deep. The plutonium content of each of these pits was not well known but was estimated to amount to kilogram quantities (Symonds, 1985). A second cleanup codenamed 'Operation Brumby' was conducted in 1967. Here steps were taken to reduce the radiological hazard by turning over and mixing the surface soil to reduce the plutonium surface concentrations. The contaminated soil was ploughed to depths of several centimetres and mixed with clean soil. The area to the north of Taranaki that was ploughed in this operation became known as the 'ploughed area'. Radioactive debris was buried in 32 pits around the site, 21 of which were at Taranaki. Other debris was buried in 60 other pits and in the crater left by the Marcoo bomb. The burial pits were covered and sealed with a cap of concrete. Following this cleanup, the site was formally closed, and the whole area, which included all the major and minor test sites, was kept under varying degrees of surveillance and subject to entry restrictions. A series of fences was erected to enclose the burial pits containing significant quantities of plutonium at Taranaki and other sites.

The Consultative Group that had functioned during the TAG era was reconvened for the rehabilitation project in 1993. This group was established to serve as a forum to discuss all matters related to the site remediation and comprised representatives from the Commonwealth, South Australia, Western Australia and the United Kingdom, together with members of the Maralinga Tjarutja Aborigine people (the traditional land owners) and their legal representatives (Lokan, 1985).

Planning of the Maralinga rehabilitation project began in 1993 with the establishment of the Maralinga Rehabilitation Technical Advisory Committee (MARTAC) whose purpose was to provide advice to the Department of Primary Industries and Energy, the project manager responsible for the site. MARTAC was given the responsibility for establishing the remediation criteria.

Three sets of criteria were established concerning the levels of contamination that were permitted to remain at the completion of the remediation operation (Maralinga Cleanup, 2002; Williams, 1990; Williams et al., 1998). They were:

(i) *Soil-removal criteria*: Contaminated soil (or the cause of contamination itself) at Taranaki was to be removed where the levels of dispersed ^{241}Am exceeded 40 kBq m^{-2}, averaged over 1 ha (10,000 m^2), or where contaminated particles exceeding 100 kBq were found or where the density of particles exceeding 20 kBq was greater than 1 in 10 m^2.

(ii) *Clearance criteria*: Where soil was removed, the residual levels of dispersed contamination in the cleared area was not to exceed 3 kBq m^{-2} for ^{241}Am, averaged over 1 ha, and particulate contamination was to meet the soil-removal criteria.

(iii) *Unrestricted land-use criteria*: Permanent occupancy and unrestricted land use was only to occur where levels of dispersed contamination were less than 3 kBq km^{-2} for ^{241}Am, averaged over 3 km^2, and the particulate contamination met the soil-removal criteria.

In other words, the criteria for the removal of contaminated particles and fragments required that (i) no particles of ^{241}Am activity higher than 100 kBq and no observable contaminated fragments should have remained outside the soil-removal contour or within the rehabilitated area at the conclusion of the operation and (ii) no more than an average of one discrete particle of activity greater than 20 kBq per 10 m^2 should be present after the remediation. MARTAC did not specify any averaging criterion for particles of 20 kBq or below, but 0.1 per m^2 or 1 per 10 m^2 was not very practical. Therefore, ARPANSA interpreted this criterion as requiring that fewer than 1,000 particles exceeding 20 kBq ^{241}Am per hectare should remain at the end of the remediation. The contaminated soil and debris were to be buried in trenches excavated close to the site and covered with a minimum of 5 m of clean material. Finally, the large or unknown amounts of contaminated debris in pits had to be rendered practically inaccessible by an *in situ* vitrification (ISV) treatment.

As the most important pathway for exposure at the Maralinga Range was inhalation, the major objective of the rehabilitation was to reduce the risk arising from exposure to radiation inhalation by individuals living an outstation lifestyle to a level that was acceptable to both the Aboriginal community and the Australian government (TAG Report, 1990). To this aim, the cleanup criteria were guided by conservative principles and estimation of doses by realistic scenarios. These included the possibility of an Aboriginal group living for an entire year on the edge of the non-residential area in regions of the highest activity permitted (approximately 3 kBq m^{-2} of ^{241}Am). Considering the dose conversion factors accepted for general use at the time, and the site-specific factors applying at Taranaki, the concentration of 3 kBq m^{-2} of ^{241}Am in the surface levels of soil was expected to lead to an annual dose of 5 mSv through inhalation of contaminated dust, under conditions of continuous occupancy (TAG Report, 1990). If, more realistically, the group spent its time randomly dispersed over the Maralinga lands outside the restricted area, or even randomly around its perimeter, the average activity levels (and hence the doses) would be expected to fall by at least one order of magnitude.

At any rate, the remediation aimed at removing all three of the possible causes of incurring high doses, namely:

- the production of large amounts of dust in locations of very high plutonium concentration;
- the deliberate collection of contaminated fragments and particles that could allow a person an accumulation of considerable and dangerous quantities of plutonium (although the likelihood of this leading to significant exposure to many people was considered to be negligible);
- the potential availability of large amounts of plutonium in the burial pits (these had to be made practically inaccessible, both to deliberate seekers and to environmental factors, by ISV or other treatment).

Eventually, the rehabilitation option chosen and agreed also with the traditional owners of the territory (the Maralinga Tjarutja), consisted in:

- removing the plutonium-contaminated soil from the ploughed areas (2.1 km^2) and burying it in burial trenches at the three contaminated sites (Taranaki, Wewak and the TM sites);
- importing clean soil to cover the scraped areas and re-vegetation of those areas;
- treating the 32 pits containing contaminated debris by a process of ISV;
- sorting through the 60 or so pits which contained the non-radioactive debris, removing any contamination that might be found and disposing of it to the main burial trench; and
- erecting a fence around the contaminated area to the north of Taranaki to prevent access.

In the course of work, further studies led to some modifications of the original work plan. ISV was restricted to the 22 pits at Taranaki (the number of ISVs was eventually even lower). Moreover, it was decided that the debris from the pits at Wewak and the TM sites were to be placed at the bottom of the burial trenches (13 and 16 m deep, respectively) excavated at those sites. Finally, the lumps of uranium present on the surface at Kuli were to be collected and buried and the central area at Kuli scraped and the soil buried.

Once the cleanup criteria had been established, the Australian Radiation Laboratory could mark out which soil had to be removed and buried. This resulted in most of the old area ploughed by the British to be included for removal, in addition to some other areas that had never been ploughed. Consideration of the potential dose to an Aboriginal living a semi-traditional lifestyle on the range also guided the initial decision to locate a fence around Taranaki. However, the idea of erecting a fence was later dropped in favour of line-boundary markers carrying warning signs advising the Aboriginals that the enclosed area was not suitable for permanent residence, but was safe enough to allow hunting and transit.

The first stage of the remediation project consisted of defining the cleanup boundaries at the sites contaminated with plutonium (Taranaki, TMs and Wewak), followed by bulk removal of contaminated soil from the three sites and burial in excavated burial trenches under at least 5 m of clean rock and soil. At Taranaki, the 22 pits in which the British had disposed unknown quantities of plutonium released during the 12 Vixen B firings were also rehabilitated. Eleven of these were treated by means of ISV, while the remaining pits were exhumed and their contents reburied in another custom-built burial trench.

To facilitate control of the process, the soil-removal area was divided into Lots ranging from 2 to 5 ha. Figure 9 (MARTAC, 2003) shows this division in Lots at the Taranaki site. Each Lot then underwent a sequence of soil removal, checking by the health physicists, re-treatment if necessary and then monitoring by ARPANSA. In a small number of cases, the measurements by ARPANSA revealed the need for further treatment in order to meet the MARTAC clearance criteria.

At the end of the remediation operation, a new radiological assessment was carried out by ARPANSA. The essential purpose of this assessment was to ensure that the whole Maralinga area had been made safe following the work undertaken during the 1994–2000 Maralinga rehabilitation project. Radiation doses for the inhalation pathway were once again calculated for a range of sites (Lokan and Williams, 1995). Not surprisingly, at all the sites considered, the dose due to inhalation was dominated by ^{239}Pu (ca. 75% of total), with minor contributions from other isotopes of plutonium and ^{241}Am. The major uncertainty in the dose calculation was the occupancy factor. As it was impossible to predict with confidence future occupancy factors for the Maralinga areas by the Aboriginal communities, a value of 100% (permanent occupancy) was generally assumed.

After the soil surfaces had been cleaned and stabilised and following a few years of natural weathering, the rehabilitation effect was noticeable in the reduced Pu concentrations in air. In April–May 2000, the air flowing over the former deposition plumes at Taranaki picked up Pu concentrations of approximately $0.1\,\mu Bq\,m^{-3}$, and at TM-100 and Wewak, the Pu concentrations in air were of the order of $0.01\,\mu Bq\,m^{-3}$. The value at Maralinga Village during the same period was $0.006\,\mu Bq\,m^{-3}$, which can be compared to the $0.002\,\mu Bq\,m^{-3}$ value for a representative northern hemisphere site such as California. In conclusion, following the remediation by removal and burial at depth of contaminated surface soil, all areas at Maralinga were shown to be within acceptable limits for all envisaged land uses in terms of absorbed doses.

The restriction on permanent occupancy within the 'restricted land-use' (non-residential) boundary surrounding Taranaki was set as a purely precautionary measure as the inhalation doses for permanent occupancy of all but a few areas (essentially within the untreated plumes) were well below the 1 mSv year^{-1} limit for members of the public.

For a semi-traditional Aboriginal lifestyle, with camp sites occupying considerable area and moving regularly, it was difficult to envisage circumstances which would have led to inhalation doses, even within most of the restricted zone, above acceptable limits. Nevertheless, restricted access to Taranaki was kept, and a fence and appropriate signposts were installed. This restricted access also reduced the highly unpredictable and essentially non-assessable hazard from possible contact with any undiscovered active particles remaining in the plumes adjacent to the soil-removal areas.

Some representative dose calculations were also performed for specific scenarios including digging, driving in a vehicle following another along a dusty track and repairing a puncture. Also for all these situations, the estimated doses were found to be acceptably low. It is now impossible for casual visitors making intermittent visits to the area – tourists, geological prospectors and surveyors, who do not engage in abnormal dust raising or large-scale soil-disturbance activities – to receive a committed effective dose by inhalation of anything approaching 1 mSv. The substantial dust loadings observed during times of severe dust storms also resulted in insignificant doses. The conclusion was then reached that under ambient conditions, concentrations of plutonium in air and plutonium resuspension factors were practically the same as worldwide background values (Shinn, 2002).

Some estimates were also made concerning the near and more distant future, taking into account a number of effects which could alter the potential doses and health risks with time. Consideration was given to possible lifestyle changes with time, if the Maralinga Tjarutja people were to move towards a more European lifestyle, with extensive areas being covered by concrete, tarmac, buildings and lawns, and to live in Western-style houses in suburban settings. The conclusion was that the dust levels and hence doses were expected to become much lower. As far as the possible influence of weathering on the future doses is concerned, a climate-modelling study of the Maralinga area (Hunt and Elliott, 2001) reached the conclusion that dust resuspension is not expected to change as a result of possible climate change.

Another obvious effect leading to decreasing doses would be via radioactive decay, although this is only significant over time scales of millennia. Over hundreds of years, assuming the contamination stays in its present location, the dose will remain approximately the same due to the long half-life of ^{239}Pu. Nevertheless, as plutonium moves deeper into the soil with time, the inhalation dose to a potential inhabitant should further decrease.

ACKNOWLEDGEMENT

This Chapter is a compilation of information and material that appeared in a number of articles and reports authored by many other scientists listed in the references and under the copyrights of several publishers that could not always be identified. Full credit is given to both these scientists and the publishers holding these copyrights for their excellent publications and reports.

REFERENCES

Bennet, B. G. (2000). Worldwide panorama of radioactive residues in the environment. *Proceedings of International Symposium on Restoration of Environments with Radioactive Residues* (Arlington, VA, 29 November–3 December 1999), IAEA, Vienna, pp. 11–24.

Burns, P. A., M. B. Cooper, P. N. Johnston, L. J. Martin, and G. A. Williams. (1994). Determination of the ratios of ^{239}Pu and ^{240}Pu to ^{241}Am for nuclear weapons test sites in Australia. *Health Physics*, 67, 226–232.

Church, B. W., D. R. Davy, D. Dervell, K. H. Lokan, and H. Smith. (1990). *Rehabilitation of Former Nuclear Test Sites in Australia*, Report by the Technical Assessment Group, Australian Government Publishing Service, Canberra.

Greenhouse, N. A., P. P. Miltenberger, and E. T. Lesard. (1980). Dosimetric results for the Bikini population. *Health Physics*, 38, 845–851.

Hamilton, T., D. Hickman, C. Conrado, T. Brown, J. Brunk, A. Marchetti, C. Cox, R. Martinelli, and S. Kehl. (1982). Department of Energy. In: *Enewetak Radiological Support Project* (Ed. B. Friesen). Final Report NVO-213. Nevada Operations Office, Las Vegas, NV.

Hamilton, T., D. Hickman, C. Conrado, T. Brown, J. Brunk, A. Marchetti, C. Cox, R. Martinelli, S. Kehl, K. Johannes, D. Henry, R. T. Bell, and G. Petersen. (2001). Individual radiation protection monitoring in the Marshall Islands: Enewetak Island resettlement support, May–December 2001, *Brief History of Nuclear Testing in the Marshall Islands*, Report UCRL-LR-149601. Livermore, CA.

Hamilton, T. F. (2004). Linking legacies of the cold war to arrival of anthropogenic radionuclides in the oceans through the 20th century. In: *Radioactivity and the Environment* (Ed. H. D. Livingston). Elsevier Science, Amsterdam, pp. 30–87.

Harrison, J. D., A. Hodgson, J. W. Haines, and J. W. Stather. (1993). The biokinetics of plutonium-239 and americium-241 in the rat after subcutaneous deposition of contaminated particles from the former nuclear weapons site at Maralinga: Implications for human exposure. *Human and Experimental Toxicology*, 12, 313–321.

Haywood, S. M., and J. G. Smith. (1992). Assessment of potential doses at the Maralinga and Emu test sites. *Health Physics*, 63, 624–630.

Hunt, B. G., and T. L. Elliott. (2001). *Potential Impact of Climate Change at Maralinga*, CSIRO Atmospheric Research Report.

IAEA. (1996). *International Basic Safety Standards for Protection Against Ionizing Radiation and for the Safety of Radiation Sources*. Safety Series No. 115. IAEA, Vienna.

IAEA. (1998a). *Radiological Conditions at Bikini Atoll: Prospects for Resettlement*. Radiological Assessment Report Series ST1/PUB/105. IAEA, Vienna.

IAEA. (1998b). *Characterization of Radioactively Contaminated Sites for Remediation Purposes*. IAEA-TECDOC-1017. IAEA, Vienna.

Johannes, D., R. T. Bell, and G. Petersen. (2002). Individual radiation protection monitoring in the Marshall Islands: Enewetak Island resettlement support, May–December 2001, *Brief History of Nuclear Testing in the Marshall Islands*, Report UCRL-LR-149601. Livermore, CA.

Johnston, P. N., K. H. Lokan, and G. A. Williams. (1992). Inhalation doses for Aboriginal people reoccupying former nuclear weapons testing ranges in South Australia. *Health Physics*, 63, 631–640.

Kehl, S. R., M. E. Mount, and W. L. Robison. (1995). *The Northern Marshall Islands Radiological Survey: A Quality Control Program for Radiochemical and Gamma Spectroscopy Analysis*, Report UCRL-ID-120429. Lawrence Livermore National Laboratory, Livermore.

Lessard, E. T., R. P. Miltenburger, S. H. Cohn, S. V. Musolino, and R. A. Conrad. (1984). Protracted exposure to fallout: The Rongelap and Utirik experience. *Health Physics*, 46, 511–527.

Lokan, K. H. (1985). *Residual Contamination at Maralinga and Emu*, 1985, ARL/TR070, April.

Lokan, K. H. (2000). Remediation of the Maralinga test site. *Proceedings of International Symposium on Restoration of Environments with Radioactive Residues* (Arlington, VA, 29 November–3 December 1999), IAEA, Vienna, pp. 321–333.

Lokan, K. H., and G. A. Williams. (23 February 1995). *Submission to Parliamentary Standing Committee on Public Works*, Maralinga Rehabilitation Project, Official Hansard Report of Public Submissions and Meetings at Ceduna.

Maralinga Cleanup. (2002). www.arpansa.gov.au/pubs/basic/maralinga.pdf

Marsh, K. V., T. A. Jokela, R. J. Eagle, and V. E. Noshkin. (1978). *Radiological and Chemical Studies of Ground Water at Enewetak Atoll, 2. Residence Time of Water in Cactus Crater*, Report UCRL-51913, Part 2. Lawrence Livermore National Laboratory Report, Livermore.

Marshall Islands Program. (2006). Marshall Islands Dose Assessment and Radioecology Program, https://eed.llnl.gov/mi, last modified 4 June 2006.

MARTAC Report. (2003). *Rehabilitation of Former Nuclear Test Sites at Emu and Maralinga (Australia)*; Report by the Maralinga Rehabilitation Technical Advisory Committee, Department of Education, Science and Training, Australia, Commonwealth Government.

McEwan, A. C. (2000). Remediation and rehabilitation programmes in the Marshall Islands. *Proceedings of International Symposium on Restoration of Environments with Radioactive Residues* (Arlington, VA, 29 November–3 December 1999), International Atomic Energy Agency, Vienna, pp. 311–319.

Miltenberger, R. P., N. A. Greenhouse, and E. T. Lessard. (1980). Whole body counting results from 1974 to 1979 for Bikini Island residents. *Health Physics*, 39, 395–507.

Miltenberger, R. P., E. T. Lessard, and N. A. Greenhouse. (1981). Cobalt-60 and cesium-137 long-term biological removal rate constants for the Marshalles population. *Health Physics*, 40, 615–623.

Nelson, V., and V. E. Noshkin. (1973). *Marine Program*, Enewetak Survey Report, NVO-14Q, Vol. 1. US Atomic Energy Commission, Nevada Operations Office, Las Vegas, NV.

Noshkin, V. (1980). Transuranium radionuclides in components of the benthic environment of Enewetak Atoll. In: *Transuranic Elements in the Environment* (Ed. W. C. Hanson). US Department of Commerce, National Technical Information Service, Springfield, VA. DOE/TIC 22800, pp. 578–601.

Noshkin, V. E., R. J. Eagle, K. M. Wang, and T. A. Jokela. (1981). Transuranic concentrations in reef and pelagic fish from the Marshall Islands. *International Symposium on the Impacts of Radionuclide Releases into the Marine Environment*, International Atomic Energy Agency, Vienna, IAEA-AM-248/146, pp. 293–231.

Noshkin, V. E., and W. L. Robison. (1997). Assessment of a radioactive waste disposal site at Enewetak Atoll. *Health Physics*, 27, 234–247.

Noshkin, V. E., K. M. Wong, R. J. Eagle, and C. Gatrousis. (1974). *Transuranics at Pacific Atolls, I. Concentration in the waters at Enewetak and Bikini*, Report UCRL-51612. Lawrence Livermore National Laboratory, Livermore.

Noshkin, V. E., K. M. Wong, R. J. Eagle, and C. Gatrousis. (1975). Transuranics and other radionuclides in Bikini lagoon: Concentration data retrieved from aged coral sections. *Limnology and Oceanography*, 20, 729–742.

Noshkin, V. E., K. M. Wong, R. J. Eagle, G. Holladay, and R. W. Buddemeier. (1976). Plutonium radionuclides in the groundwater at Enewetak Atoll. *Proceedings of IAEA Symposium on Transuranium Nuclides in the Environment.* IAEA SM-199/33, Vienna, pp. 517–543.

NRC. (1982). *Evaluation of Enewetak radioactivity containment.* National Academy of Science, National Research Council. National Academy Press, Washington, DC.

Parkinson, A. (2000). Maralinga Rehabilitation Project. National Conference-MAPW Australia, http://www.mapw.org.au/conferences/mapw2000/papers/parkinson.html

Ristvet, B. L., E. L. Tremba, R. F. Couch Jr., J. A. Fetzer, E. R. Goter, D. R. Walter, and V. P. Wendland. (1978). *Geological and Geophysical Investigations of the Eniwetok Nuclear*

Craters, Report AFWL-TR-77-242. Kirtland Air Force Base, NM: Air Force Weapons Laboratory.

Robison, W. L., K. T. Bogen, and C. L. Conrado. (1997a). An updated assessment for resettlement options at Bikini Atoll US nuclear test site. *Health Physics*, 73, 100–114.

Robison, W. L., C. L. Conrado, K. T. Bogen, and A. C. Stoker. (2003). The effective and environmental half-life of ^{137}Cs at coral islands at the former US nuclear test site. *Journal of Environmental Radioactivity*, 69, 207–223.

Robison, W. L., C. L. Conrado, R. J. Eagle, and M. L. Stuart. (1981a). *The Northern Marshall Islands Radiological Survey: Sampling and Analysis*, Summary Report UCRL-52853, Part 1. Lawrence Livermore National Laboratory, Livermore.

Robison, W. L., M. E. Mount, W. A. Phillips, C. L. Conrado, M. L. Stuart, and A. C. Stoker. (1982). *The Northern Marshall Islands Radiological Survey: Terrestrial Food Chain and Total Doses*, Report UCRL-52853, Part 4. Lawrence Livermore National Laboratory, Livermore.

Robison, W. L., M. E. Mount, W. A. Phillips, M. L. Stuart, S. E. Thompson, C. L. Conrado, and A. C. Stoker. (1980). *An Updated Assessment of Bikini and Eneu Islands at Bikini Atoll*, Report UCRL-53225. Lawrence Livermore National Laboratory, Livermore.

Robison, W. L., V. E. Noshkin, C. L. Conrado, R. J. Eagle, J. L. Brunk, T. A. Jokela, M. E. Mount, W. A. Philips, A. C. Stoker, M. L. Stuart, and K. M. Wong. (1997b). The northern Marshall Islands radiological survey: Data and dose assessments. *Health Physics*, 73, 37–48.

Robison, W. L., V. E. Noshkin, W. A. Philllops, and R. J. Eagle. (1981b). *The Northern Marshall Islands Radiological Survey: Radionuclide Concentrations in Fish and Clams and Estimated Doses via the Marine Pathway*, Report UCRL-52853, Part 3. Lawrence Livermore National Laboratory, Livermore.

Robison, W. L., W. A. Phillis, and C. S. Colsher. (1977). *Dose Assessment at Bikini Atoll*, Report UCRL-51879, Part 5. Lawrence Livermore National Laboratory, Livermore.

Robison, W. L., and E. L. Stone. (1992). The effect of K on the uptake of ^{137}Cs in food crops grown on coral soils: Coconut at Bikini Atoll. *Health Physics*, 62, 496–511.

Robison, W. L., and C. Sun. (1997). The use of comparative ^{137}Cs body burdens estimates from environmental data/models and whole body counting to evaluate diet models for the ingestion pathway. *Health Physics*, 73, 152–166.

Shinn, J. H. (2002). *Studies of Plutonium Aerosol Resuspension at the Time of the Maralinga Cleanup*, Report UCRL-ID-155063. Livermore, CA.

Simon, S. L., and J. C. Graham. (1995). Radiological Survey of Bikini Atoll, Marshall Islands: Nationwide Radiological Study, Republic of the Marshall Islands, Majuro.

Simon, S. L., and J. C. Graham. (1997). Findings of the first comprehensive radiological monitoring program of the Republic of the Marshall Islands. *Health Physics*, 73, 66–85.

Simon, S. L., and W. L. Robison. (1997). A compilation of atomic weapons test detonation data for US Pacific Ocean tests. *Health Physics*, 73, 258–264.

Stover, B. J., and W. S. S. Jee. (1972). *Radiobiology of Plutonium*. The J.W. Press, University of Utah, UT.

Stradling, G. N., J. W. Stather, S. A. Gray, J. C. Moody, M. Ellender, M. J. Pearce, and C. G. Collier. (1992). Radiological implications of inhaled ^{239}Pu and ^{241}Am in dusts at the former nuclear test site in Maralinga. *Health Physics*, 63, 641–650.

Sun, L. C., C. B. Meinhold, A. R. Moorthy, E. Kaplan, and J. W. Baum. (1997). Assessment of plutonium exposure in the Enewetak population by urinalysis. *Health Physics*, 73, 127–132.

Symonds, J. L. (1985). *A History of British Atomic Tests in Australia*. Australian Government Publishing Service, Canberra.

TAG Report. (1990). *Rehabilitation of Former Nuclear Test Sites in Australia*, Report by the Technical Assessment Group (TAG). Australian Government Publishing Service, Canberra.

UNSCEAR. (2000). *United Nations Scientific Committee on the Effects of Atomic Radiation*, Sources and Effects of Atomic Radiations, Report to the General Assembly with scientific annexes, UN, New York.

USDOE. (1982). *Enewetak Radiological Support Project*, Final Report NVO-213 (Ed. B. Friesen). Nevada Operations Office, United States Department of Energy, Las Vegas, NV.

USDOE. (2000). United States Department of Energy, United States Nuclear Tests: July 1945 through September 1992, Nevada Operations Office, Las Vegas, NV, DOE/NV-209-REV.

Williams, G. A. (Ed.). (1990). *Inhalation Hazard Assessment at Maralinga and Emu*, Technical Report ARL/TR087. Yallambia, Victoria, Australia.

Williams, G. A., M. B. Cooper, and L. J. Martin. (1998). Plutonium contamination at Maralinga-Dosimetry and clean-up criteria. *Proceedings of AIOH-98 17th Annual Conference of Australian Institute of Occupational Hygienists (AIOH)*, December 1998, Canberra, Australia, pp. 117–126.

CHAPTER 6

Remediation as Part of the Decommissioning of Nuclear Facilities

W. Eberhard Falck[1,*], Roger Seitz[2], Mike Pearl[3], Mark Audet[4], Peter Schmidt[5] *and* Horst M. Fernandes[6]

Contents

*Corresponding author. Tel.: +31-72-531-42-36
E-mail address: wefalck@wefalck.eu

[1] Consultant, Cornelis Pronklaan 102, NL-1816NR, Alkmaar
[2] Savannah River National Laboratory, 773-43A, Room 217, Aiken, SC 29808, USA
[3] UKAEA, The Manor Court, Chilton, Oxfordshire OX11 0RN, UK
[4] Atomic Energy of Canada Ltd. (AECL), Chalk River Laboratories, Environmental Technologies Branch, CRL Site Investigations and Groundwater Monitoring, Chalk River, Ont., Canada K0J 1J0
[5] Wismut GmbH, Jagdschänkenstraße 29, D-09117 Chemnitz, Germany
[6] International Atomic Energy Agency, NEFW, Waste Technology Section, P.O. Box 100, A-1440 Vienna, Austria

Radioactivity in the Environment, Volume 14
ISSN 1569-4860, DOI 10.1016/S1569-4860(08)00206-4

1. Introduction

The term 'remediation' for the purpose of this chapter is defined as the process of transforming a site to an acceptable condition suitable for its intended future use, by adopting appropriate remediation techniques during and after decommissioning of nuclear facilities. The remediation process applies to both nuclear and non-nuclear contaminants that have the potential to affect human health and the environment. Remediation encompasses *inter alia* site characterisation, identification of remedial action alternatives, implementation of a remedial action and ongoing monitoring to assure the confinement or containment of residual contamination.

The term 'decommissioning' is defined by the IAEA as the administrative and technical actions taken towards the removal of some or all of the regulatory controls from a nuclear facility (IAEA, 2004a). Decommissioning is the process by which a nuclear facility is taken out of operation after final safe shutdown and includes the administrative and technical actions towards dismantling of all systems, structures and components as stipulated by the regulatory requirements (IAEA, 1999a, 1999b, 2003a, 2006a). The release from regulatory control is the final stage in a decommissioning process and is also the final stage of a practice (IAEA, 2006d). The use of the term 'decommissioning' implies that no further use of the facility for its existing purpose is foreseen. The actions taken will need to ensure the long-term protection of the public and the environment, and typically include reducing levels of residual radionuclides in all materials so that they can be safely recycled, released and reused.

Decommissioning activities create radioactive waste that needs to be appropriately managed. The decommissioning activities are to be conducted in a timely and cost-effective manner and undue delays are to be prevented (IAEA, 2001a).

The IAEA has developed a set of documents as part of its safety standard series that focus on safety during decommissioning (IAEA, 1999a, 1999b, 2001b). Further, a Safety Guide has been developed that provides guidance to the regulatory body and the licensees for the remediation of sites undergoing decommissioning (IAEA, 2006e). It is recommended that the reader refers to the latter document for a detailed discussion of the radiation protection criteria and other radiation protection aspects of the process of decommissioning and remediation, in particular the concept of constrained optimisation (see also Chapter 1).

Decommissioning activities and remediation activities at nuclear sites are subject to some common driving forces. In particular, the regulatory requirements to meet radiological exposure limits and the selection of an acceptable end state suitable for the site's future use influence the choice of technologies and overall configuration of controls adopted. Additional factors that influence the activities include cost efficiency, available resources and stakeholder concerns (IAEA, 2002a). Because remediation and decommissioning are subject to the same driving forces, the careful identification of synergies between remediation and decommissioning may be helpful in optimising the use of available resources to achieve results faster and at lower cost (IAEA, 2004c).

There are thousands of nuclear facilities worldwide, generally licensed and non-orphan, that will ultimately require decommissioning. They range from large nuclear power reactors (IAEA, 1999a) and complex processing facilities to small research laboratories, nuclear research establishments (IAEA, 1999a), uranium and thorium mines (IAEA, 1994), conversion plants, storage facilities and manufacturing plants (IAEA, 2001b). Hence, the tasks associated with decommissioning a nuclear facility vary greatly. They may include large-scale decontamination efforts, demolition of massive concrete structures or placing the facility in a safe enclosure condition so as to allow the radioactivity to decay naturally to acceptable levels (IAEA, 1999c) or final and interim storage of waste. At the other extreme, laboratories in which radionuclides have been used may be decommissioned after some modest cleaning and decontamination activities. In all cases, the decommissioning process must be well planned and sufficient resources must be available. Over time, without proper arrangements being made for decommissioning, shutdown facilities deteriorate and ultimately constitute a radiological hazard in their vicinity from direct external exposure to radiation to the public and from potential release of radioactive material to the environment. Structural deterioration may also result in conventional hazards.

It should be noted, however, that to date radioactive release of sites has occurred only in a limited number of decommissioning projects – most notably uranium mining and milling facilities as well as some defence sites – and that decommissioning is a mature practice only in very few countries (NEA, 2006a).

The remediation goals are usually based on risk. Largely for cost reasons, the remediation is usually aimed at achieving 'restricted reuse' or 'fit-for-purpose reuse' as opposed to 'unrestricted reuse', though the latter would be the ideal goal. The remedial objectives for any restricted reuse are then based on meeting appropriate risk criteria associated with the intended land reuse. Where these risks remain acceptably low, regardless of the future uses of the land, the site can be released from any controls (IAEA, 2006e). It has to be kept in mind, however, that material originating from a released site needs to comply with the national requirements for radiation protection for materials outside the scope of regulatory control (IAEA, 2004e).

These aspects highlight the importance of implementing a forward-looking planning process. Although, in the past, decommissioning and remediation activities have often been carried out as independent activities, optimisation of effort, cost, impacts and risk can be achieved by carrying out decommissioning and remediation activities in an integrated manner. These two different approaches are referred to in this document as the 'traditional approach' (independent activities) and the 'integrated approach' (integrated activities), and the differences in these approaches are illustrated in Figure 1.

The integrated approach requires a change in thinking. Under the traditional approach, decommissioning is considered in isolation from remediation stages of a site's life cycle. This may result in decommissioning end points that have ignored the overall aims of site remediation – particularly with respect to the potential impacts on human health and the environment from any residual contamination after the facility is decommissioned. These oversights can be costly in terms of site remediation – particularly with respect to the ability to

- deal with subsurface contamination whilst the decommissioning work-force is still mobilised;
- use existing infrastructure, such as liquid effluent treatment works;
- realise potential revenue by reusing parts of the site early by remediation to a 'fit-for-purpose' end point at the time a particular facility is decommissioned, as opposed to waiting for all facilities to be decommissioned before the site can be reused.

The objective of this chapter is to provide background information about important aspects of remediation of sites that are undergoing or have undergone decommissioning. Key strategic planning issues are being outlined and site remediation activities that reduce the duplication of effort

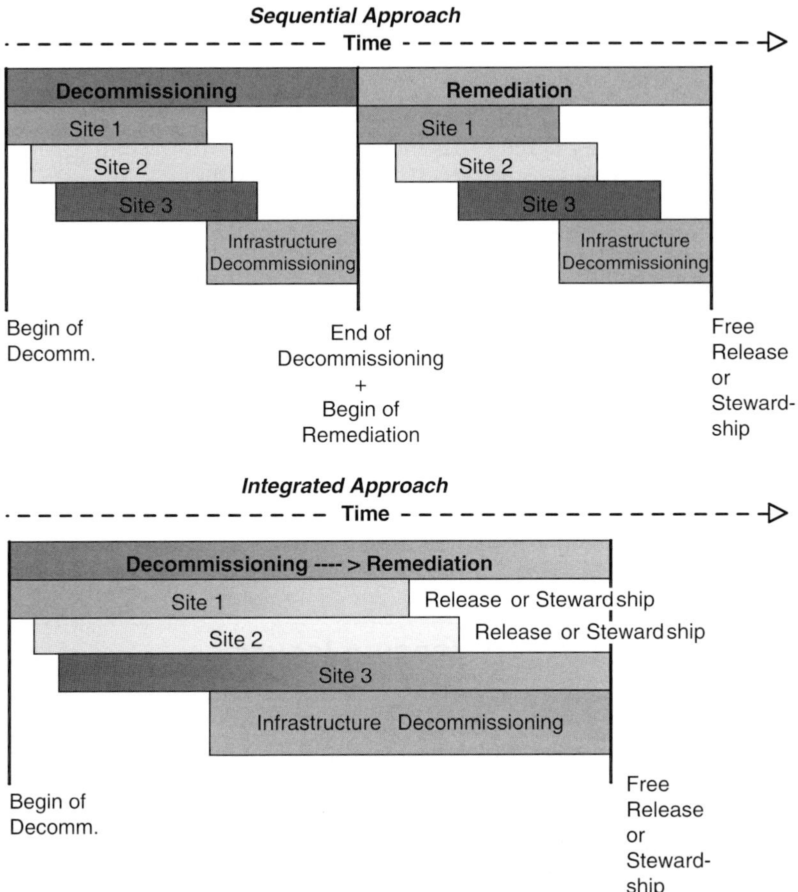

Figure 1 Sequential versus integrated approaches to decommissioning and remediation.

by various parties and that minimise adverse impacts on human health, the environment and costs are being encouraged.

2. Life Cycle Management

The nuclear community gradually becomes more aware of the need to move away from a 'technology-driven' approach to the design and management of nuclear facilities (including fuel cycle facilities) and to dealing with nuclear legacies from past practices and accidents towards a more integrated, life cycle management approach. The sustainability of the management of radioactive waste and legacies from past practices, accidents and non-fuel cycle residues can be further enhanced by moving away from *ex post facto* handling of liabilities to a more proactive and holistic approach (IAEA, 2006c).

Life cycle management in the present context involves incorporating into the decision-making process positive feedback and integrating lessons learned. It is generally agreed that incorporating decommissioning and site remediation requirements into the design of new facilities or the remaining life cycle of existing nuclear facilities can help to avoid or minimise future liabilities. The life cycle management concept does not treat each stage in the life of a facility or site as an isolated event. Instead, each phase in its overall life is viewed as having an influence on the future phases.

The long-term and life cycle management of radiological liabilities requires certain provisions and institutions. In recent years, the term 'stewardship' has been coined to describe the various activities associated with the long-term management of sites with radiological liabilities (USNRC, 2003). Generally, 'long-term stewardship' indicates the technical, societal, economic and environmental management measures needed to ensure the long-term protection of people and the environment (IAEA, 2006c).

3. Long-Term Stewardship Issues

Site reuse and provisions for long-term stewardship are closely related aspects in the life cycle of a site. Thus, the identification and selection of reuse options is a central issue that decision makers confront in making the transition from operation to decommissioning and remediation. The preferred future use for a site determines its end state for the post-commissioning phase. Establishing effective strategic planning provides a way to optimise decommissioning and the transition to stewardship (IAEA, 2006c). As a result, the successful design and implementation of decommissioning and remediation projects requires attention to a common set of tasks. Moreover, executing a successful decommissioning or remediation plan is dependent on the technical, political and economic feasibility of a desired end state (IAEA, 2002a).

Where remediation planning is not successful in identifying an optimal or sustainable end state, subsequent remedial measures may be required, or the optimal end state may be rendered unfeasible. Adopting a holistic approach early in the planning of decommissioning and remediation activities can not only result in a sustainable end state being achieved, but the ultimate optimisation case is where the end state generates revenues, which could not only (partially) offset costs but could also make public acceptance easier to achieve.

Sites, or areas of land on a site, where protection of human health and the environment requires ongoing management and control move into a post-decommissioning stewardship phase, or period of institutional care. During this period no active remediation is carried out, only surveillance (monitoring) and maintenance of engineered systems are done. In addition, land-use controls will be implemented to restrict particular reuse of the land so that risks to human health and the environment remain acceptably low. On some sites, because of radioactive decay, the stewardship period may last decades or centuries, after which the site could be unconditionally released. On other sites, particularly where there is residual activity that is long-lived, some form of stewardship in perpetuity may be necessary. Further details on issues and the requirements for long-term stewardship can be found elsewhere (IAEA, 2006c).

Controlled reuse of a site could in fact be advantageous in preventing or minimising misuse that might jeopardise either the institutional controls or the longer-term objectives for site reuse. Where residual contamination is left in discrete areas, environmental monitoring of the site will need to be an ongoing process, and it is possible that in some cases the technical measures will degrade (fail) or not perform as expected. A mechanism, therefore, needs to be in place to re-evaluate at intervals a site's status to assure that evolving technical, managerial and regulatory boundary conditions are met, which are the main objectives of stewardship (IAEA, 2006c).

4. Reuse Options

In general, there are two tiers of options to consider for a given site: broad options that essentially establish boundary conditions (constraints), such as whether any conditions will remain after remediation is complete; and specific options that define the specific end-uses of the site.

The redevelopment potential of the land, however, will depend on whether it can be remediated to levels compatible with the intended use. It also depends on the cost of remediation – bearing in mind that contaminated land on nuclear sites will need to be managed to ensure that at least minimum environmental risk criteria are met – even if this is by restricting access to potential receptors. The (re-)drawing of site boundaries and the disposition of certain features, such as impoundments for contaminated residues, will also have a strong influence on the usability and the redevelopment potential of a site.

Reuse may come in a number of guises, for example housing, new industry (industrial park), recreational uses (golf course, multi-recreational resort), museums or even an authorised disposal facility. Aspects to consider in evaluating alternative reuse scenarios include ease of access, convenient shape of plots, connection to services and other infrastructure, such as roads, railways, sewage systems, drinking water supply, electrical grid and so on.

Involving the people who are actually gaining benefit from the reuse may foster vested interest, and records management improves as the chances of continuity are greater. The objective is to create 'ownership' in the use scenarios that are compatible with the stewardship requirements.

The development potential of a site is often dependent on one or two key assets left over from the site's operating life. These assets can provide an important catalyst to a particular development or serve to improve the attractiveness of the site as an investment proposition for potential developers. Therefore, an important step in exploring the redevelopment potential of a site is to identify these potential key assets and assess their relevance to future development scenarios. Once identified, these assets should be protected from deterioration during the remaining life of the site. In the case of nuclear facilities, these infrastructure assets can include

- high-quality electricity grid supply connections;
- airstrips, road, rail or sea access with offloading facilities;
- sewage, district steam heating, potable water systems and other piping networks;
- a local workforce with a high level of technical skill;
- office space, in particular prestigious old buildings, perhaps with historical significance;
- support services (catering, public transport, etc.);
- (non-radioactive) machine shops, workshops and general production facilities, especially with large machinery, stocks of spare parts, consumables and/or
- large flat sites suitable for a substantial manufacturing investment or for a smaller investment whilst still retaining the future potential for contiguous expansion.

Recognising that financial constraints may prevent the unconditional release of the site in the near term, some of the reuse options that make use of existing infrastructure include:

- reusing (structurally sound) redundant buildings that had been used for nuclear purposes, particularly where the reuse might relate to other nuclear activities;
- leasing 'clean' buildings within a nuclear site to tenants in order to generate revenue;
- enabling other nuclear-based industries to develop on the site;
- promoting business uses that require the enhanced security infrastructure that traditionally exists at nuclear sites;

- using site infrastructure for the remediation activities, that is use of liquid effluent or solids waste conditioning treatment plants, or use of radioactive waste disposal arrangements; and
- preserving a nuclear site for its potential future use as a site for the next-generation nuclear power generation.

Of these options, the most practical reuse of a major nuclear site would be the reuse for another similar nuclear facility. In the case of a nuclear power plant, it is expected that only the reactor building, including the systems and components, would be radioactive. Hence all other buildings and equipment could be put to fit-for-purpose reuse after evaluating the structural soundness and radioactive cleanliness.

The reuse potential for a site can also be evaluated in terms of options that enable parts of the site to be conditionally reused and thus result in no or little cost (or very low maintenance) only. In this case, restrictions would ensure that any residual radioactivity cannot cause adverse effect on human health and the environment during the period before site closure. After site closure these restrictions may need to apply only if the site enters a period of long-term stewardship.

Although the cost-neutral/low-maintenance-cost options appear to generate no revenue, they may enable other costs associated with more intensive remediation to be offset, such as those associated with the generation, conditioning and disposal of large volumes of relatively short-lived radioactive waste.

Other broad options where the site infrastructure is not retained include creation of nature reserves, use for a recreational facilities or establishment of industrial (nuclear) heritage sites.

5. Synergies Between Decommissioning and Remediation

5.1. Identifying synergies

Careful consideration of the life cycle of a site through strategic planning may assist in linking the decommissioning plan and the remediation plan to future use. This will increase the likelihood that synergies between operations, decommissioning and post-decommissioning are maximised as indicated in Figure 2.

The term 'synergy' refers to the concept that working together or cooperating in a combined effort by sharing information and resources to accomplish some project tasks can produce more benefits than are achieved through independent and consecutive efforts. Synergies are possible between remedial activities and decommissioning activities because each

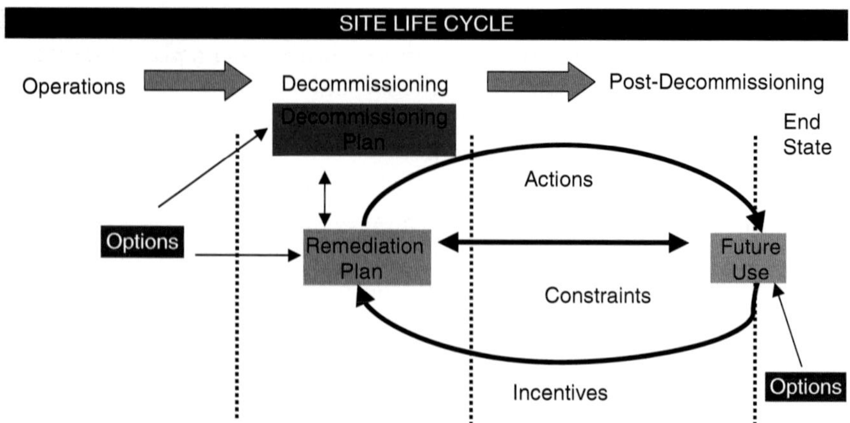

Figure 2 Synergies between decommissioning and post-decommissioning.

effort is based on a common objective. As a result, the successful design and implementation of remediation and decommissioning projects involves a number of common tasks including

- project management,
- site characterisation,
- risk assessment,
- materials and waste management,
- records management,
- quality management (QM) programme,
- measures to ensure occupational safety and health,
- administrative measures to ensure regulatory compliance,
- measures to ensure stakeholder involvement.

Identifying potential synergies in each of these activities will allow a completion of projects in a more cost-effective manner and, perhaps, in a shorter time.

Typically these issues are managed using a strategic planning process that involves first defining the preferred site end state, then identifying, integrating and scheduling all decommissioning and remediation activities to meet the site end state objectives. This high-level, strategic planning is another common activity that includes synergies of importance.

For a multi-facility site, strategic planning is the co-ordination of multiple projects to decommission the buildings and remediate the building sites. It is through this strategic planning process that all decommissioning and remediation activities (projects) can be identified and the steps taken to develop the supporting common tasks in a manner that efforts and costs are minimised and impacts and risks are optimised.

In general, decommissioning/remediation planning addresses such issues as workers safety and environmental protection, safety assessments, working procedures, time schedule, training and other technical and administrative aspects. For extensive projects, it is helpful to identify important technical, operational and administrative aspects when defining individual management requirements, even if the same personnel eventually manage these aspects. It will be preferable, particularly in the early stages of the project, to include, in the decommissioning/remediation team, persons who were involved in the operation of the facility/plant/site with institutional knowledge.

5.2. Site characterisation synergies

Site characterisation work primarily includes radiation field and contamination surveys (buildings, soils) as detailed in Chapter 2 (IAEA, 1998, 2000a; CIRIA, 2000, 2002). Assessment of historical records provides a qualitative characterisation of the site and may help to identify previous problematic activities and incidences, thus helping to target the physical site characterisation process. Synergies include

- common records associated with facility histories, building and land usage, environmental compliance monitoring, unusual incidents;
- common records associated with defining radioactive and hazardous chemical inventories;
- common records associated with facility drawings;
- a holistic view of the whole of the decommissioning task and an ability to ensure that the decommissioning and remediation surveys are sufficient to support the risk management measures;
- stakeholder confidence that the ongoing management of the site clearly demonstrates an intention to control and manage potential risks;
- integration of long-term monitoring arrangements for any residual subsurface contamination into the site's life cycle - taking account of those decommissioned areas that are at an intermediate site end point.

5.3. Risk management synergies

Risk assessment is one of the central elements of remediation and decommissioning planning because it can aid decision makers in determining an acceptable end state suitable for future use. In addition, risk assessment is used as a planning tool to identify and manage possible occupational safety and health risks to workers involved in implementing the remediation/decommission project by identifying significant exposure pathways.

The results of risk assessments will be used to design safety plans that minimise the risk of accidents during remediation/decommissioning. The application of risk assessment to determine potential risks that may result from waste management activities, including on-site or off-site transportation, can aid in the identification of risk-minimising alternatives for managing the residues from remediation or decommissioning. Ecological risk assessment will help to identify potential adverse effects on the environment.

Separate risk assessments of each activity may lead to risk 'displacement' by the transfer of risk to other activities rather than overall risk reduction. For example, soil removal may reduce on-site risk but creates off-site risk associated with transportation, treatment and disposal. Integrating the risk across the full spectrum of the life cycle of nuclear facilities ranging from operation, to decommissioning and remediation, to post-decommissioning can help to reduce the overall risk and to optimise radiation protection.

5.4. Materials and waste management synergies

Decommissioning of a nuclear facility may generate large amounts of surplus materials and varying volumes of waste that are quite different from the normal operational wastes. Through careful planning and sequencing of dismantlement operations, most of the waste can be segregated into inactive materials and (low-level) waste. Dismantling of activated and contaminated components would generate radioactive waste and the volumes of this can be kept low through appropriate strategies and technical options. The techniques selected for dismantlement need to be properly evaluated based on overall primary and secondary wastes that would be generated (IAEA, 2000b).

Significant reduction in volumes of wastes generated can be achieved through a well-formulated decontamination programme, appropriate dismantling techniques, contamination control and suitable radiological and administrative control measures. Reuse and recycle strategies can substantially reduce the amount of material (mainly metals and concrete) that has to be classified as low-level radioactive waste (LLW).

In many cases, decommissioning and remediation are purposely undertaken in stages with considerable time delays between each phase. This allows short-lived radionuclides to decay, in order to reduce the dose to the workers and also the amount of radioactive waste produced. At the end of each stage an interim end point will be reached where any remaining radioactivity is in a form that cannot cause unacceptable impacts on human health and the environment.

Soil remediation may generate large quantities of soil that was contaminated by abnormal events or chronic leakages of pipe work or storage facilities. Minimising remediation waste generation by applying risk management approaches, for example segregation, is likely to reduce waste management costs. Recycling and reuse of releasable materials from remediation may further reduce remediation costs. Taking measures to limit subsurface contamination and to minimise the footprint during remediation/decommissioning is an additional synergy. Appropriate timing can be an additional option, as some contaminants decay or degrade through the period of institutional care, thereby removing the need to manage *ex situ* potentially contaminated materials or soils.

5.5. Occupational safety and health synergies

The dismantling of structures or the deployment of remediation technology will have certain health and safety risks associated with them (Travis et al., 1993). Such risks may also result from accidents that occur during the handling of residues including their on-site or off-site transportation. Health risks may result from workers being exposed to radionuclides or hazardous chemicals during remediation or decommissioning.

Potential synergies may be obtained by having remediation projects collaborate with the related decommissioning projects in collecting information from ecological studies, medical surveillance programmes, epidemiological or toxicological studies, regulatory requirements and reference sources. Working together to design and collect necessary occupational safety and health data will avoid duplication.

5.6. Records management synergies

Identification and rationalisation of common data, logical structuring of the information management system so that it mimics the remaining life cycle of the nuclear facility and consideration of the audience for the record is likely to result in considerable synergies.

The sources, types of data and form of data are generally disparate (IAEA, 2002b) and the purpose of collection can be completely different for a current or future use. This is particularly true of historic data. Identification of common types of data and their rationalisation relative to the requirements of a number of potential users, therefore, enables new data to be collected once only, with the intended audience in mind, rather than many times, with only the specific job-in-hand in mind.

Another important synergistic aspect of managing a large data store is to consider how all the different users might access these data. As much of the data relate to facilities or areas of ground, access typically would be provided through a Geographical Information System (GIS) (Coppins et al., 2003).

5.7. Stakeholder involvement

The term 'stakeholder involvement' refers to the activities conducted during the design and implementation phases of the remedial action and/or decommission project that attempt to determine the needs and concerns of various parties including elected officials, interested citizens, workers, businessmen and environmentalists (IAEA, 2002a). The goal is to foster a dialogue to help create positive relationships between project managers and stakeholders.

Although project managers often think about individual projects, stakeholders may have a more general perception of the site that does not necessarily distinguish between remediation and decommissioning. Active involvement of stakeholders during the design phase of projects may help in the identification of end points, definition of priorities and selection of technologies. Using a shared staff of public involvement experts may reduce manpower costs and facilitate broader awareness of stakeholder perceptions about acceptable and unacceptable risks and end states for the site. Such insights may help project managers reduce or avoid misunderstandings, especially when some elements of the overall effort at the site are controversial. Stakeholder involvement from the very beginning may also help to create a sense of 'ownership' in the chosen paths and final site uses, thus facilitating the stewardship requirements (IAEA, 2002a).

6. Project Management

Careful planning and management are essential to ensure that decommissioning/remediation is accomplished in a safe and cost-effective manner (IAEA, 2004c). Many management-specific needs arise during the course of decommissioning or remediation projects. Site-specific management issues will differ at each facility. Factors such as schedule, work progress, outcome of meetings and regulatory issues may influence planning. A major benefit is dialogue with the regulatory authorities to minimise inconsistencies both in implementation of the decommissioning/remediation plan and achieving unrestricted release criteria.

Various lessons have been learned from the commercial decommissioning/remediation projects to date that will help future projects become more successful. The complexity and uncertainty of the radiological (and non-radiological) contamination aspects have the most impact on schedule and cost of decommissioning and remediation. Some of the key elements of safe and effective management planning are as follows:

- A good management system that works from day one is put in place.
- Managers study the best practices elsewhere and keep the lessons learned in mind before planning for decommissioning/remediation and set solid up-front expectations for the workforce.
- Integrated task planning is essential.

- Co-ordination between the various disciplines involved is a big challenge; effective co-ordination can only be carried out from a central place.
- It is beneficial to carry out safety-related tasks through well-written, reviewed and approved step-by-step procedure.
- All disciplines, including crafts, health physics, industrial safety and others need to be involved when developing and scheduling job evolutions.
- Taking the time to evaluate and understand the level and experience of the workforce and individual awareness of the importance of safety will pay dividends.
- It usually proves more efficient and cost-effective in the long run to use highly skilled workers for the higher risk tasks to avoid the cost and liabilities associated with training or the consequences of using unskilled workers.
- It can be a mistake to ignore the advice of long-time plant employees who understand and know plant characteristics and history.
- Detailed pre-job briefings are vital to ensure hazards are avoided during job performance.
- Work-in-progress briefings given to work crews at critical junctures during complex high-risk tasks ensure that all job requirements are understood and implemented.
- Because the regulatory process for licence termination is continuing to develop, it is critical to understand the roles of each organisation involved.

In order to ensure a safe and effective project, it is essential to establish clearly stated, verifiable end states. This is probably the most important single factor. The end states must be derived from the goals and objectives of the organisation responsible for the tasks and also be acceptable to the organisation taking over the facility at the end of decommissioning and remediation and any remaining stakeholders. The end state must be independently measured and reported. If iteration is needed to achieve goals, then this must be taken into account.

Many sites have an 'operation team' style of organisation to begin with and it is recognised that their expertise and knowledge is vital. Experience, however, has shown that it is not effective to use an 'operation' style organisation because decommissioning and remediation are more like construction than operation, with the team facing new tasks constantly.

Flexible, short-term planning horizons for projects will make them more tractable. Decommissioning/remediation can be full of surprises and often techniques are being tried for the first time on a large scale. Experience has shown that a high-level overall schedule backed up by a detailed planning process for relatively near term activities works the best. This type of planning coupled with an effective data gathering and analysis programme will allow for effective and flexible use of resources.

7. Residues Management

7.1. Basic considerations

The decommissioning and remediation activities will result in different types of residues or waste streams that each requires adequate management (IAEA, 2004d). The two principal options are recycle/reuse and disposal. It should also be kept in mind that decommissioning and remediation can result in the contamination of additional materials or media as the respective technologies are deployed (IAEA, 2004b). These materials also need to be managed appropriately. In general, however, a large proportion of the residues arising will be inactive, which means that they will be available for unconditional release.

7.2. Decontamination residues

In the case of nuclear power plants and research reactors, most activated materials are contained within the reactor vessel and its internal components, as well as in the biological shielding that surrounds the vessel. Typically, these components contain materials such as stainless steel, alloy steel, aluminium, reinforced concrete, graphite, zirconium alloys and so on. The source of radioactive contamination is mainly neutron activation. Process equipment and components used to contain process materials, whether a reactor coolant or reprocessing plant process streams, become contaminated with fission products, activation products and some transuranic (TRU) isotopes.

Facilities from the front end of the fuel cycle, such as mines and ore processing, and enrichment and fuel fabrication plants would be contaminated by naturally occurring radionuclides. Other parts of a facility may be contaminated if there had been any liquid, gaseous or particulate leaks. The waste arising out of dismantlement of active structures' systems and components is known as primary waste.

Radioactively contaminated liquids can also result from the decommissioning of a facility, for example the liquid wastes arising from the decontamination or flushing of systems. The types of radioactive contaminant in the liquid are dependent on the type of facility being decommissioned and the exact location in the process where the waste stream is being generated. These are generally known as secondary wastes.

Appropriate segregation and decontamination processes will reduce the volume of radioactive material requiring treatment significantly. Typically, non-radioactive solid materials include items such as piping, pumps, valves, tanks and duct work, and structural and electrical equipment. Inactive liquids and solid materials can be disposed of or recycled in accordance with applicable regulations and using conventional methods.

Gaseous effluents and aerosols would be finely dispersed radioactive materials resulting from cutting and abrasive surface cleaning methods. Some cutting and cleaning methods produce large volumes of toxic smoke and fumes. Contamination control coupled with filters in the ventilation stream is required to prevent spreading of contamination.

Worldwide many countries have developed comprehensive treatment strategies for waste arising from decommissioning projects. The starting point of these strategies is the preparation of an inventory of the radionuclides present, as these will dictate the operational, transportational and disposal practices to be employed during decommissioning, depending on the availability of on-site or off-site near-surface disposal facility.

7.3. Remediation strategies

Remediation methods for contaminated land are generally based on the following options (IAEA, 1999d, 1999e, 2004a, 2004b):

- *Natural attenuation* – where the natural properties of the ground are utilised to retain or destroy the contaminants by physical, chemical or biological processes (IAEA, 2004b). This is usually the 'baseline option', but is not equal to a 'do nothing option'. Careful site assessment and monitoring is required to assure the feasibility of this option. The scope and applicability of monitored natural attenuation has been discussed in detail in a recent report (IAEA, 2006b).
- *Removal* – where the contaminated material is removed off-site to an engineered disposal site. Removal techniques include for instance excavation of contaminated soil or the pumping of contaminated groundwater. The contaminated materials often require treatment and conditioning before emplacement at the disposal site and of course a suitable site must be available and licensed to receive the respective residues. For a recent review of relevant techniques see IAEA (1999d, 2004b).
- *Immobilisation* – where contaminated materials and media are treated to reduce the mobility and hence the source term of radionuclides. Typical methods include cementation or vitrification. Immobilisation can be applied *ex situ* (but perhaps on-site) as part of a removal and disposal scheme (IAEA, 2003b) or *in situ*, that is underground.
- *Containment* – where contaminated ground is isolated from the surrounding environment thereby preventing or reducing the migration of radionuclides away from the source. Typical measures are liners, cappings, injection curtains, etc. Containment may be combined with *in situ* immobilisation to give a defence in depth, or with removal to provide extra time for carrying out this operation. In managing groundwater contamination, 'funnel-and-gate' schemes in combination

with for example 'reactive walls' are a variant of these themes. For a detailed review of these techniques see IAEA (2004b).

- *Destruction* – where the contaminants are destroyed or converted to less harmful substances by chemical, biological or thermal processes. These methods would not be applicable to radionuclides *per se*, but to organic contamination that may be associated with it. Techniques such as incineration can also be used to reduce residue volumes, but one must be aware of the possible liberation of volatile nuclides such as ^{210}Po. Again, a critical review of the various techniques is provided in IAEA (2006a).

The different techniques and strategies naturally result in different types of waste streams. The materials to be managed usually include top-soils, soils, groundwaters and man-made materials such as concrete used in foundations.

The ultimate choice of waste management processes will depend on a wide variety of technical and non-technical considerations (IAEA, 2002a).

7.4. Integrated residues management

An integrated and holistic approach to decommissioning and site remediation affords the opportunity to assess these waste streams in terms of possibilities to combine them for more efficient management. Their fit-for-purpose reuse within the project or opportunities to segregate the wastes such that they can be disposed and maintained in possibly less expensive on-site facilities help in avoiding the cost and risks of transport to an off-site licensed waste management facility. Integrating measures might include

- retention of existing hard standing areas (e.g. base slabs from demolished buildings) to cap a soil contamination;
- use of existing, redundant hard standing areas with simple covers to store contained, unconditioned very low level radioactive waste;
- use of areas of immobilised contamination (going in hand with improved geotechnical and civil engineering properties) to support site remediation infrastructure, for example waste stores, treatment centres or treatment facilities;
- substitution of 'new' construction materials with suitable or suitably conditioned (very) low-active residues;
- segregation of very low active materials, particularly crushed concrete for reuse
  - in foundations or base layers for building foundations (the upper layers may need to be clean material);
  - as backfill for voids between containerised LLW in a repository or
  - in liner or cap/cover systems for LLW repositories.

8. Quality Management

Developing a QM programme early in the planning process will help to ensure that all regulatory requirements are met with regard to safety of the public and workers and protection of the environment during decommissioning and remediation. Such a programme would ensure that an adequate safety culture is inculcated in all concerned because the highest level of safety is achieved only when all are dedicated to the common goal.

The QM programme is expected to ensure a systematic approach to all activities affecting safety and quality, including, where appropriate, written verification and certification that each task has been performed in accordance with prescribed limits, regulation and approved procedures. Also, this would ensure that necessary effective corrective actions are taken. The QM programme would also establish an internal audit system to ensure that all safety-related activities are performed as per requirement stipulated by regulatory authorities. The QM programme must provide adequate flexibility while maintaining consistency with accepted quality standards.

A site-specific QM manual will detail all requirements of the QM programme, covering *inter alia*

- management functions including establishment of an organisational structure of trained and certified personnel with functional responsibility and levels of authority with clear lines of communication;
- all decommissioning/remediation activities to be performed after adequate planning and with well-prepared, reviewed and approved procedures;
- a verification programme to ensure that decommissioning/remediation activities are carried out in a safe manner and as planned and intended;
- internal audit to identify non-compliance, if any;
- corrective actions to be taken in case of any failure or deficiency noticed during implementation/subsequent monitoring;
- documentation of all required information during decommissioning/ remediation and storage in a retrievable manner for future reference and verification.

Since decommissioning and remediation of a nuclear facility involve handling of radioactive material on a large scale, the QM programme should take into consideration the following aspects for proper accounting of radioactivity:

- All information and records relating to radioactivity in the nuclear facility are collected and documented prior to decommissioning and remediation.
- For remediation to restricted reuse the records pertaining to location, configuration, quantities and types of radioactivity remaining at site

during prolonged period are periodically updated and properly accounted for.

- All radioactive materials that were present at the site at the commencement of decommissioning/remediation are properly accounted for till their ultimate destination is identified.

9. Practical Experience in Remediation and Site Reuse

Decommissioning and remediation activities are an opportunity to achieve reuse objectives consistent with site release criteria. Because national standards differ for site release criteria, various examples are available illustrating experiences in decommissioning and site reuse. This section provides more detailed descriptions of specific projects drawn from a diverse set of nuclear facilities.

9.1. Environmental remediation at the Idaho National Laboratory site

9.1.1. Introduction

This section summarises waste retrieval activity that was conducted as part of remediation of a radioactive waste disposal facility located at the Idaho National Laboratory (INL) site in the United States. The information provided in this section has been obtained from a number of different sources. However, the primary sources were a remedial investigation and baseline risk assessment report (Holdren et al., 2006) and a United States Department of Energy (USDOE) report summarising the Glovebox Excavator Method Project (USDOE, 2004). This section is divided into three parts: a background summarising general information about the site, an overview of the Glovebox Excavator Method and a discussion of implementation of the Glovebox Excavator Method.

9.1.2. Background

The INL Site, formerly the Idaho National Engineering and Environmental Laboratory, originally established in 1949, is a USDOE-managed reservation that historically has been devoted to energy research and related activities. The varying missions that have been undertaken have resulted in a large amount of wastes being disposed over the lifetime of the site. The Radioactive Waste Management Complex (RWMC) has been used for management and disposal of radioactive and mixed wastes for more than 50 years since the Laboratory was established.

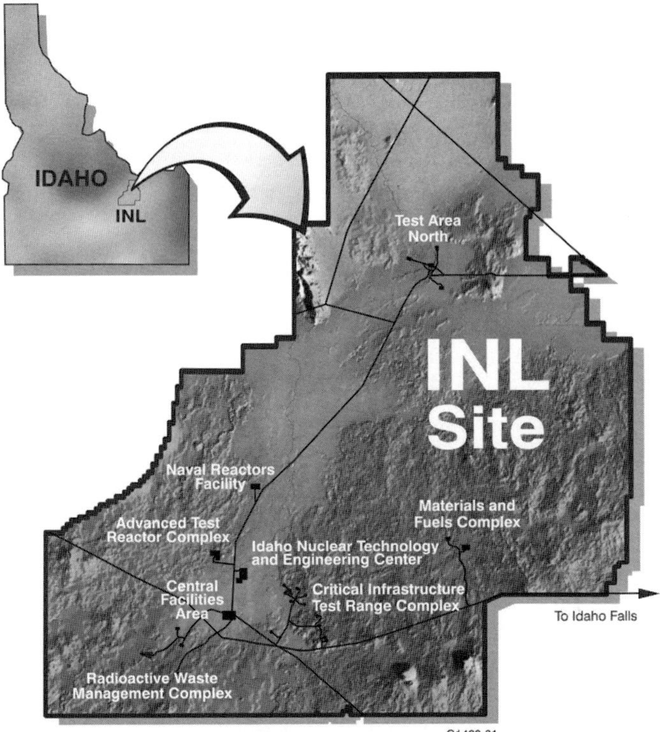

Figure 3 Idaho National Laboratory Site.

Past disposal operations have included disposal of some long-lived LLWs that required retrieval and deep geologic disposal at the Waste Isolation Pilot Plant (WIPP) as part of remediation of the historic disposal facility.

The INL Site is located in southeastern Idaho (Figure 3) and occupies 2,305 km^2 in the northeastern region of the Snake River Plain. Regionally, the INL Site is nearest to the cities of Idaho Falls and Pocatello. The INL Site extends nearly 63 km from north to south, is about 58 km wide at its broadest southern portion and occupies parts of five southeastern Idaho counties.

The RWMC covers 72 ha (Figure 4), including the operations and administration area (approximately 9 ha), the Subsurface Disposal Area (39 ha) and the Transuranic Storage Area (23 ha). Burial of radioactive waste in the Subsurface Disposal Area since the early 1950s has resulted from building and operating a wide variety of reactor types at the INL Site, other research and military activities, and accepting disposal of radioactive and hazardous waste from outside facilities (primarily from the Rocky

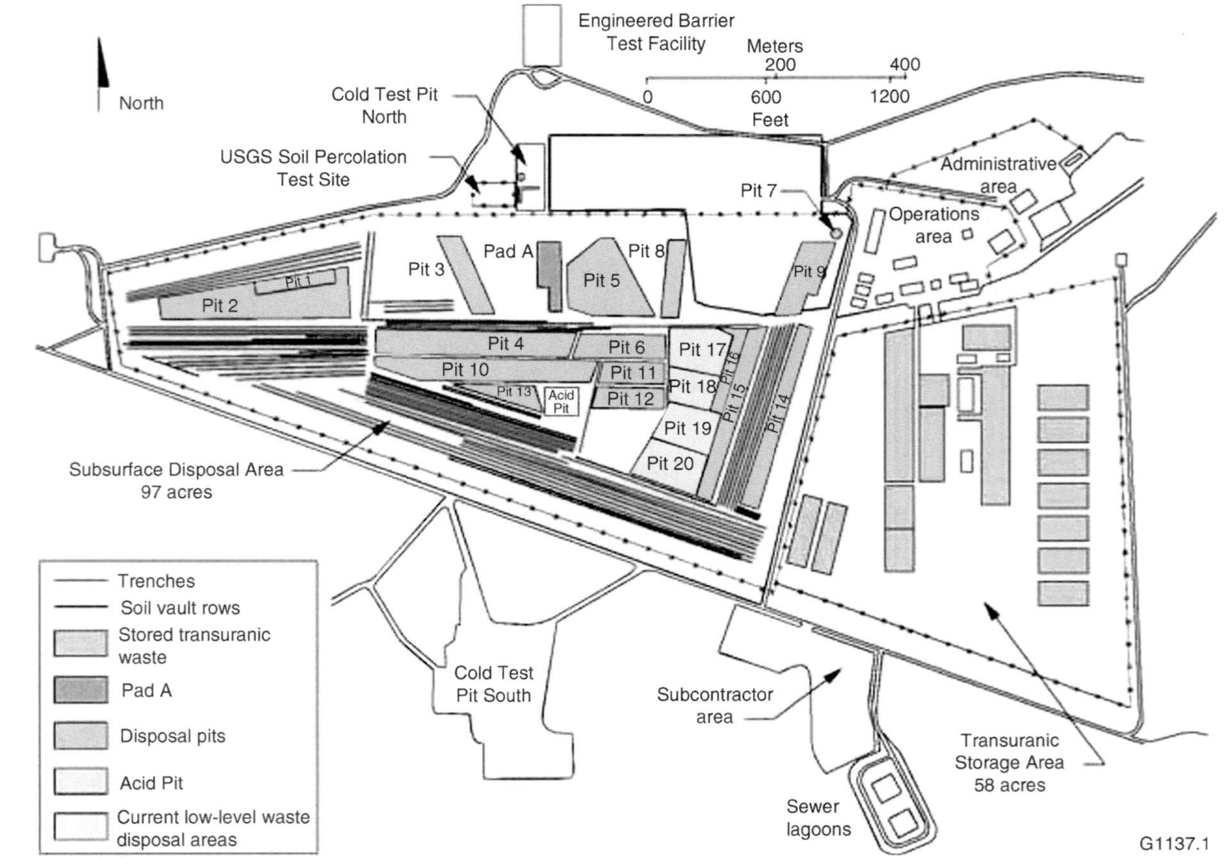

Figure 4 The INL Radioactive Waste Management Complex.

Flats Plant). Current environmental management activities at the INL Site include the following:

- treating, storing and disposing of waste;
- removing or deactivating facilities that are no longer of value;
- cleaning up historical contamination that presents risk to human health or the environment;
- providing long-term stewardship.

The INL Site region is classified as arid to semiarid because of low average rainfall of 22.1 cm $year^{-1}$, with somewhat less precipitation more recently. The RWMC has no permanent surface water features; however, the local depression tends to hold precipitation and to collect additional runoff from surrounding slopes. Surface water from episodes of rain or snowmelt eventually either evaporates or infiltrates into the vadose zone (i.e. unsaturated subsurface) and the underlying aquifer.

The RWMC was sited in a location that had sufficient surficial deposits to provide a reasonable burial depth. Undisturbed surficial deposits within the RWMC area range in thickness from 0.6 to 7.0 m. Irregularities in soil thickness reflect the undulating surface of underlying basalt flows. Below the shallow surficial sediment is a thick sequence of basalt flows intercalated with thin sedimentary interbeds. In the past, waste was typically disposed at depths starting at the top of the basalt, but the currently operating disposal pits were blasted deeper into the basalt to provide increased disposal capacity.

The regional subsurface consists mostly of these layered basalt flows with a few comparatively thin layers of sedimentary interbeds. Because subsurface formations are unsaturated most of the year, they are characterised as a vadose zone; however, ephemeral lenses of perched water have been detected in association with interbeds.

The Snake River Plain Aquifer underlies RWMC at an approximate depth of 177 m and flows generally from northeast to southwest. The aquifer is bounded on the north and south by the edge of the Snake River Plain, on the west by surface discharge into the Snake River near Twin Falls, Idaho, and on the northeast by the Yellowstone basin. The aquifer consists of a series of water-saturated basalt layers and sediment.

Land within the INL Site is administered by DOE and is classified by the United States Bureau of Land Management as industrial and mixed-use acreage. Approximately 98% of land on the INL Site is open and undeveloped. Due to the inhospitable nature of the INL Site, especially near the RWMC, there have been no significant residential developments in spite of the presence of hundreds to thousands of jobs for more than 50 years. No permanent residents live within the boundaries of the INL Site and no cities are located within several miles of the INL Site boundary.

Large tracts of land are reserved as buffer and safety zones around the boundary of the INL Site, whereas portions within the central area are reserved for INL Site operations. Future land use (and aquifer use) is expected to remain essentially the same as current use – research facility within the site boundaries, with agriculture and undeveloped land surrounding the INL Site.

The RWMC was established in the early 1950s as a disposal site for radioactive waste. Radioactive waste has been buried at the Subsurface Disposal Area within the RWMC in underground pits, trenches, unlined soil vaults, concrete vaults and one aboveground pad. Disposal of long-lived LLW (TRU) occurred in the Subsurface Disposal Area from 1952 to 1970. Since 1970, TRU waste has been placed on asphalt pads in interim storage at the Transuranic Storage Area (see Figure 4) and is currently being processed and then shipped for disposal at the WIPP. Acceptance of TRU waste for storage from off-site generators was discontinued in 1988. Disposal operations are currently limited to containerise LLW with the highest-activity LLW disposed in concrete vaults.

9.1.3. Glovebox Excavator Method Project

Pit 9 is a specific area within the Subsurface Disposal Area that was used for disposal of radioactive waste between 1967 and 1969. Pit 9 is located in the northeast corner of the Subsurface Disposal Area (see Figure 5). The Record of Decision (USDOE, 2008) specifies the preferred alternative of retrieval of TRU waste from Operable Unit (OU) 7-0 (the clean-up designation for Pit 9). On 1 October 2001, the INL published the Waste Area Group 7 Analysis of OU 7-0 Stage II Modifications (INEEL, 2001), which identified a feasible approach for retrieving waste from Pit 9. The overall objectives for the Glovebox Excavator Method Project were as follows:

(1) Demonstrate waste zone material retrieval.
(2) Provide information on contaminants present in the underburden.
(3) Characterise waste zone material for safe and compliant storage.
(4) Package and store waste on-site, pending final disposition.

The project was designed to safely conduct a waste zone material retrieval demonstration in a selected area of Pit 9. The volume to be retrieved was planned to be roughly between 75 and 125 m^3. The project processes consisted of excavation and retrieval; sampling, packaging and provisional storage; shutdown; deactivation, decontamination and decommissioning; and environmental monitoring. Project facilities include a Weather Enclosure Structure, Retrieval Confinement Structure (RCS), excavator, ventilation system and other supporting equipment. The packaged waste zone material retrieved by the project was transferred to an on-site facility for temporary storage, pending final disposal.

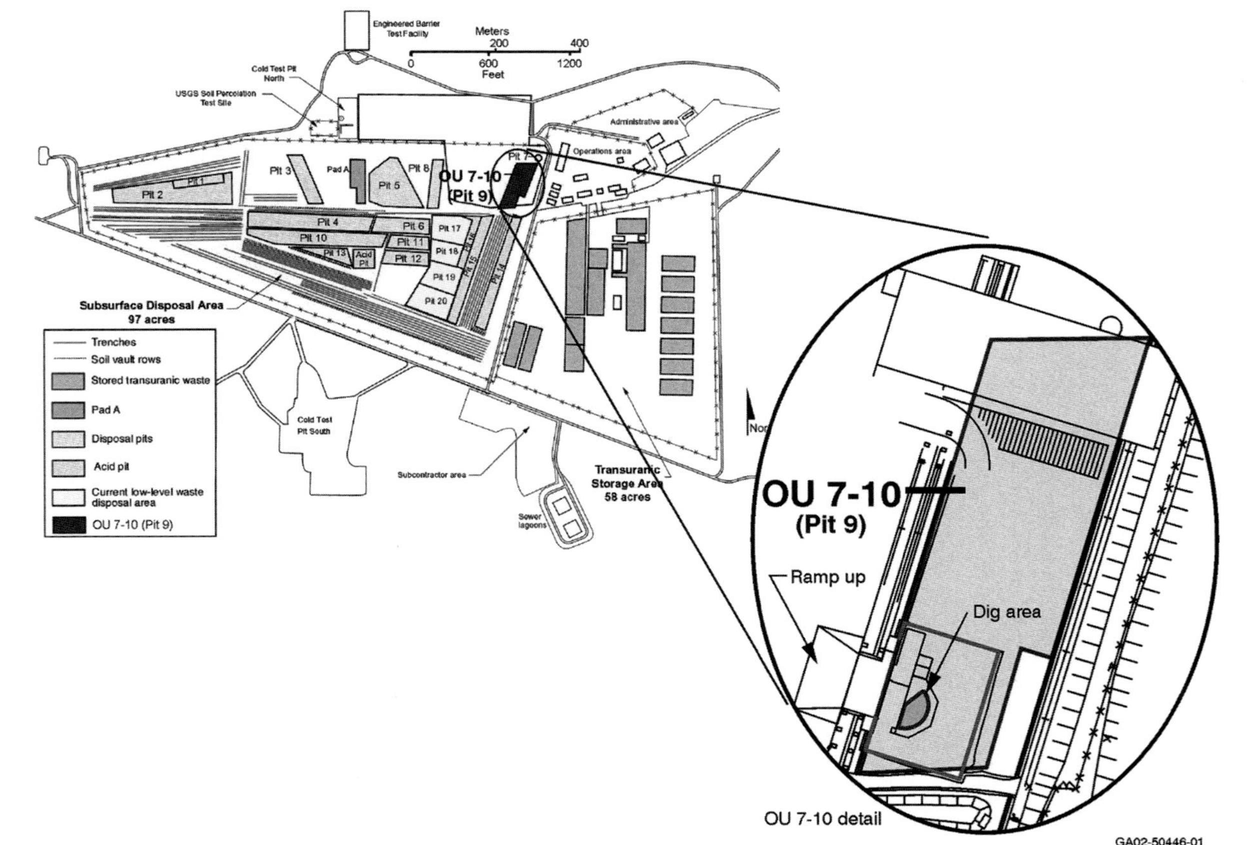

Figure 5 Diagram of the Radioactive Waste Management Complex with an expanded view of the OU 7-10 Glovebox Excavator Method Project area.

The Glovebox Excavator Method Project focused on a location in the southwest end of the Pit 9 area. It is defined by a fan-shaped area with a 6 m radius and the angular extent of 145°. Figure 6 presents the plot plan of the Pit 9 area showing infrastructure and the project location. The area addressed by the Glovebox Excavator Method was approximately 115.5×38.7 m.

Inventories of waste in Pit 9 and the Subsurface Disposal Area pits and trenches have been generated using existing and available historical records. The inventories contain uncertainties about various items including exact locations of waste inside the pit, extent of contaminant migration, specific isotopic information and chemical form, and valence state of the contaminants.

The waste in Pit 9 resulted from Rocky Flats Plant's weapons production operations and INL nuclear reactor testing activities and includes a variety of radionuclides, organic and inorganic compounds. The inventory of materials disposed in Pit 9 is summarised in the Record of Decision (USDOE, 2008).

The depth of the pit from ground surface to the bedrock is approximately 6 m. The soil cover or overburden has been estimated to be 1.2–1.8 m thick.

The Rocky Flats Plant waste contains radiological and non-radiological contaminants. The material shipped to Pit 9 from the Rocky Flats Plant included weapons-grade plutonium and uranium isotopes. Weapons-grade plutonium contains ^{238}Pu, ^{239}Pu, ^{240}Pu, ^{241}Pu and ^{242}Pu. Uranium isotopes shipped to the RWMC included ^{235}U and ^{238}U. Also included in the waste shipments were ^{241}Am and ^{237}Np, which are daughter products resulting from the radioactive decay of ^{241}Pu. In addition to the ^{241}Am produced by the decay of the inventory, ^{241}Am removed from weapons-grade plutonium during processing at the Rocky Flats Plant was also disposed of in Pit 9. This extra ^{241}Am is a significant contributor to the total radioactivity in Pit 9.

The primary organic chemicals known to be in Pit 9 include carbon tetrachloride, trichloroethene, 1,1,1-trichloroethane, tetrachloroethene, lubricating oils, Freon 113, alcohols, organic acids and versenes (ethylenediaminetetraacetic acid (EDTA)). Examples of inorganic chemicals known to be in the waste include hydrated iron, zirconium, beryllium, lead, sodium nitrate, potassium nitrate, cadmium, dichromate, potassium phosphate, potassium sulphate, silver, asbestos and calcium silicate.

9.1.4. Implementation of Glovebox Excavator Method Project

The project facility operated 24 h a day, 7 days a week. Operations personnel were divided into four crews, each covering a sliding 12-h shift (i.e. 4 days on followed by 4 days off).

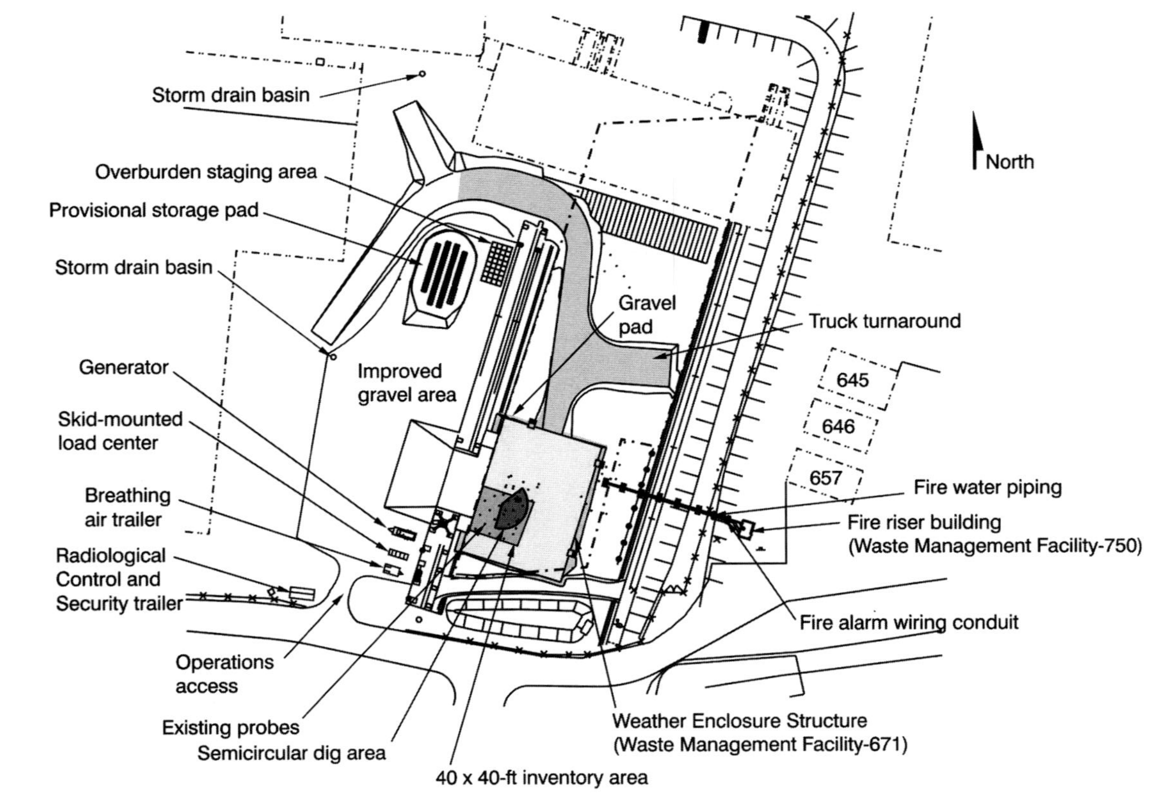

Figure 6 Facilities in the OU 7-10 area and the project site (fan-shaped area).

9.1.4.1. Overburden removal. Overburden removal commenced on 12 December 2003, upon receipt of DOE authorisation to proceed with project operations. Personnel, wearing a single set of personal protective equipment (PPE) with respirator, removed approximately 1.1 m of soil from the excavation area in the RCS. Operators used dust suppression equipment to minimise the amount of fugitive dust generated during this activity. The overburden soil dust suppression system functioned adequately during overburden activities. Operations personnel primarily used the excavator to remove soil and place scoops of overburden soil into soil sacks staged within the RCS. Operations personnel also entered the RCS and used shovels to manually remove overburden soil from around a probe cluster that had been installed to inspect the waste prior to retrieval. Once filled, the soil sacks were closed, surveyed for contamination and removed from the RCS and Weather Enclosure Structure. These sacks were later disposed of in the Subsurface Disposal Area as LLW.

Three soil sacks were filled, removed from the RCS and Weather Enclosure Structure and staged along the west side of the project interim storage area during the first day of operations. Overburden operations continued for a week, during which time a total of 39 soil sacks were removed. An average of just under five soil sacks were removed per day, with a peak rate of 12 sacks removed during 1 day of operation. The overburden removal process and procedures were refined as operations personnel gained experience working in the pit. The soil sack design volume allowed for up to 1.8 m^3 of material to be placed in each sack. The estimated sack fill volume for the overburden removed was approximately 80%. Therefore, approximately 57 m^3 of overburden was removed from the excavation site (39 sacks $\times$ 0.8 $\times$ 1.8 m^3 per sack).

Completion of overburden removal occurred on 19 December 2003, after removal of overburden soil from across the excavation area to a depth of 1.1 m below ground surface. In preparation for waste retrieval operations, personnel entered the RCS following overburden completion and cleaned and removed overburden removal-unique equipment. Waste retrieval equipment not previously staged in the RCS was also placed into the RCS and other tasks required to prepare the RCS for waste retrieval activities were completed.

9.1.4.2. Waste retrieval. Waste retrieval operations commenced on 5 January 2004, following receipt of concurrence to proceed from the Department of Energy. Excavator operators took scoops of waste zone materials (Figure 7) and placed these materials in transfer carts at one of three gloveboxes. Glovebox operators moved the transfer carts into the gloveboxes, segregated the waste zone material (Figure 8), separated and measured suspect fissile material and packaged the waste in appropriate storage containers (i.e. 55 gal drums) in a safe and compliant manner

Figure 7 Excavator working during project digging operations.

(Figure 9). Once the drums were filled, operators changed out drums and transferred them for assay measurement and subsequently to interim storage in WMF-628.

During the early waste retrieval effort, excavator operators removed only soil from the excavation site. Waste was not encountered until approximately 1.8 m of overburden soil had been removed. After removal of the overburden soil, operators identified waste materials and recorded this information along with the drum identification number where the waste was packaged. When operators suspected fissile material in the waste, the suspect material was placed in a separate bucket and moved for measurement and subsequent placement in an appropriate drum, ensuring that criticality limits were never exceeded. Once waste was packaged in drums, operators transferred the drums for full drum assay, followed by placement in the storage facility.

During waste retrieval, drum-generation rates averaged approximately 9.5 drums per day, with a peak rate of 27 drums in a 24 h period. A total of 454 drums were filled during the retrieval effort, most containing approximately 0.14 m^3 of waste materials, which met a project objective of removing more than 57 m^3 of material (Figure 9). As experience was gained over hundreds of hours of operation, operating procedures were refined and streamlined, increasing overall effectiveness and throughput while

Figure 8 Glovebox operators segregating waste zone material.

maintaining safety. Of note during this refinement process was the reducing of PPE required during drum change-out activities.

Monitoring by Radiation Control Technicians of contamination levels in the drum-load out area found no contamination releases during drum change-out operations, allowing a modification in the PPE required for operators. The modification permitted operators to work more effectively, thus increasing the number of drums removed daily while maintaining a safe working area. Both Industrial Hygiene and Radiological Control (RadCon) personnel monitored their respective areas of concern throughout operations.

Operations were curtailed for short times when high levels of radon (from atmospheric inversions) were measured. Operators followed RadCon recommendations for addressing this condition, and operations continued after radon levels were reduced. The ventilation support system functioned well during operations, maintaining required negative pressures

Figure 9 Waste being packaged in a 55 gal drum.

(i.e. air flow into the RCS from the Weather Enclosure Structure). Likewise, the RCS itself maintained its integrity throughout the operating period.

All three packaging gloveboxes were used during waste retrieval and packaging operations. Transfer carts within the gloveboxes operated adequately; however, the carts began to experience excessive vibration towards the end of the operations campaign, requiring the operating speed to be reduced. Continued operations significantly beyond achieving project objectives would have necessitated repair or replacement of the drive system components. Gloveport gloves were changed safely multiple times during operations, maintaining confinement and providing operators with functional and safe access to waste materials. Drum-loading enclosures used for drum change-out activities underneath each glovebox also functioned adequately. No contamination was detected in the enclosures during initial operations, which allowed for opening of the enclosures, thereby increasing drum throughput.

Waste drums found in the pit had little structural integrity due to corrosion. However, plastic bags and plastic containers had retained much of their integrity. Some bags were less pliable and more brittle, but most were in extremely good condition. Writing and markings on plastic containers and labels protected by plastic were often still very clear and legible.

Probes placed into the excavation area during Stage I activities (Figure 10) were successfully and safely moved within the excavation area during retrieval activities. Movement of the probes allowed for greater access to

Figure 10 Graphite drum in the midst of P9-20 probes.

the pit by the excavator. Probe-puller caps installed on the probes before commencement of waste retrieval greatly facilitated the backhoe in grasping and removing the probes.

*9.1.4.3. **Waste drum storage.*** Waste excavated from the project excavation site filled 454 55-gal drums. These drums were surveyed for radiological contamination and assayed to determine radiological isotope activities and fissile mass. Composite samples were analysed to support application of hazardous waste numbers. Drums were then transported to the storage facility.

9.1.5. Conclusions

At completion of the waste retrieval phase, the facility was placed in a warm standby condition. The purpose of the warm standby condition is to maintain the facility in a safe, stable condition that reduces surveillance and maintenance costs. Actions taken to achieve warm standby permit the facility to be restarted for future use, with the exception of the excavation area that was grouted after retrieval efforts were completed.

The Glovebox Excavator Method Project was a successful demonstration of a safe method for retrieving buried radioactive waste in a manner that minimises worker exposure and potential for contamination. Lessons learnt from this effort helped to form the basis for the final Record of Decision that documents the formal regulatory agreement on remedial actions to be completed within the Subsurface Disposal Area (USDOE, 2008).

9.2. Remediation of a radioactively and chemically contaminated site at Harwell, UK

9.2.1. Background

The Southern Storage Area (SSA) comprises an area of land covering some 7.2 ha, situated approximately 1 km south of the UKAEA Harwell, Oxfordshire. The SSA shared a boundary fence with Chilton Primary School. There are also nearby residences and a farm.

The SSA was used by the Royal Air Force (RAF) until 1945 as an ammunition store. From 1946 the site was used for a variety of waste storage and handling operations and for the 'permanent' landfill burial of mixed chemical, beryllium and LLW – these wastes being stored in a number of burial pits.

Typical operations included flask storage, decontamination and sea dump drum packing. Physically the site consisted of open ground, small huts, concrete track ways and many large earth mounds surrounding the original bomb storage bays. A preliminary clean-up of the site was carried out during 1988–1990, to eliminate the need for the site to be licensed under the Nuclear Installations Act (as was required for the main Harwell site). However, this remediation was not sufficient to allow unrestricted access to the site, which therefore remained secure. The objectives of the remediation were twofold:

(1) Physical: To clean up the land to a condition suitable for unrestricted public access. Put simply, to remediate the site such that it would be safe for children to play on. This implied a risk target and the determination of concentrations of contaminants that would be acceptable to be left on the site after clean-up.
(2) Psychological: The SSA was a sensitive site with some local controversy. It was considered necessary not only to make the site suitable for public access but also to be seen to have done so in a transparent and acceptable manner. A second objective was therefore the removal of doubt.

The project came about as a necessary part of UKAEA's mission to restore the environment. In the 1990s an opportunity arose to link the remediation to the development of a neighbouring area for housing. A combination of the need to regularise the authorisation status of the site, the commercial opportunity and UKAEA's ongoing mission resulted in a project for complete clean-up starting in 1999.

9.2.2. Selection of the remediation strategy

The characterisation of the site began with a review of historical records for the site, relating to munitions processing and material storage/disposal operations. Although records of the SSA lacked detail, they provided an overall scope and were used to plan the safety design of the characterisation phase. Interviews with ex-staff proved useful. Following a review of available historical information, a phased characterisation of the site was undertaken, including initial walkover surveys and trial pit investigations, followed by more detailed characterisation using a range of methods such as soil gas surveys, soil sampling, core sampling, trial pit sampling, probe surveys, groundwater monitoring and geophysical methods. In order to determine the best practicable environmental option (BPEO) for contamination present at the site, a formal environmental assessment process was undertaken in collaboration with the National Radiological Protection Board. The BPEO was defined by the Royal Commission on Environmental Pollution in 1988 as 'the option that provides the most benefits or least damage to the environment as a whole, at acceptable cost, in the long term as well as the short term'.

Options were generated prior to consultation with regulators, local authorities and the public. A large number of options for end condition and remediation technology were created using workshop methods covering the range of

- do nothing,
- capping the wastes,
- complete removal.

Using information from the site characterisation and best practice models, pollution linkages were developed for the site. Two linkages emerged as important:

1. human health impact (inhalation, ingestion, contact, radiation exposure ...) and
2. groundwater impact.

A consultation workshop with regulators and the local authority was used to provide input into the assessment of options and their scoring. In parallel, a programme of public communication was used to test opinion on the preferred option. The BPEO emerged to be complete removal of all wastes from the site.

As a precursor to the implementation of the preferred option, risk assessment was used to create a series of risk-based clean-up levels (RBCLs)

Table 1 Summary of RBCLs and clean-up targets for radionuclides generated by BPEO.

Radionuclide	RBCL ($Bq\,g^{-1}$)	Background 95th percentile value ($Bq\,g^{-1}$)	Clean-up target ($Bq\,g^{-1}$)
^{210}Pb	0.59	0.1[a]	0.7
^{226}Ra	0.04	0.045	0.09
^{228}Ra	0.06	0.05[a]	0.11
^{137}Cs	0.1	0.017	0.12
^{60}Co	0.02	0.03[a]	0.05
^{241}Am	0.33	0.015[a]	0.35
^{234}U	2.5	0.018	2.52
^{235}U	0.56	0.03[a]	0.59
^{238}U	1.4	0.016	1.42
^{238}Pu	0.25	0.002[a]	0.25
$^{239+240}Pu$	0.44	0.003	0.4
^{241}Pu	13	0.04[a]	0.4
^{228}Th	0.04	0.022	0.06
^{230}Th	0.63	0.024	0.65
^{232}Th	0.25	0.023	0.27
^{90}Sr	6.4	0.01[a]	0.4

[a]Analytical limit of detection.

for the chemicals and radionuclides of potential concern. These were designed to achieve a risk target of 1.0×10^{-6} per year. For the radioactive substances, this equated to 20 μSv per annum dose to the public after remediation (in addition to local background). For comparison a typical background exposure for the United Kingdom is 2,200 μSv s from natural sources. Table 1 gives a summary of the RBCLs and clean-up targets for radionuclides generated by this process.

When the informal consultations were complete, the environmental assessment was published in detail and in summary. The published version was subject to external independent peer review by consultants working under contract to the local authority. The peer review and response was built into a final version of the assessment and also publicly distributed. The final environmental assessment was submitted in support of conditions relating to the Town and Country Planning Application for the neighbouring housing development and formally consulted.

9.2.3. Implementation and validation of the remediation strategy

After the BPEO had been accepted, further stakeholder communication took place, the business case was finalised and contract specifications were prepared that carried forward the promises made to date. Contractors were selected competitively with safety and environmental competencies as a top priority. In parallel with the environmental assessment process, the UKAEA safety management system was implemented and a series of safety cases produced. The principal contractor drew up a detailed remediation plan and this became the definitive statement of how the works were to be completed. After the contractors had completed off-site design and documentation, a detailed topographic survey of the site was undertaken, in addition to baseline surveys of noise and ecology. Background radioactivity and chemical surveys were also undertaken. The contractor set up and commissioned the safety controls and other support systems required for the works. Commissioning included some rehearsal of emergency plans with the local services. The works were then completed over a 2-year period. Validation of the works was carried out at the end of each phase and at the end of the overall project. After demobilisation of the contractor the site was landscaped and a clean layer of topsoil added. The contractor produced a post-remediation report, a health and safety file and a land quality statement. Throughout the works UKAEA maintained close supervision and controlled the implementation through a series of process systems and staged approvals. For example, every waste shipment leaving the site was controlled by the contractor and additionally signed off by UKAEA. It maintained a full-time site team of supervisors and employed further support contractors to carry out independent audit, surveying, monitoring and validation.

9.2.4. Clean-up method: general areas

The general land areas of the SSA were of low hazard. Typical contamination was $<10\,\mathrm{Bq\,g^{-1}}$ of ^{137}Cs in patches up to a few square metres. Chemical and radioactive contamination was predominantly at the surface, but could also occur randomly throughout the depth of made ground. The approach adopted was to dig over every part of the SSA down to base geology in layers 300 mm deep. Before digging in an area the next layer was surveyed and sampled using a full range of methods. Any 'hotspots' were removed from the layer using targeted digging and then the entire layer was bulk dug and placed elsewhere on the SSA for later reuse as clean fill. Waste was put into $1\,\mathrm{m^3}$ woven bulk bags and over-wrapped. The wastes were then sent to a dedicated waste assay facility on the SSA. Some very low hazard wastes and scrap were bulk dug and held in stockpiles for later shipment off site in bulk waste containers. All drains, roads, buildings, foundations and other structures were surveyed and removed as waste or processed for reuse. All bagged wastes were over-packed in reusable containers for shipment to landfill. The over-wrapped bulk bags were disposed of intact to a landfill cell and immediately capped.

9.2.5. Clean-up method: burial pits

One of the principal risks to workers and the public arose from the presence of significant quantities of beryllium and beryllium oxides in the burial pits. Beryllium is a light, strong metal used in the nuclear industry because of its material and neutron performance properties. Some of the forms of oxide as a respirable dust can be very toxic to sensitive individuals and beryllium has a very low maximum exposure limit (MEL). All of the pits were dug inside a ventilated enclosure designed to ensure inward leakage. Exhaust ventilation air was filtered through high-efficiency filters, monitored and discharged via stacks. The enclosure was contained within an outer weather shelter. Workers used a hierarchy of protective measures including damping down, careful digging, monitoring, hygiene precautions, health surveillance and finally personal and respiratory protective equipment. Waste was segregated at the workface and put into 1 m^3 polypropylene woven bulk bags. A representative sample for analysis was taken during filling. The bags were posted out of the enclosure into an over-wrap and taken to a dedicated waste assay facility. Large wastes were compacted at the workface and put into bags or wrapped individually and exported from the enclosure. The pits were dug either side-on or from the top-down and were over-dug to remove any materials that had migrated from the sides or base. In general, chemicals and radionuclides had moved <1 m into the surrounding geology (the main exception being the known chlorinated hydrocarbon groundwater pollution arising from the chemical pits). After removal of all materials above the clean-up targets the pit surfaces were sampled and validated. The remediated pit holes were backfilled with low-permeability clay and geotextile layers at the request of the Environment Agency.

9.2.6. Waste management

All bagged wastes were processed through a waste assay facility on the SSA. A representative sample from each bag was taken during filling and the wastes were monitored during filling, according to the Radioactive Substance Act and its Exemption Orders. Samples or combinations of samples were analysed for metals, organic chemicals (gas chromatography-mass spectrometry), volatile organic compounds and poly-chlorinated biphenyls (PCBs) and screened total alpha/beta activity. The use of total alpha/beta radioactivity screening provided a sensitive indication of potential radioactivity, but was unreliable as an indication of absolute level when radioactivity above background was present. Bags suspected of being LLW were rotated on a turntable in front of a calibrated high-resolution gamma-spectrometry system designed to police the limits necessary to consign the waste correctly under the Radioactive Substance Act and Exemption Orders. The combination of these measurements was used to decide the appropriate waste route. Fingerprinting was not utilised formally because of the heterogeneity of the wastes, but it was possible to group waste types using experience gained as the work proceeded. The decision methodology utilised was agreed with the Environment Agency prior to use. Duplicate quality assurance samples were taken for 5% of the total.

9.2.7. Verification monitoring

Throughout the remedial works a verification programme was undertaken to confirm compliance with the remediation objectives. The programme involved continuous sampling and monitoring, culminating in a final survey. The programme was designed to demonstrate the following:

- soil reused on site met the remediation targets;
- the final surface (and the base of pits prior to infilling) met the required remediation targets;
- soil at the final surface before radionuclide activity concentrations in the topsoil met the remediation targets;
- any soil left on site containing material that exceeded the RBCL still met the risk target.

This was demonstrated by refining the risk analysis reported in the environmental assessment by using actual monitoring and other site-specific information. In total approximately 10,000 samples were taken for all sampling criteria, including waste categorisation, with almost 19,000 separate analyses performed.

9.2.8. Technology performance

9.2.8.1. Technical. The SSA land remediation is a leading example of the clean-up of a radioactively and chemically contaminated site. The clean-up targets were achieved and the land is now suitable for unrestricted public access. The verification process revealed that only 57 results out of a population of more than 13,000 measurements exceeded the RBCLs for reasons other than limitations of the measurement device (i.e. the limit of detection exceeded the RBCL) or natural background levels. Thus, more than 99.5% of the measurements were directly compliant. For the remainder a revised risk estimate was derived using the refined site-specific risk assessment methodology. The key achievements of the remedial works include the following:

- 14,000 m^3 of wastes that exceeded a clean-up target disposed off site mainly to licensed landfill (a small quantity went to high-temperature incineration);

- 230 m^3 of LLWs disposed of to the BNFL Drigg repository;
- 4,500 m^3 of scrap and unsuitable inert materials removed from the site to landfill or for recycling;
- 250,000 m^3 of soils sorted through;
- in total 11 landfill pits and 7 ha of land totally remediated to base geology;
- land returned as suitable for unrestricted public access;
- land re-profiled and designed for use as an amenity area;
- no significant dose or other exposure to workers;
- no off-site release distinguishable from background.

A significant issue arose because some of the old degraded oils in the burial areas contained PCBs at low concentrations. In many cases this complicated or prevented use of landfill waste routes and led to increased costs. This had not been identified by the characterisation to a sufficient degree. The project was completed prior to the introduction of the Landfill Regulations 2002. These regulations significantly complicate finding cost-effective routes for hazardous chemical wastes.

9.2.8.2. Workability. The project took place over two winter periods and operations continued throughout. Over one winter the outer weather shelters suffered wind damage in severe storm conditions. This did not give rise to any environmental concerns but did cause delay. The project experienced a period where rain levels exceeded '1 in 10 year' levels and this caused some delay.

UKAEA always expected to find some munitions left behind from the Second World War, and munitions surveys were built into the remediation design. The extent of munitions found was beyond expectations, however, and gave rise to cost and delay. Some 1,200 live practice bombs, 13,000 small arms munitions, 30 landmines and many other odd items were discovered. Three munitions burial pits were found and munitions were otherwise scattered randomly across the site. Several large German bomb casings were uncovered (500 and 750 kg). The first of these to be discovered led the Harwell site emergency standby arrangements to be enacted and to the evacuation of the school, garden centre and nearby residences. The event lasted for a few hours until RAF Bomb Disposal was able to confirm that the device had been previously defused.

9.2.8.3. Stakeholder interests. The sensitivity of the SSA and the project objective to build confidence in the clean-up led to stakeholder involvement being a key enabling part of the project. A formal system was adopted which involved

- systematic development of a project policy (openness, transparency, truthfulness, effectiveness, timeliness ...);

- development of goals (no surprises, confidence generating ...);
- the identification of relevant stakeholders;
- the selection of communication, consultation and participation methods for training the project staff;
- monitoring and reviewing of success through local attitude survey.

Many techniques were used in parallel. Particular attention was given to the local community and the school. Some of the techniques used were as follows:

- local liaison committee briefings
- public meetings
- local media
- site visits
- talks to the parish council
- a regulators forum
- project-specific newsletters
- one-to-one dialogue

9.2.8.4. Technology risk management. Formal risk management tools were used to support the business case for the remediation and to focus risk management actions. Key risks were distributed between the contractor and UKAEA, allocated to that party which was in the best position to manage the risk. Some risk events occurred that were either entirely unexpected or that exceeded upper expectations.

Overall conventional waste volumes were 50% higher than expected from the estimates produced using the characterisation information. Put simply, the characterisation missed a number of waste patches that fell between sampling points. This is an unfortunate reality of land remediation projects where the waste is buried too deeply for surface surveying and is widely distributed. Land remediation is necessarily carried out in 'three dimensions' and consequentially waste volumes can increase quickly on the discovery of unknown areas.

Counterbalancing the above risk, the project achieved a 100% reduction in LLW waste volumes compared to predictions. The use of careful segregation at source and assay based on sensible averaging volumes proved practicable and cost-effective.

Some areas of the general land were found to contain fibrous asbestos wastes. This was easily dealt with using specialist sub-contractors and required mini-containment operations in some cases. Cement-based asbestos materials were ubiquitous in parts of the site and required hand picking.

9.2.9. Conclusions

The wastes in the SSA at Harwell were not causing any immediate environmental impact, although the site did have a hydraulic containment system to contain groundwater contaminated with chlorinated hydrocarbons. The project was mainly driven by redevelopment of a part of the site as part of the wider redevelopment and divestment of the Harwell campus, and the potential to develop an income stream from what would otherwise have been a 'sterile' part of the site.

Careful and considered stakeholder dialogue played an important part in the success of the project – not only at the project definition stage but also as part of continuous feedback during implementation phase.

9.3. Characterisation of the Dounreay Castle site (UK) using the Groundhog system

9.3.1. Background

Dounreay Castle is located at the mouth of the Mill Lade at the northern boundary of the UKAEA Dounreay site. As a result of past operations at the UKAEA Dounreay site, the castle environments were affected by radioactive contamination. The two sources of contamination identified were effluent dispersion experiments carried out in the mid-1950s and leakage of the low-radioactive drainage system.

9.3.2. Site history

The castle dates back to the 16th century and was last occupied in 1863. It is today in a ruinous state, unroofed and overgrown. The structure is based on a tower house of L-shaped plan that is normally associated with the lowlands of Scotland. A 19th-century cottage abuts the castle's most easterly wall. This is one of the last buildings that once formed part of an extensive post-medieval settlement of the area. The castle has been granted scheduled monument status.

During the mid-1950s a series of dispersion characteristics experiments were carried out at the Dounreay site. The purpose of the experiments was to provide data for the design of a sea discharge system for effluent containing radioactivity. The experiments involved the discharge of a mixed fission product liquor into the Pentland Firth. As a result of leakage and spillage in the castle grounds, fission products contaminated the courtyard of the castle.

The contamination of the castle drain (combined sewer) and the foreshore occurred due to the migration of fission products and actinides from the low-radioactive drainage system to the non-radioactive drainage system.

9.3.3. Surveys

Groundhog was used for a baseline gamma flux survey. The survey was carried out before the remediation. This was done to identify the areas of highest external radiation dose. Once identified, this contamination was removed so that restrictive working practices and increased dose were ameliorated in the short term.

9.3.4. Operational constraints

Groundhog does not identify the depth of the contamination. A high degree of PPE may be required on highly contaminated sites to carry out the initial study. The degree of PPE depends on the mechanism deployed, that is if the site is suitable, the system can be vehicle mounted and will reduce the need for PPE as long as appropriate filters are in place on the vehicle. In terms of the capability of the system, there is a depth limit for accurately monitoring beta mm for gamma and at the surface on hard standing, smooth soil or short grassed areas for beta. The beta system is slower than the gamma system and can only be used in fair weather. The system accuracy can also be affected by 'shine' generated by adjacent nuclear facilities.

9.3.5. Results

Groundhog highlighted the areas of highest external dose to be located within the castle courtyard (Figure 11). Activity concentrations of up to

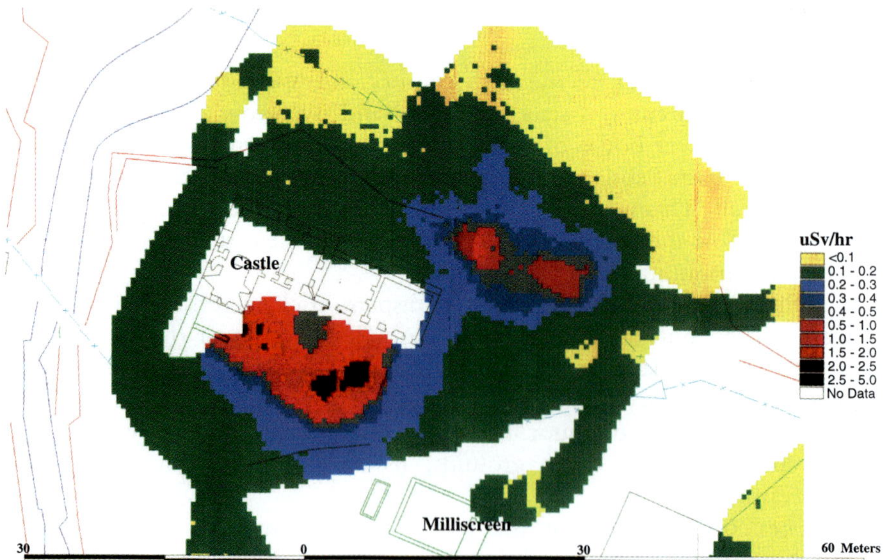

Figure 11 Groundhog pre-remediation survey at Dounreay Castle (SAFEGROUNDS, n.d.).

2,000 Bq g^{-1} ^{137}Cs were recorded in the upper layers of the soil profile. The contamination was shown to extend into the cottage but not the tower house of the castle.

The foreshore area was contaminated with up to 35 Bq g^{-1} ^{137}Cs, 18 Bq g^{-1} $^{239+240}$Pu, 8.6 Bq g^{-1} ^{238}Pu and 7.4 Bq g^{-1} ^{241}Am. Monitoring of boreholes installed upstream of the castle environments showed that migration was insignificant. The remediation involved the excavation of an area of 900 m^2 to a maximum depth of 3 m. The excavation was backfilled with clean material as well as excavated material that was assessed as being below the target limit (1 Bq g^{-1} artificial alpha and 4 Bq g^{-1} beta). Following the remediation a survey of the site was carried out using the *Groundhog* system. The results showed that the levels were less than 0.3 μSv h^{-1} across the site (Figure 12).

9.3.6. Assessment of technology performance

9.3.6.1. Technical. The Groundhog system provided a fast way of assessing the extent of the radioactive contamination. It also allowed the areas of highest external dose to be identified and remediated so as to overcome the need for restrictive working practices.

9.3.6.2. Financial. Due to the rapid nature of the Groundhog system, sites can be surveyed very quickly. This reduces the cost of surveying.

Figure 12 Groundhog post-remediation survey of Dounreay Castle (SAFEGROUNDS, n.d.).

Also, the identification of the most highly contaminated areas can potentially reduce the overall remediation costs.

9.3.6.3. Legal/regulatory. The Ionising Radiation Regulations 1999 require contaminated sites to be designated. Groundhog results allowed a more detailed risk assessment to be conducted to support production of the safety case, before works were carried out on the site. Remediation commenced in the castle courtyard, the area of highest external radiation dose identified by Groundhog. The reason for this was to remove the contamination that resulted in elevated doses, so that restrictive working practices and increased dose uptake could be removed in the short term, in accordance with the ALARP (as low as reasonably practicable) principle.

9.3.6.4. Workability. The Groundhog system is easily transported around the site. It can be either vehicle mounted or carried by a person. The information is then downloaded into a GIS, which gives a visual representation of the site data and allows integration with health physics and other data.

9.3.6.5. Technology risk management. The Groundhog system only identifies beta^{137}Cs. Care needs to be taken with the geology of the site, as the presence of different soils and rock types can affect readings. The Groundhog system does not identify the depth of the radioactive contamination.

9.3.6.6. Remediation. In close co-operation with a specialist archaeological contractor, the ground around the castle was carefully excavated in slices of 20 cm depth. This method allowed the safe removal of contaminated material and a detailed recording of the site archaeological history. The initial stage of the remedial works is shown in Figure 13.

Remediation works completed during 1998 allowed open access to the site for the first time in 40 years (Figure 14).

9.3.7. Conclusions

It is often forgotten that remediation of a site is more than just consideration of the means by which contaminants are removed, or contained. In the case of the Dounreay Castle, the sensitivity of the historical setting and architectural heritage of this part of the site played an important role in the planning and implementation of the remediation scheme.

The remediation targets for the site took into account the remote location of the site and relatively restricted access. This is a particularly significant aspect of managing sites undergoing decommissioning integrated with remediation as the target concentrations for clean-up required for a

Figure 13 Beginning of the remediation of Dounreay Castle (SAFEGROUNDS, n.d.).

Figure 14 Dounreay Castle after remediation (SAFEGROUNDS, n.d.).

site which is being managed, and where land use can be controlled, may well be less stringent than for a site which is unconditionally released.

9.4. Canadian experiences in remediation of decommissioned nuclear sites

9.4.1. Introduction

The Canadian nuclear industry comprises a wide range of industry sectors including radium and uranium mining and milling, refining and fuel fabrication, power production, research and development, nuclear medicine, and various industrial applications of nuclear technologies. These industries are supported by a waste management industry, which comprises waste processing, storage and disposal facilities. Decades of nuclear operations within these sectors have resulted in a wide range of nuclear sites and facilities of varying age, design and operational history, including facilities of relatively primitive design that were operated in the early decades of development of the nuclear industry, that is the 1930s to 1960s. In some cases, these earliest facilities had been simply abandoned without remedial measures taken (in particular the radium refining facilities), and in some cases residential developments had been established over the contaminated sites. These early facilities and sites are now subject to decommissioning and remediation programmes in varying stages, as are the more recent facilities that have been shut down (Figure 15).

Figure 15 Sites of Canadian Decommissioning and Remediation Projects.

Contemporary management of modern nuclear facilities typically includes up-front life cycle planning to accommodate future decommissioning and remediation requirements, with the planning basis including potential future reuse options. The older 'legacy' and 'historic' nuclear sites that are subject to decommissioning and remediation now were not designed and operated with a view towards reuse, or the early assumptions for reuse have since changed to options that reflect current thinking in environmental protection. Therein lies a key challenge in the decommissioning and remediation programmes in Canada – how to cost-effectively remediate the older nuclear sites.

Among the industrial sectors mentioned, the most significant decommissioning and remediation challenges exist within the sectors that were active in the earliest years, when

- nuclear safety regulations and oversight processes had not been established;
- radiological hazards were not fully understood (e.g. primitive contamination control and waste management practices);
- the design of nuclear facilities did not employ multiple containment systems or other protective features required in modern nuclear facility designs (resulting in numerous accidental releases of radioactivity to the environment) and
- radioactive effluents were routinely released to the environment with limited controls (e.g. dispersals of low-activity liquid effluents into soils).

These earliest sectors that have left the most significant radioactive contamination and waste legacies include mining and milling, refining and research and development.

In applying modern standards for radiological safety and environmental protection, the contaminated sites have largely been identified, and programmes put in place to restore the sites for different scales of reuse, whether restricted or unrestricted. In some cases, institutional controls may be needed for perpetuity (e.g. tailings management facilities). Among the different decommissioning and remediation programmes or projects underway in Canada, the following represent a cross section of the progress, accomplishments and challenges in Canada:

- remediation of historic waste sites from early radium and uranium mining, milling and refining facilities (e.g. Fort McMurray, Surrey, Port Hope, Port Granby, Scarborough sites);
- remediation of the Elliot Lake uranium mining and milling complex;
- decommissioning of small radioisotope laboratories and calibration and irradiation facilities;
- decommissioning and remediation programmes at Atomic Energy of Canada's (AECL's) national nuclear research and development sites, Chalk River Laboratories (CRL) and Whiteshell Laboratories (WL) and

- decommissioning of prototype research, isotope production and power reactors (e.g. Douglas Point, Gentilly-1, NPD, NRX, ZEEP, PTR, WR-1, SDR – see Figure 15).

9.4.2. Remediation of early refining and industrial sites

The earliest Canadian contaminated sites originated from mining, milling, refining, processing and manufacturing activities involving radium and uranium that were carried out in the 1930s to 1950s. Referred to as 'historic waste sites', these sites are generally characterised as relatively small areas of low-level contamination left in soils, with the radiological contaminants being those associated with the radium and uranium processing activities. The initial processing activities were in support of medical and industrial applications of radium, but after the advent of the Canadian nuclear power programme, uranium mining and refining activities were prevalent.

Most of the historic waste sites occur at the sites where the radium ore was handled or processed, leaving a series of contaminated areas from the point of origin, which was the Port Radium mine in the Northwest Territories, to the processing and manufacturing end points, which were refining and conversion facilities in southern Ontario (e.g. Port Hope and Port Granby). These sites and other similar contaminated sites are subject to remediation programmes managed by AECL's Low-Level Radioactive Waste Management Office (LLRWMO), which is a key component of AECL's overall Waste Management and Decommissioning Programme. Additional contaminated sites are discovered from time to time in Canada, but the larger historic waste sites managed by the LLRWMO include:

- Northern Transportation Route, Northwest Territories and Alberta;
- Fort McMurray, Alberta;
- Port Hope and Port Granby, Ontario;
- Scarborough, Ontario; and
- Surrey, British Columbia.

Northern Transportation Route: From the 1930s until the 1950s, a water transportation route 2,200 km in length was used to transport uranium and radium ore from the Port Radium mine on Great Bear Lake, Northwest Territories, to Waterways (now Fort McMurray), Alberta, for onward rail shipment to Port Hope, Ontario. Characterisation work in the early 1990s lead to the identification of 47,000 m^3 of uranium-contaminated soils at several sites (i.e. rail sidings where the ore was handled and spilled). These sites are currently being remediated by soil removal. The remediation criteria being applied will enable the lands to be released for unrestricted reuse.

Fort McMurray, Alberta: Uranium-contaminated soil and building materials were found at an unused warehouse in Fort McMurray, Alberta, in 1992. Subsequent investigations identified several other contaminated sites in the city. Clean-up of these sites was carried out (soil removal), resulting in roughly 31,000 m^3 of contaminated soil being disposed of as industrial waste in a designated landfill area in Fort McMurray. The remediated sites have been released for unrestricted reuse.

Port Hope and Port Granby, Ontario: Operation of radium refining facilities starting in 1932 in the Town of Port Hope. These facilities were eventually dismantled and removed as production shifted from radium to uranium refining and a uranium refining plant was opened. Initially, processing residues were retained on the plant site, but before long, residues were placed at several other locations in/around Port Hope. Other sites in Port Hope became contaminated through a variety of ways including spillage during transportation; unrecorded, unmonitored or unauthorised diversion of contaminated fill and materials; and erosion. After a period of time, the waste management operations were shifted to the Port Granby area. A number of areas of low-level soil contamination in and around residential developments resulted from these activities. Following from extensive radiological surveying of soils and buildings, roughly 120,000 m^3 of contaminated soils and building materials were recovered. Initially, the recovered soils and materials were shipped by rail to one of the radioactive waste management facilities at AECL's CRL for long-term management, but the excessive volumes encountered led to the remediation work being suspended, then resumed after interim soil storage facilities were established in the Port Hope area. Following from an extensive public consultation programme, these soil deposits are planned to be consolidated into an improved long-term storage facility. The development of this storage facility will enable additional remediation work to be carried out in the Port Hope area, with the planned end state for the contaminated sites being unrestricted reuse. The soil storage facility will require long-term management, hence institutional controls will apply.

Scarborough, Ontario: The radium-recovery operations and other activities on a farm in the mid-1940s resulted in radium-contaminated soils. In the mid-1970s, urban expansion of the nearby city of Scarborough resulted in the farm property being developed into a residential area without knowledge of the history of the site. In 1980, radium contamination was discovered in one residential area and additional contamination was discovered in another nearby area in 1990. With the assistance of the community, a proposal was developed to excavate the soil and take it to a soil-sorting and interim storage site in an industrial area where it was sorted. Following from comprehensive surveying, radium-contaminated soils were removed from more than 60 residential and commercial properties.

The remediation criteria that were applied enabled unrestricted reuse of the lands. The soils identified from the soil-sorting process as containing low-level radioactive contamination were transferred to one of the radioactive waste management facilities at CRL (50 m^3). The mildly contaminated soils resulting from the soil-sorting process (about 16,000 m^2) were placed in an engineered storage mound in Scarborough. The nuclear regulatory agency, the Canadian Nuclear Safety Commission, is currently assessing the regulatory requirements for the storage mound.

Surrey, British Columbia: Operation of a niobium smelter (producing niobium which is used as an alloying agent in a variety of metals) resulted in thorium contamination in and around the industrial property. Niobium ore was imported and smelted on the property during the 1970s. As an accessory mineral in the ore, the ore contained naturally occurring radioactive thorium, which remained in the slag by-product from the smelting process. Some of the slag was inadvertently mixed with sand and gravel and used as fill on a nearby site, and a small volume of the contaminated material was also used in a nearby railway yard. Remediation of the contaminated sites involved soil and slag removal following radiological surveying. Roughly 5,000 m^3 of thorium-contaminated soil and slag were removed from one site and shipped to commercial facility in the nearby U.S. state of Oregon. In addition, a small volume of contaminated slag from the rail yard (less than 100 m^3) was shipped to one of the radioactive waste management facilities at CRL.

9.4.3. Decommissioning and remediation of the Elliot Lake uranium mine and milling complex

Uranium was discovered in the Elliot Lake area of Ontario in the early 1950s, leading to the development of several mines and associated milling facilities and mill tailings management areas (tailing ponds) in the mid-1950s. In total, there were roughly 100 mine openings (e.g. pits, shafts, ventilation raises) and 10 tailings management areas. Similar to the situation with the radium refining facilities, the initial uranium mining activities pre-dated the establishment of nuclear safety regulations and oversight, hence the first decade of uranium production was not controlled to the extent after 1970 when the Atomic Energy Control Act (now the Nuclear Safety Control Act) was enacted and the Atomic Energy Control Board (now the Canadian Nuclear Safety Commission) was established. By the time that the NSCA was enacted, uranium mining in the area had significantly declined as a result of a decrease in demand from the USA. This resulted in many of the smaller mining operations shutting down, and the mine sites (including the tailings ponds) being abandoned. This initial phase of uranium mining left a legacy of uranium-contaminated soils, rock and mining infrastructure. The demand for uranium in support of the Canadian nuclear power reactor (Canada Deuterium Uranium (CANDU)) programme resulted in resurgence in the mining operations, enabling several mining companies to remain in operation into the 1990s. By the early 1990s, however, depleted reserves and low uranium prices caused the last mines in the area to close. These last closures resulted in a substantial range of facilities to be decommissioned and lands to be remediated, many with mining companies to undertake but others without.

Decommissioning of the mine and mill sites involved the dismantling of all infrastructure and the release of metals, where possible, for recycling. The materials that could not be released because of the residual contamination levels were placed either in some of the underground workings (underground landfilling) or in surface landfills along with other contaminated materials, waste rock and tailings. Contaminated soils at the mine and mill sites were collected and placed either inside the mine workings with the building rubble or in the surface landfills. Following conventional practice in decommissioning mine sites, all mine openings were capped to prevent intrusion, and the underground workings were then allowed to flood. Comprehensive environmental monitoring programmes were then applied to enable the monitoring of conditions around the mine sites.

The tailings ponds created from the milling operations represented the most significant contamination hazards at the mine sites. However, dealing with such problems is discussed elsewhere in this volume.

The socio-economic impacts of the mine closures on the nearby city of Elliot Lake warrant some discussion; because of proactive effort from Federal, Provincial and Municipal governments and community members, the impacts to the community were greatly mitigated, and today the community enjoys the economic stability that many mining towns and cities are unsuccessful in attaining.

Elliot Lake was established as a planned community for the uranium mining after the discovery of a significant uranium ore-body in the area. The community planning was carried out in the mid-1950s by a special agency (Planning and Development Department of the Ontario Ministry of Housing) created by the Provincial Government to ensure the development of Elliot Lake as a viable community. This proactive planning effort provided the basis for ensuring that Elliot Lake would not turn into a shack town after the mines closed. The city experienced several boom-and-bust cycles over the decades with the fluctuating price of uranium, which is typical of single-industry communities. The population peaked around 26,000 in the 1960s and is currently stable at roughly 12,000. By the early 1990s, depleted reserves and low prices caused the last mines in the area to close. With the closures in sight, existing Federal and Provincial programmes and services were utilised to identify economic development opportunities in the region. A diversification plan was developed and implemented, diversifying the economic base of the community and stabilising the economy. A few of the diversification successes include a treatment centre for substance abuse, a thriving tourism industry, forest products harvesting and a successful retirement living programme that has attracted over 2,000 families from all over Canada, the United States and Europe. Other innovative ways to diversify the economy included establishing in the city

- a Nuclear and Mining Museum (http://www.elliotlake.com/nuclearmuseum/) and the northern home of the Canadian Mining Hall of Fame (http://www.halloffame.mining.ca/halloffame/);
- a field study facility for Laurentian University (the Elliot Lake Research Field Station, which facilitates the study of environmental radioactivity) and
- an analytical services laboratory.

9.4.4. Decommissioning of small radioisotope labs and calibration and irradiation facilities

A wide range of small radioisotope laboratories and calibration and irradiation facilities exist throughout the Canada. These facilities are very small in comparison to the types of nuclear facilities used in other industry sectors, and the quantities of radioactive materials used in the facilities are usually relatively small, as are the hazards. Nonetheless, the small radioisotope laboratories and calibration and irradiation facilities operate under individual nuclear licences, and therefore require de-licensing before the facilities can be decommissioned and released for reuse.

The decommissioning of these facilities is typically relatively simple, as usually there is very little contamination generated from operations. Traditional approaches are applied in identifying the contamination sources from historical information and surveying. The characterisation work can extend beyond the physical boundaries of the facility, but soil and groundwater contamination typically does not occur at the small radioisotope facilities. Radioactive wastes generated from facility decontamination operations are often shipped to the radioactive waste management facilities at AECL's CRL for long-term management.

The planning for reuse of the facilities can be driven by the immediate business needs for the property. Often the facilities are located within larger buildings that are renovated for reuse in new industrial applications. Where the building structure is unsuitable for the planned reuse (whether too small, or inappropriate construction, or too old), the building is demolished and a new building constructed on the site.

9.4.5. Decommissioning and remediation of large nuclear research and development sites

AECL, as Canada's national nuclear research establishment, has operated two research sites for several decades: Chalk River Laboratories and Whiteshell Laboratories.

9.4.5.1. Chalk River Laboratories. AECL's Chalk River Laboratories (Figure 16) were the birthplace for the Canadian nuclear programme. Construction of the CRL site started in 1944, and its development and operating history includes the construction and operation of seven research reactors and numerous associated/supporting nuclear laboratories. The CRL site consists of a 70 ha developed (industrial) site located within a larger undeveloped area (supervised area – 3,700 ha) that serves primarily as an exclusion zone. The developed area, or inner area, includes over 100 buildings and facilities. The development and operation of the site can be broadly characterised in two phases. The initial phase was oriented towards the production and recovery of plutonium and ^{233}U (i.e. defence role), hence the facilities constructed in the early years included, in addition to one laboratory-scale test reactor (ZEEP) and a larger research reactor (NRX), facilities for processing irradiated uranium and thorium and packaging the recovered products, as well as development laboratories, administrative buildings and facilities for key site-support services.

Figure 16 AECL's Chalk River Laboratories.

The second phase started in 1954 when the research focus shifted to include the application of nuclear technology for electrical power generation based on the natural uranium fuelled, heavy water moderated concept, subsequently dubbed CANDU. By the late 1950s, the defence role ended. To support the new mandate, existing facilities were used and new facilities were constructed, for example facilities for fuel development, fabrication, testing and post-irradiation examination of fuels and reactor components. Also, engineering programmes were initiated to support the development of prototypes for the CANDU nuclear power reactor and advanced reactor concepts. Support facilities and services such as machine and instrument shops, analytical laboratories, engineering, computation, stores, radiation protection, environmental and biological research, nuclear materials and waste management, administration, cafeteria and so on were constructed as required. Largely through the research programmes and operations carried out at CRL, AECL continues to fulfil national objectives, such as producing about half of the world's medical isotopes, providing instruments to the IAEA for safeguards inspections, and recent testing of mixed-oxide fuel for CANDU reactors.

Another important aspect of the CRL site is the presence of nine different waste management areas (WMAs, Figure 17) that were operated in succession over the years. Some of these WMAs present significant challenges with regard to the decommissioning and remediation work required. In particular, the oldest waste areas were developed at a time when a different level of protection to the environment was afforded, for example dispersals of low-active liquids into soils, burial of drummed higher-level radioactive liquids without secondary containment or direct burial of waste objects (without containment). The CRL WMAs receive waste not only from site operations, but in providing a national waste management service, the waste management facilities also service the needs of most non-utility Canadian nuclear facilities (e.g. medical facilities, universities and industry).

Although research and isotope production operations will continue for decades at the CRL site, a site-decommissioning programme has been established in the last decade in order to decommission the buildings and facilities employed during the initial phase of site operation. Accordingly, a well-developed decommissioning programme exists at CRL, and progress is being made in decommissioning older redundant buildings, in parallel with ongoing site operations and refurbishment activities. The decommissioning programme is predicated on the assumption that the current nuclear operations at the site (and ongoing decommissioning activities) will continue over a 100-year operating period, but with a decline in site operations towards the end of the period. It is planned that over this period of time, one or more waste disposal facilities will be constructed at the site, for example a shallow rock cavity (SRC) for low- and intermediate-level wastes and/or an intrusion-resistant underground structure (IRUS) for the near-surface burial

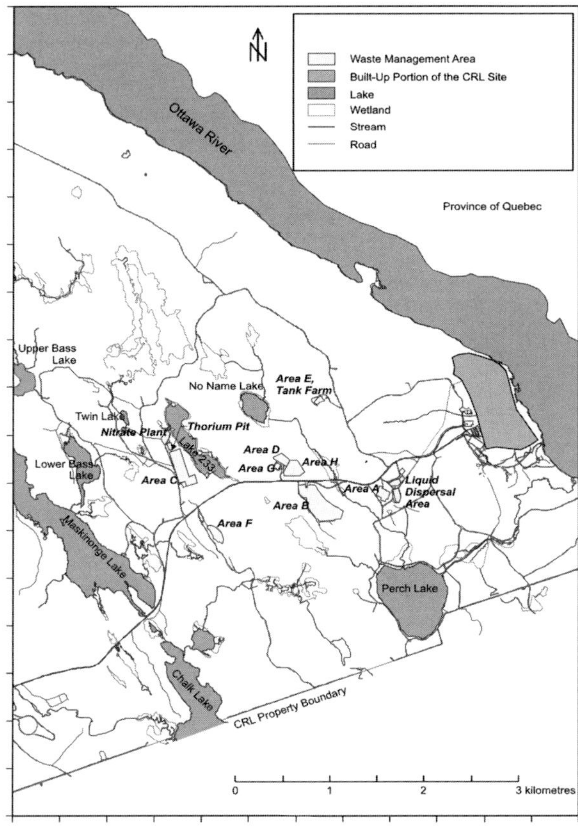

Figure 17 CRL waste management areas.

of LLWs. In order to enable the decommissioning and remediation activities required for bringing the site to the desired end states (discussed below), a number of support facilities are planned to be constructed over the 100-year operating period (with some being designed and constructed at the present time). These decommissioning 'enabling facilities' include, for example waste handling and processing capabilities such as bulk waste monitoring (clearance), detailed waste characterisation, liquid waste consolidation, liquid waste immobilisation, improved waste storage, incineration, high-level waste retrieval, high-level waste inspection and repackaging (shielded facilities). These facilities will facilitate decommissioning of not only the buildings and nuclear facilities at the CRL site, but also the removal of structures and materials buried at the WMAs. The last tier of decommissioning at the end of the site operating period will be to remove the enabling facilities. Closure of the waste disposal facilities would also occur at this time.

As the buildings, nuclear facilities and WMAs at the site are decommissioned, a combination of active and passive land remediation will be applied to address soil contamination resulting from releases from the facilities. Active remediation will entail soil removal and the application of one or more technologies to remove or reduce the radiological contaminants. Passive remediation will entail allowing the residual radioactivity to diminish by radioactive decay. From the monitoring of groundwater plumes on the CRL site, it has been observed (as is the case at many other nuclear R&D sites established in the 1940s and 1950s) that the radiological contaminants of concern (e.g. actinides) are relatively immobile, and therefore active remediation work can be applied in relatively small areas of the plumes. Through pathway modelling and dose assessment, the residual activity levels will be assessed to determine the scope of active remediation required to meet dose targets related to the decommissioning end states. The length of the institutional control (IC) period (300 years) was selected in order to enable passive remediation to be effective for the radionuclides with half-lives shorter than 30 years, such as ^{3}H, ^{90}Sr and ^{137}Cs, the first two of which are the principal contaminants in the CRL groundwater plumes.

Through the decommissioning and active remediation work carried out throughout the 100-year operational period, most hazard sources will be removed and many areas of the site, including many of the contaminated land areas, will have been stabilised and qualified, for either unrestricted use or industrial reuse. It is expected (planned) that there will be residual levels of activity remaining in only a few areas of the site that will require the application of institutional controls to prevent unacceptable exposures to the residual hazards while the hazards passively reduce. At the end of 300-year IC period, all areas of the site will have been qualified for reuse with minor, if any, restrictions.

The CRL decommissioning programme is being executed in a planned, logical sequence designed to optimise cost, safety and environmental risk (e.g. waste retrievals will be deferred until the necessary waste processing and repackaging and new storage facilities area are available). Operation of the CRL site in parallel with execution of the decommissioning programme also presents constraints that must be addressed in the planning. With the wide range of decommissioning activities to be completed, optimisation is assessed using a 'prioritisation review' that is applied every 2 years. The evaluation process involves priority-ranking of all of the decommissioning and remediation activities, with the prioritisation (scoring) categories being HSSE (health, safety, security and the environment) needs, business needs (e.g. regulatory requirements, or site refurbishment programme needs). One key aspect of the HSSE evaluation is the review of potential implications of the sequence and deferral of decommissioning activities on the scope of the associated remediation activities. The decommissioning activities that have large gains in averting large remediation requirements are given the highest priority, which then results in high priorities being assigned to the associated enabling facilities.

It is therefore recognised in the decommissioning planning process that careful consideration of remediation requirements is an essential component of optimising cost, safety and environmental risk.

The fact that some decommissioning and remediation activities must be deferred in the site decommissioning plan drives the requirement for mitigation systems to control the potential environmental impacts. Three plume interception and treatment systems are in operation at CRL (one passive and two active), and a fourth (passive) system is being designed.

Progress has been made in recent years at CRL in decommissioning redundant buildings and nuclear facilities, and in addressing the higher-priority waste storage issues. Waste retrievals are being carried out, and two waste management sites have been equipped with infiltration barriers. A comprehensive integrated environmental monitoring programme and environmental management system continues to be operated, providing key information into the assessment process, which in turn enables optimisation of the decommissioning and remediation activities.

*9.4.5.2. **Whiteshell Laboratories.*** AECL's Whiteshell Laboratories (Figure 18), located in the Province of Manitoba, were established in the early 1960s to facilitate expansion of the nuclear research programmes initiated at CRL.

The WL research programmes were also oriented towards the application of nuclear technology for electrical power generation, and also included other

Figure 18 AECL's Whiteshell Laboratories.

nuclear research programmes that were unique to the site, which included a small reactor development programme. The larger nuclear facilities located at the site included the WR-1 research reactor (a 60 MWt organic liquid-cooled reactor), the Slowpoke Demonstration Reactor (a 10 MWt pool reactor), a Van de Graaf accelerator and a neutron generator facility. Supporting nuclear facilities included research and analytical laboratories, a hot cell facility, a liquid waste treatment centre and a small WMA.

Similar to CRL, the WL site comprises a small developed area (220 ha) surrounded by an undeveloped exclusion zone (roughly 4,375 ha). The developed area consists of approximately 50 buildings and structures.

Starting in the mid-1990s, AECL began consolidating its research and development programmes to CRL, and in 1998 AECL received government concurrence to proceed with the planning actions necessary to close the WL site and initiate decommissioning. Unlike the situation at CRL, the buildings and infrastructure at WL were in reasonably good condition, and the site was located in reasonable proximity of a major city, making the site eligible for reuse as an industrial park. The industrial park concept formed the basis of the decommissioning plan developed for the site, and the initial phase of this park included retaining a licensed nuclear section (necessary to support the decommissioning operations). Therefore, as sections of the property were cleared (determined to be free of radioactivity), they were released for non-nuclear industrial use. The decommissioning and remediation activities are currently ongoing.

With the WL site being developed with the benefit of considerable operating experience at CRL, and involving only a few nuclear facilities of more modern in design philosophy, there were few occurrences where radioactivity leaked from the facilities to the environment over the years of operations. Accordingly, although significant efforts are required to complete the planned decommissioning activities for the site, the remediation requirements for the site are limited.

Decommissioning of the WLs began in 2002 and is now well underway. To gain experience safely, initial decommissioning consisted of removing a series of relatively simple and uncontaminated outbuildings. Small nuclear facilities like the Van de Graaf accelerator and neutron generator were then removed and sold for reuse. Now, more challenging facilities are being decommissioned, including shutdown and decontamination of half of the site's shielded facilities (hot cells). Many of the manipulators and some of the hot cell windows have been removed and sent to CRL for reuse. Upcoming tasks include immobilising several hundred litres of highly active liquid reprocessing waste, and finishing the shutdown and decontamination of the main radiochemical laboratories, then demolishing them.

At the time that the WL decommissioning plan was developed, a large percentage of the 4,375 ha property surrounding the main site was believed to be unaffected by site's nuclear operations. This was confirmed through the following staged approach. After including a suitable buffer zone around the utilised lands, AECL designated the balance of the property ($\sim$3,000 ha) as the 'Unaffected Area' (UA), then conducted a radiological survey to demonstrate that this was the case. The approach was to evaluate lands both on and off the WL site for radionuclide concentrations in the environment and demonstrate that any radioactive material identified in the UA was indistinguishable from levels present in the background reference areas. The survey methods employed included an airborne gamma scan survey over the site and over an off-site background reference area and ground gamma scan surveys along the roads, trails and power line rights-of-way that access the UA. Soil samples and *in situ* gamma spectroscopy readings were obtained within the UA and in off-site background reference areas. The survey results demonstrated that there had been no impact on the UA from WL site operations. With this determination, the lands will be released for unrestricted reuse.

Low levels of contamination were known to exist at the WL process water outfall in the nearby Winnipeg River. A radiological survey of the river sediments was conducted in order to obtain sufficient data for a preliminary estimate of the inventory to support an assessment of the impact of contaminated sediments on biota in the river and on humans. This assessment was performed to verify that the operation of WL within its regulated release limits had led to no significant impact in the river sediments and that no remedial action was required. The survey was conducted by scuba divers equipped with two gamma detectors configured for underwater use and a global positioning system (GPS) to log data collection locations. The survey included an upstream background reference area for comparison to the two study areas. It was determined from the impact assessment that, even with extremely conservative dose estimation methods, the doses would be below accepted guidelines, and therefore remediation of the riverbed sediments was not warranted.

The WL WMA is the only other area on the research site where ground contamination exists. The plan for decommissioning this part of the site entails removing the higher-hazard wastes, components and affected soils, then qualify the low-hazard wastes and residual contamination for *in situ* disposal. The approach will involve the application of infiltration controls and ongoing environmental monitoring to ensure that the environmental performance supports the disposal case. After an IC period designed to allow sufficient radioactive decay, the planning assumption is for the WMA to be released for unrestricted reuse.

9.4.6. Decommissioning and remediation of redundant reactor facilities

The development of the Canadian nuclear power reactors of the CANDU series involved the construction and operation of three prototype reactors;

the Nuclear Power Demonstration (NPD) reactor at Rolphton, Ontario (22 MWe), the Douglas Point reactor at Tiverton, Ontario (206 MWe), and the Gentilly-1 reactor at Bécancour, Québec (250 MWe). Because these reactors have served their purpose, they have been shut down and the initial phase of decommissioning completed (hazard removal and preparation for storage with surveillance). In addition to these early power reactors, there are three small experimental reactors and two larger research reactors at the two AECL nuclear research and development sites at different stages of decommissioning:

- Zero Energy Experimental Pile (ZEEP) at CRL (zero energy),
- Pool Test Reactor (PTR) at CRL (2 kWt),
- Slowpoke Demonstration Reactor (SDR) at WL (10 MWt),
- National Research Experimental (NRX) at CRL (42 MWt) and
- Whiteshell Research (WR-1) at WL (60 MWt).

Of the eight shutdown reactor facilities, only two have resulted in releases of radioactivity to the environment (both resulting from leaks in fuel storage bays), resulting in the requirement for remediation, and of these two, only one is significant (NRX, discussed below).

The remediation requirements (remediation targets) for the reactor facilities relate to whether or not the facilities are co-located with other operating nuclear facilities, and whether any releases from the shutdown reactors have resulted in plumes at the sites. Where the redundant reactor facilities are located on licensed sites, where other nuclear facilities remain in operation (which is the case with seven of the eight reactor facilities), the ongoing operations can also constrain the scope of remediation in several ways. First, it may not be possible to fully remediate contaminated soils or groundwater, where the contaminant plume extends beneath neighbouring operating nuclear facilities. In this case, active remedial measures can be applied to accessible areas, and remediation of inaccessible areas of the plume must be deferred until the operating facilities are decommissioned. A similar situation exists with site-support services (e.g. buried piping and electrical lines) – the scope of initial remedial measures must be evaluated against the site impacts from interruptions in key support services. The evaluation must, however, also include estimating the potential consequences of deferring the remediation work. Where the future remediation requirements could potentially be significantly increased (i.e. the remediation requirements are sensitive to the timing and/or scope of early remediation), priority should be assigned to increased initial scope. The second constraint pertains to the target end state of the reactor facility and the surrounding licensed site. It may be impractical to remediate the redundant reactor to an unrestricted reuse (green field) target when the nuclear site is operated as an industrial site (i.e. resulting in a green field island within a brown field area). What may be more practical is to remediate the reactor facility in two phases – active remediation to the release target for the site (whether green field or brown field), followed by passive remediation. The remediation strategy must be developed with close attention to the potential for deferred activities to result in significantly escalated remediation requirements during the second phase of remediation.

This is the case with the NRX reactor at AECL's CRL, where ongoing leaks from fuel storage bays have resulted in a tritium and ^{90}Sr plume extending across part of the site. When the concrete bay structure is removed, it is expected that other contaminants will be found in the soils, but the contamination should be localised and amendable to soil removal. After this active remediation is completed, the remainder of the plume will be remediated passively over the operating life of the site (100 years) and/or the ensuing IC period (300 years).

The other reactor facilities are not expected to have resulted in significant ground contamination, that is to the extent that remediation would be limited to only local soil removal.

9.4.7. Conclusions

Development of the nuclear industry in Canada first started in the 1930s, with radium mining and refining operations being the initial focus. Similar to other countries, the Canadian nuclear programme rapidly diversified during 1940s to 1960s, resulting in a wide range of nuclear sites and facilities of varying age, design and operational history. The earliest nuclear facilities were of relatively primitive design, and present the most significant decommissioning and site remediation challenges today.

Decommissioning and remediation programmes have been established in the last decade, and progress is being made in remediating the nuclear sites to different standards for reuse. At the smaller sites, soil removal has proven to be feasible, enabling the contaminated lands to be released for unrestricted reuse. The same is true of the small radio-laboratories and calibration facilities, where often the buildings housing the facilities can be reused without constraints. In the case of mine and milling sites, where the requirements for active remediation would be extensive, stabilisation, mitigation (groundwater treatment) and passive remediation has been applied. Although at these sites it will be possible to eventually shut down the groundwater treatment plants, institutional controls will likely be necessary indefinitely. The modern nuclear facilities and power plants are designed with a view towards decommissioning and the operating processes are effective in preventing the release of radioactivity and other contaminants to the environment. Because of this, the future remediation requirements are expected to be relatively minor. Waste disposal technologies continue to undergo development in Canada, complementing the decommissioning and remediation programmes.

9.5. Remediation of a uranium mining/milling site under decommissioning: the Wismut case

9.5.1. Background

In 1947, the Soviet occupation forces in Germany established the state-run stockholding company (SAG) Wismut. Run by the Soviet military, the companies' sole aim was the exploitation of the East German uranium deposits to serve their nuclear programme. During the early, 'wild' years, uranium mining in Saxony and Thuringia was characterised by complete disregard for the environment of the densely populated areas and a reckless exploitation of natural and human resources. From 1954 on, the new bi-national Soviet-German company (SDAG) Wismut continued uranium mining with a workforce of up to 120,000 employees. However, in the wake of the German reunification in 1990, more than 40 years of intensive uranium mining and milling came to an end. During that period, the Wismut company had produced a total of 231,000 tons of uranium. In global terms, this ranks Wismut in post-war uranium production as number three, after the United States and Canada.

In 1991, the Federal Republic of Germany took over 100% of the shares of SDAG Wismut, and assumed the responsibility for the nuclear legacy. The assets were transferred into a company under German corporate law: Wismut GmbH and the Wismut Remediation Project were launched. Remediating the vast radioactively contaminated uranium production sites in East Germany has been one of the largest ecological and economic challenges facing the united Germany till today. In terms of complexity and size, the Wismut Project is unique, even by international standards. The project involves remedial activities at sites located at considerable distance away from the Wismut headquarters in Chemnitz, for example the Aue site 40 km away or the even 100 km away Königstein site (Figure 19).

In addition to the contaminated areas listed in Table 2, numerous industrial structures and facilities needed decommissioning after termination of the uranium production, including four underground mines (Aue, Pöhla, Ronneburg, Königstein and Gittersee) with 56 shafts and a total length of the shafts, underground tunnels and galleries of about 1,470 km, two large processing plants (Crossen and Seelingstädt, see Figure 20), 13 ore loading stations, 32 storage bunkers and 63 km of railway tracks.

The amount of material of different contamination levels arising from the dismantling and demolition of the structures and facilities was estimated to include:

- Concrete/reinforced concrete: 250,000 m^3
- Timber: 16,000 m^3
- Masonry: 100,000 m^3
- Scrap metal: 262,200 t

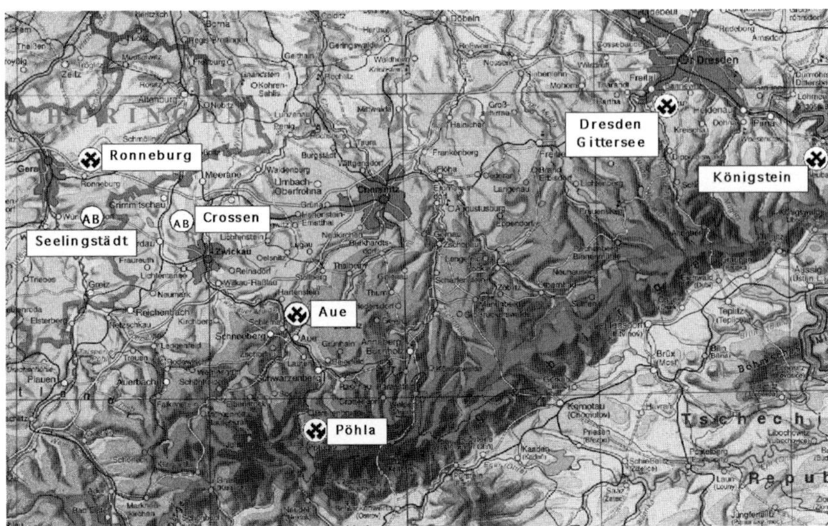

Figure 19 The Wismut mining and milling sites.

Table 2 Contaminated areas and legacies at Wismut sites in East Germany.

	Sites			
	Aue and Pöhla	Königstein and Gittersee	Ronneburg	Seelingstädt and Crossen
Operation	Mining	Mining, underground leaching	Mining	Milling
Industrial area (km^2)	5.7	1.4	16.7	13.1
Mine dumps				
Number	20	3	16	9
Area (km^2)	3.7	9.4	6.0	5.3
Volume (m^3)	47,000,000	4,500,000	188,000,000	72,000,000
Tailings ponds				
Number	1	3	3	7
Area (km^2)	0.035	0.046	0.09	7.1
Volume (m^3)	300,000	200,000	250,000	160,000,000
Open pit mine				
Number			1	
Area (km^2)			1.6	
Volume (m^3)			84,000,000	

Figure 20 The former uranium ore processing plant at Seelingstädt in 1991; the facility has been demolished and the site reclaimed since.

9.5.2. Typical patterns of contamination

Depending on the production history, four typical contamination patterns are found at the Wismut areas, representing contamination by

1. scattered ore (specific activities of the key nuclide ^{226}Ra are around 0.3–2 Bq g^{-1} in near radioactive equilibrium with the other nuclides of the ^{238}U decay chain);
2. scattered tailings material (specific activity around 1–10 Bq g^{-1} for ^{226}Ra as the key nuclide, not in radioactive equilibrium);
3. lost uranium concentrate (specific activity up to some kBq g^{-1} for ^{238}U as the key nuclide) and
4. naturally leached uranium, for instance at sites from which waste dumps were removed (specific activity up to some Bq g^{-1} for ^{238}U as the key nuclide).

Surface contaminations on machinery, equipment, metal scrap and so on also do reflect the production history. In principle, the same four nuclide vectors are encountered. The surface contamination on equipment and scrap arising from the decommissioning and demolition of mine ventilation structures are dominated by the long-lived radon daughter nuclides ^{210}Pb and ^{210}Po. Typical surface activities range from around 1 Bq cm^{-2} for contaminations by scattered ore to around 100 Bq cm^{-2} where chemical leaching processes resulted in the concentration of radionuclides in residues

Table 3 Typical nuclide vectors normalised to the key nuclide (italic).

Material	^{238}U	^{230}Th	^{226}Ra	^{210}Pb
Waste rock material	0.95	0.95	*1*	0.91
Uranium concentrate	*1*	0.002	0.001	0.0007
Naturally leached uranium	*1*	0.08	0.1	0.1
Tailings	0.05	0.65	*1*	0.95
Ventilation air precipitation	0.03	0.03	0.03	*1*

and uranium concentrate. Table 3 presents the five typical nuclide vectors at Wismut sites, normalised in each case to the dominating nuclide.

During demolition and clean-up at the processing sites, radioactivity has been found not to be homogenously distributed and to reach down to depths of some metres below ground. At former ore storage sites, naturally leached radioactivity penetrated into the ground less than 1 m and with a horizontally homogenous contamination pattern.

9.5.3. Strategy for the demolition and site remediation

Objectives and scope of the Wismut Project follow from the legal requirements of the German 'Federal Mining Law', which stipulates the owner's obligation to abate public hazards and to mitigate damages caused by mining, as well as to prevent future hazards after mine closure. The 'Ordinance for provision of radiation protection for waste rock dumps and industrial settling ponds' (HaldAO) regulates the radioactivity aspects of remediation, whereas the 'Water Resources Management Act' ensures the protection from contamination of surface and groundwater.

The extent of remedial measures to be undertaken is derived by object-specific investigations and remediation feasibility studies, rather than by applying uniform standards. The remediation workflow, unlike common civil engineering projects that have a linear succession of tasks, is an iterative process. Within the workflow, a conceptual site model guides the optimisation of designs and investigations, while both operational exposures and environmental base lines are monitored. The acquired data are collated and analysed at corporate level to provide decision-making support for senior management.

At the present advanced stage of the Wismut remediation project, the reutilisation of reclaimed areas and objects is receiving increased attention. Remediation of the Wismut legacy is viewed as an opportunity to return the affected land to productive use, thus enhancing the revitalisation of the former mining areas in East Germany.

By definition, there are no legal restrictions on the utilisation of areas that have received a complete clean-up. That is why the remediation goal is

Figure 21 Spa garden Schlema (in the background recontoured and partly covered waste rock dumps).

(whenever feasible) to maximise the number and size of areas reclaimed for unrestricted use. Large areas of sites such as those from which waste rock piles were removed cannot be completely cleaned up in each case. As a rule, they can only be released for restricted reuse. After proper capping of the contaminated parts, such areas can only be utilised for forestry, as pasture land, or for industrial and trade use. Forestry has the advantage of being a low-maintenance reutilisation option, sustainable in the long term. However, exemptions are possible, if long-term monitoring and maintenance can be ensured. Thus, a mutually beneficial integration of reclamation plans with the communal/regional development has been successfully implemented at two former mining towns, leading to rebirth of the health spa of Schlema (Figure 21) and the development of the Federal Garden and Landscape Exhibition in 2007 (Bundesgartenschau 2007) hosted by the towns of Ronneburg and Gera.

With respect to the reuse of structures and buildings, a more restrictive policy is followed by Wismut. For various reasons continued use is not feasible and the majority of the installations and buildings are destined for dismantling and demolition (Figure 22).

Many buildings and facilities are obsolete and partly also radioactively and/or chemically contaminated. Decontamination and subsequent refurbishment of the objects is neither technically possible nor can it be justified for economic reasons. The same goes for technical equipment and machinery,

Figure 22 Blasting and preparation for demolition of a former ore storage and loading station (Königstein site).

like hoisting and conveying equipment, power plants and ore processing installations. Exceptions are buildings outside of actual production sites, such as staff social building complexes and the former site management buildings. Whenever possible, Wismut intends to sell off these assets. Thereby care has to be taken with regard to possibly enhanced radon indoor concentrations, since in the first years SDAG Wismut used crushed waste rock as aggregate in building materials or to construct the basements of buildings.

9.5.4. Waste management

The huge amount of waste and the vast range of materials having different types and levels of contamination require an appropriate waste management strategy. Thus, Wismut takes advantage of the open space available in not yet flooded mines and at sites where contaminated material were disposed of *in situ*. The places for disposal of contaminated material include

- underground storage sites ('dry' mine galleries, i.e. those above groundwater level);
- waste rock dumps (construction of cassettes, placement of the waste, coverage);
- beach areas of tailings ponds (same technology as for dumps) and
- engineered hazardous waste disposal facility (constructed on top of a waste rock pile).

Figure 23 Disposal of 'big bags' filled with immobilised water treatment residues at a waste dump.

Special attention is given to the operational waste that results from the application of modern remediation technologies for the treatment of contaminated material. Wismut operates modern treatment plants at six sites (Schlema, Königstein, Pöhla, Helmsdorf, Ronneburg and Seelingstädt). The contamination is concentrated in residues, reaching specific activities of up to 1,000 Bq g^{-1} ^{238}U and 20 Bq g^{-1} ^{226}Ra (Schlema site). Special techniques for the immobilisation of these residues need to be applied. The disposal of the Schlema residues, which are immobilised by mixing with cement and fly ash, filled into 'big bags' and emplaced into a specially constructed facility on a waste dump, is shown in Figure 23.

However, despite of the space available for disposal, it is the policy of Wismut to separate waste according to the type and level of contamination. Thereby, it is intended to recycle materials free of radioactive contamination. In some cases, this approach may also be applied for low-contaminated waste. Salvaged uncontaminated metallic scrap and scrap with a total alpha surface activity below 0.5 Bq cm^{-2}, for instance, is sold for smelting (see also the following section). Demolition rubble is crushed before being reused within the Wismut Project in a variety of applications, such as for interim cover layers on tailings ponds.

9.5.5. Release procedures

For the reuse of sites, buildings and materials contaminated by uranium mining activities, the German Commission on Radiological Protection recommended in 1991 a primary level of effective dose of 1 mSv per annum, based upon the variation bandwidth of natural exposure. This level refers to additional exposure caused by contamination with natural radionuclides from uranium production, that is it would be applied on top of the background radiation exposure. From this primary level, secondary standards were derived for regulating restricted reuse, namely:

- When the specific activity A_i of the dominating nuclide is below $0.2\,\mathrm{Bq\,g^{-1}}$, a site can be released without restrictions.
- The same applies to the reuse of equipment with a total surface alpha activity below $0.05\,\mathrm{Bq\,cm^{-2}}$.
- When A_i is below $1\,\mathrm{Bq\,g^{-1}}$, the site can be released with restrictions, for instance no building permits for new residential accommodation, recreation centres, kindergartens, sports and recreation facilities and the like are granted. Forestry and agricultural use as grassland would be allowed. In addition, a cover of clean soil is required in order to limit direct exposure dose rate to $<0.3\,\mathrm{nSv\,h^{-1}}$.
- Metal scrap can be released for smelting when the total surface alpha activity does not exceed $0.5\,\mathrm{Bq\,cm^{-2}}$.

Decommissioning and remediation of uranium production sites entail the monitoring of considerable waste streams. The contamination in many cases is not homogenous over the site, which requires an effective measurement strategy for a representative determination of radiological quantities such as A_i or the total surface alpha activity. To this end, Wismut developed the concept of 'intelligent combination' of laboratory analyses (mainly gamma and alpha spectrometry) and field measurements (dose rate measurements, *in situ* gamma-spectrometry, alpha/beta measurements using hand-held contamination monitors, etc.). The flow chart of this approach is given in Figure 24.

An example for the application of this approach is the release of scrap with the total surface alpha activity lower than $0.5\,\mathrm{Bq\,m^{-2}}$. For the representative determination of the total surface alpha activity of scrap, Wismut developed a measurement technique based on *in situ* measurements of beta count rates (N_β) using hand-held surface contamination monitors. Derivation of the total surface alpha activity from N_β is based on the proportionality between beta and alpha activities through a nuclide vector specific calibration factor k_β ($\mathrm{Bq\,cm^{-2}}$ per cps). Special calibration pads were set up for the Wismut-typical nuclide vectors (see Section 9.5.2). Quality assurance was carried out by 'scrape-off' surface samples. The samples are analysed in the laboratory using low-level gamma spectrometry. The total surface alpha activity can only be determined point-wise by hand-held contamination monitors, that is not over the total surface area of a scrap metal piece. For this reason, statistical methods were used to average

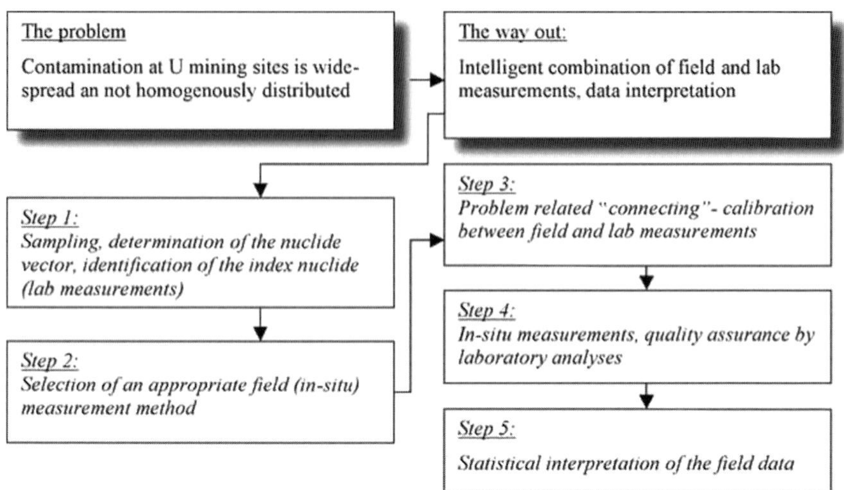

Figure 24 Flow chart for the combination of field and laboratory measurements for the release of materials.

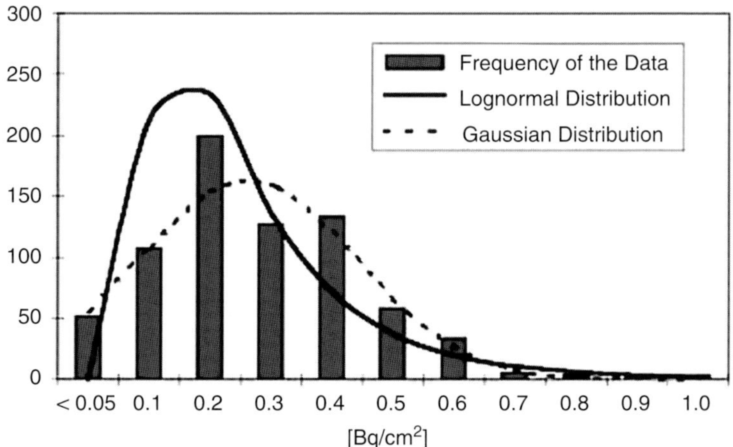

Figure 25 Measured total surface alpha activity values at a scrap metal heap at Wismut; typical distributions pattern of natural radionuclides at contaminated uranium mining and milling sites.

the distribution of total surface alpha activity values. Typically, log-normal distributions were observed (Figure 25).

Agreement was found between Wismut and the radiation protection authorities, that a representative, but not covering the complete surface, total surface alpha activity of a scrap heap is considered to be below the release limit of 0.5 Bq cm^{-2}, when the upper limit of the confidence interval of the statistic distribution of all measured data (generally log-normally distributed) falls below this level. A similar approach was used for the release of contaminated areas.

9.5.6. Final remarks on returning uranium mining legacies to productive use and conclusions

Considerable efforts were made over the past years for the benefit and the protection of the population living in the Wismut impacted region. The significant reduction of adverse environmental impacts and the reutilisation of vast surface areas for industrial use, for housing construction or as a nature reserve have put revitalisation of these regions on track. The idea 'New Horizons through Remediation' became more than a political slogan, it became a programme.

Integrated state-of-the-art technologies from around the world were applied to conduct decommissioning and remediation of the radioactive legacies left by the uranium mining in a efficient and cost-effective way, leading to sustainable solutions. However, the sustainability of the remediation solutions and post-remedial stewardship is best guaranteed if, in addition to applying high-technology, the reclaimed land and building objects can be put to a productive use, either restricted or ideally unrestricted. To achieve this goal a consideration must be given to the socio-economic effects of environmental remediation, particularly its contribution to the regional revitalisation and development, in addition to the mere containment of health and environmental risks and the future use of the remediated areas/objects should be specified prior to remediation.

The Wismut Project provides proof that value-added results can be achieved with no additional (or at reimbursable) costs, if remediation is undertaken with a well-defined reutilisation in mind. Vice versa, successful utilisation goals can only be developed in co-operation with the envisaged future user (i.e. municipality or developer), the community and the regulatory authorities. If consensus is achieved with the stakeholders prior to remedial works, a 'remediation by objectives' becomes practicable, that is the objectives for the individual remedial steps can be determined, consistent with the ultimate utilisation goal.

9.6. Remediation of the uranium production facilities at Poços de Caldas

9.6.1. Introduction

The uranium industry in some emerging economies, such as Brazil, sometimes has the task to above all provide the fuel for the respective domestic nuclear power plants. As a result, the development of the industry is very much controlled by the needs of national strategic planning. One of the consequences is that costs based on market values are often disregarded and mining of very low grade ores is undertaken. Such mining may lead to large areas in need of remediation, as the dominating aim of the project would have been the exploitation of the uranium deposit with no consideration of future environmental remediation costs. Once the ore is exhausted and mining and milling operations are terminated, operators (typically state-owned companies) may not be able to afford the burden of remediation programmes. The necessary expenditures will have to compete with other social needs and remediation projects will suffer delays in their implementation. Meanwhile, however, the cost for emergency measures such as effluent treatment will run up. The lack of experience in planning and implementing decommissioning and remediation programmes may exacerbate the problem. On the other hand, various countries with a number of completed decommissioning/remediation projects in the nuclear field have been able to mature this practice. The experience gained in the decommissioning/remediation of sites in developed countries may not be, however, readily transferable to developing countries.

The following describes the remediation of the first uranium mining and milling facility in Brazil, which ceased operation in 1997. The approach adopted is reviewed and technical issues related to the remediation of the respective industrial sites are discussed.

9.6.2. The mining site of Poços de Caldas

The Brazilian economy grew by 7.4% per year on average between 1950 and 1980. This growth slowed down in the 1980s to some 1.6%. There were a number of reasons for this, one of them being the dramatic increase in World oil prices. Brazil was then forced to reduce oil imports and the Government encouraged the use of alternative primary fuels. The Brazilian Nuclear Programme was part of this strategy. It was decided to build the first uranium mining and milling facility at Poços de Caldas, in the southeast of the country, which would feed the nuclear power plants then under construction.

The Poços de Caldas deposit is associated with a circular volcanic structure that intruded into the bed rocks of granites and gneisses during the upper Cretaceous about 87 million years ago and evolved in successive steps until about 60 million years ago.

The uranium enrichment is related to hydrothermal events (primary mineralisation) and to later weathering processes (secondary mineralisation).

The mine covered an area of about 2.5 km^2 and has been divided into three mineralisation units, designated as ore bodies A, B and E for mining purposes. The mining and milling facilities began commercial operation in 1982, but the originally intended production of 500 tons of U_3O_8 per year was never reached. The uranium deposits are of low grade

(675–1,700 mg kg^{-1}) and the uranium occurs mainly in the form of pitchblende. By 1995, in total 1,172 tons of U_3O_8 had been produced. During the development of the mine 44.8×10^6 m^3 of rock was removed. Of this, 10 million tons were used as construction material (roads, ponds, etc.). The rest was disposed of onto two major rock piles. Presently, acid drainage waters are collected in the mined out pit and pumped to a neutralisation plant before discharge into a local stream, whereas the solid treatment sludge is disposed of in the tailings dam. More recently, due to the exhaustion of disposal capacity in the tailings dam, the precipitates have been deposited in the mined out pit (Figure 26). The effluent from the tailings dam is treated with $BaCl_2$ to remove radium isotopes from the solution. The precipitates are collected in two holding tanks and the overflow is discharged into streams.

In Brazil, a uranium mining and milling project needs to obtain the environmental permits from the Brazilian Institute of the Environment (IBAMA) and the Nuclear Licensing of the Brazilian Nuclear Energy Commission (CNEN). The decommissioning of a mining facility requires the following actions:

- backfilling with mine debris and sealing of all wells, holes, galleries or any other excavation for research or ore recovery, in the surface or subsurface, to prevent the occurrence of accidents;

Figure 26 Open pit with the deposition of the slurry from the acid water treatment.

- actions to limit the potential risks to the human health and safety;
- classification of areas in the mine to prevent the release of toxic substances to the environment;
- implementation of a closure and site remediation plan, to be approved by the Regulatory Authority (CNEN), where possible future uses shall be foreseen.

These criteria are laid down in Standard CNEN-NE-1.13 (1989) (www.cnen.gov.br) that regulates the licensing of uranium and thorium mining and milling facilities. Post-closure environmental release criteria, however, are not stipulated, but it is implicit that the releases must not exceed those authorised for the operation of the installation.

As mentioned above, uranium mining and milling facilities are also regulated by the federal environmental authority (IBAMA). In this sense and to comply with the Brazilian Constitution (article 225 paragraph 2), the decree No. 97623 of 10 April 1989 demands that every existing project in the country involving mineral extraction should furnish a Site Remediation Plan within no more than 180 days after the promulgation of the decree. The decree has also established that in the case of a new project the plan must be presented during the environmental licensing procedure for the project. Economical aspects of the environment remediation are also taken into account and costs related to this activity must be part of the overall project budgeting.

9.6.3. Remediation strategy development

The plan to be presented by the operator has to be compatible with the local ecosystem, taking into account possible future land uses. It was agreed between the regulator and operator that the plan to be presented will be sub-divided into four topical parts:

- tailings dam and its area of influence,
- waste rock dumps,
- open pit and
- industrial area.

The latter covers the dismantling of buildings and equipment as well as general decontamination to specified clearance levels to be set by the regulatory authority.

For each one of the above items, remediation plans will have to be detailed, in such a way that the intended objectives will have to be specified as well as the intended strategies and expected performance. A cost-effectiveness analysis will have to be presented in addition. Project and financial schedules will have to be detailed in such a way that the regulators may monitor each of the ongoing activities.

9.6.4. Remediation plan

The Poços de Caldas site is the first of its kind in Brazil to be decommissioned and remediated. On one hand, the design of the operation benefited from a regulatory framework that strived to avoid major environmental impacts, such as those in the case of Wismut (see examples in this book), on the other hand, it failed to predict the occurrence of acid mine/rock drainage that has led to environmental impacts.

The low annual uranium production resulted in actual revenues too low to cover the environmental costs associated with the remediation works. In addition, the site received waste from a mineral sands processing facility located about 600 km away. Some of these wastes (^{228}Ra-rich solid waste) have already been disposed of in the tailings dam. Other waste (called 'Cake II' – a residue resulting from the processing of monazite ore to produce rare earth concentrates) is still being stored in barrels on-site.

The overall situation provides two of the most serious challenges a remediation programme can face, that is the lack of a good life cycle management plan for the operations and the resource constraints in terms of both manpower and financial resources. There is also a shortage of the necessary technical expertise in the country to implement and regulate such remediation activities. It is likely that if decommissioning/remediation activities had been adequately considered at least during the operational phase, if not during the licensing procedure, the current problems would be less serious.

The lack of local expertise will necessarily lead to the need of hiring international consultants and contractors to assist with the development and implementation of a remediation plan. Fees asked by such qualified companies are beyond the Brazilian market realities. Hence, innovative strategies need to be pursued. Due to the fact that the mining operator is a state-owned company (Industrias Nucleares do Brazil (INB)), it is subject to administrative and bureaucratic rules and regulations that would not necessarily apply to private companies, such as going through elaborate national and international bidding processes, which are often slow and cumbersome. Not to mention disruptions due to limited terms of office of governmental officials, who may change every 4 years following federal elections.

Taking all these considerations into account, a phased approach has to be put in place. But this also means that a substantial part of the site will have been released before the actual end of institutional control over the site as a whole.

This phased approach leads to the division of the site into the four topical areas mentioned above. Site categorisation according to the operational history and thus according to the likelihood of contamination will be of key importance in this process. One may assume in the first instance that administrative buildings, social and recreational facilities or even the plants producing sulphuric acid for the ore leaching process are not contaminated at all, or at such low levels that would allow free release according to criteria that still need to be established. On the contrary, those areas where mining and milling wastes were disposed of will require special treatment and due to the long half-lives of the radionuclides present will have to remain almost certainly under perpetual institutional control.

9.6.5. Provisional details of the remediation plan

As no overall strategy to comply with the legal requirements has been proposed yet by the operator, the Institute of Radiation Protection and Dosimetry developed a plan (though without an officially binding effect). It describes the steps needed to tackle the various legal, operational, technical and financial issues.

- The first step involves an administrative arrangement for the formation of a working group composed of the consulting company in charge of planning the remediation on behalf of the operator, the regulatory federal and state authorities (CNEN, IBAMA, FEAM), representatives of local stakeholders and knowledgeable technical institutions (National Institutes, Universities, etc.). These latter institutions would provide relatively cheap manpower to carry out field work, such as monitoring, data collection, interpretational and predictive modelling and so on. This could also effect, at least to some extent, some sort of technology transfer. Periodic meetings of the working group will allow the monitoring of the overall progress, the discussion of data gathered, providing corrective actions on the overall programme, and thus trust building between all parties concerned, reducing further delays in granting the necessary licences. Stakeholder involvement is of key importance in the process.
- Finding a sustainable long-term solution for the waste rock dumps and the tailing ponds is a key element in the overall project, but will not be discussed here.
- Meanwhile, the dismantling of the non-contaminated infrastructure can take place together with the transfer of equipment and machinery to the still active production centre of Caetite. This is motivated by the wish to increase production at the Caetite site, which produces uranium by a heap-leach process, to meet an anticipated demand increase due to planned construction of a third nuclear power plant in the country. A conventional leaching plant will be constructed in part from the material received from the dismantled Poços de Caldas facility.
- An alternative course of action in this phased approach could be to adapt the Poços de Caldas plant to process uranium from different sources, or to produce rare earth concentrates. Such strategy would take advantage of the fact that the site already stores radioactive waste. However, it faces

strong opposition from pressure groups who believe that it is simply a transfer of 'radioactive wastes' from other parts of the country to the Poços de Caldas site. The downside of this course of action would be that the construction of another tailings dam would be needed.

- Recovery of uranium from acid drainage waters could also play a role in the remediation scheme. Preliminary feasibility studies have indicated that about 30 tons year^{-1} of U_3O_8 can be recovered. With the increasing prices of uranium in the international markets, the revenue from the sales of this uranium could make a much desired contribution to financing the site decommissioning and remediation.

10. Conclusions

It is essential that decommissioning and remediation be approached as an integral part of a nuclear facility's life cycle. This enhances recognition that site reuse constitutes the final post-decommissioning phase in a site's life cycle. As a result, decommissioning and remediation activities are an opportunity to achieve reuse objectives consistent with site release criteria, free release of a site being the ideal end point. However, it would be inefficient to adopt a total dismantling policy for a nuclear facility that requires demolishing all structures and removing the materials from the site, if some buildings could be used in the future as part of the site's new use. In such instances, it may be feasible to adopt a selective decontamination and dismantling policy that retains the viability of some structures for redevelopment. Such an approach reduces the 'footprint' (area subject to decommissioning) at a site, thereby reducing waste volumes. On the other hand, it may be expedient to follow a concept of restricted reuse, whereby it may be feasible to limit soil remediation to a level appropriate for industrial redevelopment or a nature reserve and where residential reuse may have to be prohibited in the future. The buffer zones that separated the operational area of the nuclear facility from public access may be suitable for redevelopment for residential or other unrestricted uses after decommissioning. It was found, indeed, that within such zones natural habitats have developed largely undisturbed over many decades and it will be worthwhile to preserve these.

Careful selection of site reuse options will facilitate long-term stewardship measures that may be needed, where unrestricted release of a site is not feasible for technical or socio-economic reasons. Long-term stewardship is the collective term for all those measures that are undertaken to ensure that institutional control is maintained to the necessary degree. An important measure of planning and performance success is long-term liability reduction.

Many international cases provide various examples of successful decommissioning/remediation projects resulting in redevelopment of nuclear sites. There is evidence of how the reuse objectives affect the planning and management of decommissioning/remediation. But it is also noted that on several occasions the major weakness in decommissioning and remediation projects was poor or inadequate planning and management, leading to time and cost overruns. Often availability of adequate resources on time is a major constraint. A holistic planning approach, viewing the decommissioning–remediation–stewardship as respective stages in the life cycle of a site will aid in optimising and prioritising the sequence of work in a manner that resources are made available adequately and risks are reduced commensurably as the overall project moves forward.

Hence, it is important that a well-formulated strategic plan is put in place that can be sustained and keeps the end use in view, taking into account the various aspects, such as human resources, extent of work, safety, regulatory and other stakeholders concern, technology availability, cost, schedules, risk, feasibility and so on. In the case of a multi-facility site it would be advisable that from the outset a strategy to the decommissioning/remediation of the whole site be determined, from which the plan for the decommissioning/remediation of individual facilities or parts of the site could be implemented and put to reuse.

REFERENCES

Construction Industry Research and Information Association (CIRIA). (2000). *Best Practice Guidance for Site Characterisation: A Report for the SAFEGROUNDS Learning Network.* (Eds A.C. Baker, et al.). Construction Industry Research Information Association (CIRIA), London. Available through www.safegrounds.com

Construction Industry Research and Information Association (CIRIA). (2002). *Good Practice Guidance for Managing Contaminated Land on Nuclear Licensed and Defence Sites, SAFEGROUNDS Learning Network.* Construction Industry Research Information Association (CIRIA), London. Available through www.safegrounds.com

Coppins, G., M. Ayres, and M. Pearl. (2003). A data management and geographic information system (GIS) for the management of land quality on UKAEA sites. *Proceedings of 9th International Conference on Radioactive Waste Management and Environmental Remediation (ICEM '03)*, Oxford (CD-ROM).

Holdren, K. J., D. L. Anderson, B. H. Becker, N. L. Hampton, L. D. Koeppen, S. O. Magnuson, and A. J. Sondrup. (2006). *Remedial Investigation Baseline Risk Assessment for Operable Unit 7-13/14*, DOE/ID-11241, US DOE, Idaho Falls.

IAEA. (1994). *Decommissioning of Facilities for Mining and Milling of Radioactive Ores and Close-Out of Residues.* IAEA-TRS-362. IAEA, Vienna.

IAEA. (1998). *Characterization of Radioactively Contaminated Sites for Remediation Purposes.* IAEA-TECDOC-1017. IAEA, Vienna.

IAEA. (1999a). *Decommissioning of Nuclear Power Plants and Research Reactors Safety Guide.* IAEA Safety Standard No. WS-G-2.1. IAEA, Vienna.

IAEA. (1999b). *Decommissioning of Medical, Industrial and Research Facilities Safety Guide*. IAEA Safety Standard No. WS-G-2.2. IAEA, Vienna.

IAEA. (1999c). *Decontamination and Dismantling of Nuclear Facilities*. Technical Report Series No. 395. IAEA, Vienna.

IAEA. (1999d). *Technologies for Remediation of Radioactively Contaminated Sites*. IAEA-TECDOC-1086. IAEA, Vienna.

IAEA. (1999e). *Technical Options for the Remediation of Contaminated Groundwater*. IAEA-TECDOC-1088. IAEA, Vienna.

IAEA. (2000a). *Site Characterization Techniques used in Environmental Restoration Activities*, Final Report of a Co-ordinated Research Project 1995–1999. IAEA-TECDOC-1148. IAEA, Vienna.

IAEA. (2000b). *Predisposal Management of Radioactive Waste, Including Decommissioning Requirements*. IAEA Safety Requirements No. WS-R-2. IAEA, Vienna.

IAEA. (2001a). *Organization and Management for Decommissioning of Large Nuclear Facilities*. IAEA-TRS-399. IAEA, Vienna.

IAEA. (2001b). *Decommissioning of Nuclear Fuel Cycle Facilities Safety Guide*. IAEA Safety Standard No. WS-G-2.4. IAEA, Vienna.

IAEA. (2002a). *Non-Technical Factors Impacting on the Decision Making Processes in Environmental Remediation*. IAEA-TECDOC-1279. IAEA, Vienna.

IAEA. (2002b). *Record Keeping for the Decommissioning of Nuclear Facilities: Guidelines and Experience*. IAEA-TRS-411. IAEA, Vienna.

IAEA. (2003a). *Remediation of Areas Contaminated by Past Activities and Accidents*. IAEA Safety Requirements No. WS-R-3. IAEA, Vienna.

IAEA. (2003b). *Predisposal Management of Low and Intermediate Level Radioactive Waste Safety Guide*. IAEA Safety Standard No. WS-G-2.5. IAEA, Vienna.

IAEA. (2004a). *Glossary of IAEA Safety Related Publications*. IAEA, Vienna, http://www-ns.iaea.org/standards/Glossary/D.pdf

IAEA. (2004b). *Remediation of Sites with Low Levels of Dispersed Radioactive Contamination*. IAEA-TRS-424. IAEA, Vienna.

IAEA. (2004c). *Planning, Managing and Organizing the Decommissioning of Nuclear Facilities: Lessons Learned*. IAEA TECDOC-1394. IAEA, Vienna.

IAEA. (2004d). *Transition from Operation to Decommissioning of Nuclear Installations*. IAEA-TRS-420. IAEA, Vienna.

IAEA. (2004e). *Application of the Concepts of Exclusion, Exemption and Clearance*. IAEA Safety Standard Series No. RS-G-1.7. IAEA, Vienna.

IAEA. (2006a). *Remediation of Sites Contaminated by Hazardous and by Radioactive Substances*. IAEA-TRS-442. IAEA, Vienna.

IAEA. (2006b). *Applicability and Limitations of Monitored Natural Attenuation at Radioactively Contaminated Sites*. IAEA-TRS-445. IAEA, Vienna.

IAEA. (2006c). *Management of Long-Term Radiological Liabilities: Stewardship Challenges*. IAEA-TRS-450. IAEA, Vienna.

IAEA. (2006d). *Decommissioning of Facilities Using Radioactive Material*. IAEA Safety Standard Series No. WS-R-5. IAEA, Vienna.

IAEA. (2006e). *Release of Sites from Regulatory Control on the Termination of Practices*. IAEA Safety Guide No. WS-G-5.1. IAEA, Vienna.

INEEL. (2001). *Waste Area Group 7 Analysis of OU 7-10 Stage II Modifications*. Report INEEL/EXT-01-01105. Idaho National Engineering and Environmental Laboratory, Idaho Falls.

NEA. (2006). *Releasing the of Nuclear Installations – A Status Report*. NEA Publication No. 6187. OECD-Nuclear Energy Agency (OECD-NEA).

SAFEGROUNDS. (n.d.). www.safegrounds.com/pdf/groundhog_case_study.pdf, 5pp.

Travis, C. C., P. A. Scofield, and B. P. Blaylock. (1993). Evaluation of remediation worker risk at radioactively contaminated waste sites. *Journal of Hazardous Materials*, 35(3), 387–401.

USDOE. (2004). *Remedial Action Report for the OU 7-10 Glovebox Excavator Method Project.* Report DOE/NE-ID-11155. US DOE, Idaho Falls.

USDOE. (2008). *Record of Decision for Radioactive Waste Management Complex Operable Unit 7-13/14.* Report DOE/ID-11359, Rev. 0. US DOE, Idaho Falls.

USNRC. (2003). *Long-Term Stewardship of DOE Legacy Waste Sites: A Status Report.* The National Academies Press, US National Research Council (NRC), Washington, DC, http://books.nap.edu/catalog/10703.html

CHAPTER 7

Remediation Planning of Uranium Mining and Milling Facilities: The Pridneprovsky Chemical Plant Complex in Ukraine

Oleg Voitsekhovych* *and* Tatyana Lavrova

Contents

1. Introduction

This chapter provides basic information on the activities related to the current status of remediation planning at the Pridneprovsky Chemical Plant (PChP) and focuses on developing a new concept for decontamination of the former uranium extraction facilities and the need for proper management of the uranium residue in compliance with the best international practices.

Uranium mining was intensively conducted in Ukraine from the end of the 1940s to the beginning of the 1990s. Most of the uranium deposits have

*Corresponding author. Tel.: +38044 525 86 33; Fax: +38044 525 11 30
E-mail address: oleguhmi@yahoo.co.uk

Radiation Monitoring Department of the Ukrainian Research Institute for Hydrometeorology, Prospect Nauki 37, 03028 Kyiv, Ukraine

Radioactivity in the Environment, Volume 14
ISSN 1569-4860, DOI 10.1016/S1569-4860(08)00207-6

been explored in the Dnieper river basin, while some smaller deposits can be found within the basin of Ingulets river (Figure 1). In that period there were also several large uranium production facilities in the former Soviet Union; however, they were closed in 1991, the disintegration of the former Soviet Union leading to uranium production being significantly reduced (Chernov, 1998; IAEA, 2002).

The milling plant and uranium extraction facilities in Zhevti Vody are still in operation by the UkrAtomprom Industrial Consortium, and the rehabilitation programme for all uranium facilities within that site is the consortium's responsibility. The most difficult situation is to provide an optimised rehabilitation action plan for legacy sites such as the uranium tailings and other facilities situated in Dnieprodzerzhinsk town.

The former state industrial enterprise PChP was one of the largest metallurgical facilities, where uranium ores were processed from 1948 until 1991. It was one of the largest uranium milling facilities of the former Soviet Union (Korovin et al., 2001). During that time, uranium extraction was carried out using the raw ore products delivered from Central Asia, Germany and the Czech Republic. In addition to imported ores, the PChP processed uranium-bearing sludge obtained from cast iron smelting of iron ores from the uranium mines of Ukraine (Figure 1). In the early 1990s, due to the disintegration of the former Soviet Union and consequently of the

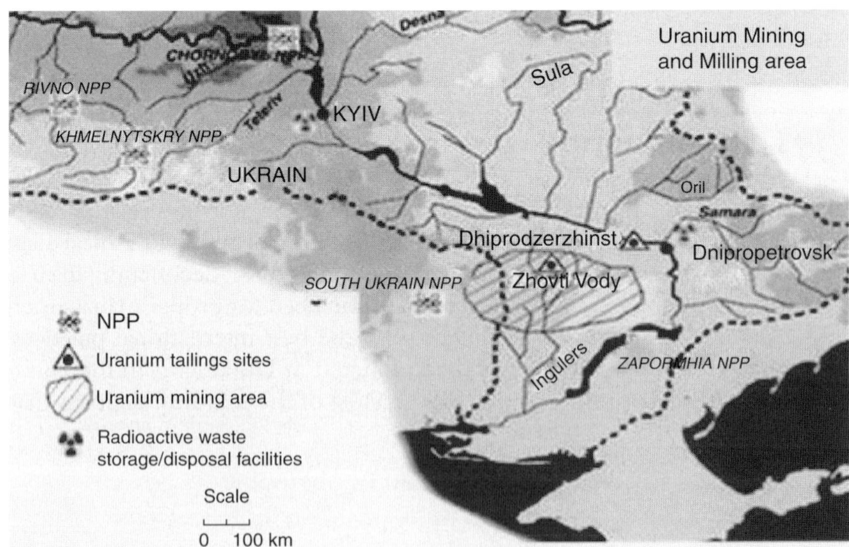

Figure 1 The sites of uranium ore deposits and locations of the main milling and uranium extraction facilities in Ukraine (Dnieprodzerzhinsk and Zhevti Vody towns).

uranium industry, the PChP was split into several separate companies and processing of uranium was stopped.

Nine tailings impoundments were created in the area containing about 42 million tonnes of uranium extraction residues with a total activity of 3.2×10^{15} Bq (86,000 Ci) (IAEA, 2002). Some of the highly contaminated equipment and scrub metals used at the facilities were deposited at storage sites within the territory of the industrial zone of Dnieprodzerzhinsk, and other residues were disposed of about 14 km to the south east of the site. Each tailings impoundment has been inventoried with information obtained from rather limited studies carried out during the recent decade under the programme initiated by the Ministry of Fuel and Energy of Ukraine. However, the quality and complexity of information available for each particular facility are still rather low and require more specific studies.

The PChP territory is enclosed by a concrete fence. The area of the former PChP is divided by the railway line into two large areas, that is into upper and lower parts. The main former uranium extraction facilities of PChP are situated in the upper part of the territory, and the largest tailings dump (Dnieprovskoe, see point 4 in Figure 2) is in the lower part, south

Figure 2 The Pridneiprovsky Chemical Plant for uranium production. 1: Zapadnoe tailings; 2: Centralny Yar tailings; 3: Yugo-Vostochnoe tailings; 4: 'D' (Dnieprovskoe) tailings with an indication of the main strategy on relocation of the tailings dumps from the industrial site to the tailings 'D' and 'S' (Sukhachevskoe) sites, located 14 km away from the PChP industrial site. The rocky pit in this area is considered a potential site for relocation of the uranium residues from the site.

of the Konoplyanka river (creek) – also referred to as the drainage canal. The upper part of the PChP territory, where the facilities lie, is much more contaminated with U–Th series radionuclides due to the higher impact of the former uranium extraction facilities. No properly engineering barriers were created for most of the tailings. After full capacity was reached, each tailings impoundment was usually covered by local soils, debris and other industrial wastes.

The distinctive feature of this former uranium legacy site and its uranium tailings is located in the populated area of Dnieprodzerzhinsk town (about 276,000 citizens). The residential area is situated rather close to the industrial zone at a distance of 1–2 km from the nearest tailings impoundment (see Figures 1, 2). Therefore, the fate of the former uranium production facilities and issues related to possible options for its remediation are very sensitive for the local population.

The termination of milling activities in the early 1990s resulted in both the need to decommission some PChP buildings and to fully convert others. According to the legal requirements which are still in use (SPLKP-91, 1991) for conversion of the former uranium milling facilities, it was necessary to carry out a full inventory of all locations where residues from uranium extraction were disposed of and other highly contaminated wastes dumped to provide a safety assessment of the former milling facilities, and to carry out some decontamination of the existing 'hot spots'. However, these requirements have some shortcomings for different reasons, the main being that the existing regulatory basis for uranium mining and processing (as for other NORM mining and processing activities) is not covered by regulations because these address other types of radioactive wastes. Therefore, until recently, there were no clear regulatory criteria developed for management of uranium residues.

The State Remediation Programme (developed in 2003) has been commissioned in 2005 and aimed to identify the first priority remedial actions focused on decontamination of the former extraction facilities, demolishing the most contaminated buildings and to make proper cover of the tailings. However, this programme has not been successfully completed because of the undeveloped legal and regulatory framework needed to serve as a basis for the proper implementation of the remediation plan. The lack of experience in remediation planning and lack of available funds also became a significant constraint, which in fact led to suspension of the programme in 2008.

According to the decision of the Ministry of Fuel and Energy of Ukraine, a new concept for the modified State Remediation Programme was developed in 2008 (Voitsekhovych et al., 2008). The new programme should take into account experience and knowledge from the best international practices (UMTRA, 2005; UMREG, 2008).

2. Current Activities in and around the PChP Territory

Despite decontamination of the legacy site not being fully implemented since 1991, about 20 companies were still in operation in the PChP territory in 2008. Most of the enterprises just make use of the area at the uranium production legacy site but are themselves not related to the former uranium processing activities. However, their workplaces are situated close to the locations of the highly contaminated tailings dumps or the former buildings used for ore milling and extraction, which may expose workers to gamma radiation, and may cause internal exposures due to Rn emanations and alpha-aerosol dispersion. Some small businesses make use of contaminated facilities, which were not decontaminated in a proper way. Other larger companies do not use the contaminated facilities, but operate in close vicinity to these. There are only a few enterprises that have formally established a radiation safety service and culture, where staff are classified as professional workers according to radiation protection criteria. The regulatory constraints and specific roles for companies in this territory are still not well developed and require urgent improvements.

Special attention towards radiation safety of the personnel will be needed during the demolition of contaminated buildings (former uranium extraction facility), some of them situated just about 20 m from the companies still in operation. This specific situation will require establishment of a comprehensive safety culture and radiation protection system for staff working in this territory, but will simultaneously and significantly constrain the economic development of the enterprise if remediation is not implemented as soon as possible.

In addition to the businesses already operating in this territory, interest to further exploit the former empty facilities of this site is high. For example, some workshops of the former hydrometallurgical plants that were used in the past for extraction of uranium concentrate have been sold to a new owner who intends to use these workshops for processing gold-containing raw materials. It is clear that such 're-profiling' of the former uranium facilities requires completely decontaminated or demolished facilities to allow the flourishing of newly emerging business developments.

Assessments have clearly shown that there is an impact on the environment from most sites with uranium tailings and wastes. The impact is mostly due to releases of radionuclides from the uranium decay series (^{238}U, ^{230}Th, ^{226}Ra, ^{210}Pb and ^{210}Po) to surface waters and the groundwater table, as well as from radon emissions and dust dispersion into the air. External exposure due to gamma irradiation also becomes relevant in cases where direct access to contaminated areas is possible.

Since 1991 regular environmental and radiation monitoring programs have not been in place until after 2003 when regular monitoring and site-specific studies were re-established in the special programs framework initiated and funded by the Ministry of Fuel and Energy of Ukraine. New research programmes were also initiated with financial support from international projects (STCU, 2007; ENSURE, 2008). The results of those projects and assessments were used as a basis for remediation planning starting from 2009 to 2015, and are presented in the following assessments.

3. Typical Patterns of Site Contamination

Typical external gamma dose rates in the territory are generally rather low: 0.15–0.30 $\mu Sv\,h^{-1}$. However, in some places, such as tailings surfaces, external gamma dose rates may reach 1–3 $\mu Sv\,h^{-1}$ and even 30–60 $\mu Sv\,h^{-1}$. In such local 'hotspots' (e.g. the 'Centralny Yar' tailings), ^{226}Ra activity in soils at the surface of tailings reach 0.1–0.2 $kBq\,g^{-1}$. The Rn exhalation in such 'hot spots' was measured to be 2–6 $Bq\,m^{-2}\,s^{-1}$. Surveillance studies showed that the surface cover at such tailings sites is not sufficient to reduce exhalation rates.

Surface contamination on machinery, equipment, metal scrap, etc. from the period of uranium production still exists, and such materials are currently kept close to the former uranium production workshops. Some of the most contaminated debris and metal constructions were dumped together with uranium residues at the tailings dumps, containing, amongst others, ^{226}Ra and long-lived radon daughter nuclides ^{210}Pb and ^{210}Po on the surfaces of the contaminated equipment and scrap. Monitoring data demonstrate that the composition of ^{238}U and ^{226}Ra in the PChP territory varies from hundreds to several thousand Becquerels per kilogram, compared with the local soils that contain only 15–30 $Bq\,kg^{-1}$ dry weight. The main long-lived nuclides in the tailings (the uranium extraction residues) are $^{234,238}U$, ^{230}Th, ^{226}Ra and ^{210}Pb ^{210}Po with activities up to $10^5\,Bq\,kg^{-1}$ and higher. The typical ranges of the environmental contamination are given in Table 1 for different natural objects and tailings sites.

Aerosol pollution is also relatively high at the legacy site in comparison with the naturally occurring background levels in the vicinity and is a result of wind re-suspension and radon-progeny radionuclide dispersion over this area.

The main contributor to radiation exposure in this region is indoor radon and outdoor contamination. Typical outdoor concentrations of radon were observed at up to $2–4 \times 10^2\,Bq\,m^{-3}$. Higher concentrations of radon were found in some buildings used by workers in the industrial premises, storage rooms, etc. and range between 10^3 and $10^4\,Bq\,m^{-3}$.

Table 1 Typical ranges of environmental contamination by radionuclides of the ^{238}U series.

Type of samples	Units	^{238}U	^{230}Th	^{226}Ra	^{210}Pb	^{210}Po
Soils in the inhabited areas	$Bq\,kg^{-1}$	20–30	50–70	20–40	50–90	40–70
Soils at the industrial site	$Bq\,kg^{-1}$	10^2–10^3	10^2–10^3	10–10^2	10^2–10^3	10^2–10^3
Soils and tailings material at the uranium residue dump sites	$Bq\,kg^{-1}$	10^3–10^4	10^3–10^4	10^4–10^5	10^4–10^5	10^4–10^5
Groundwater (pore water) from the aquifer at the tailings site	$Bq\,m^{-3}$	10^2–10^5		10–10^3	10–10^3	15–300
Surface water in a drainage system inlet to the Dnieper river	$Bq\,m^{-3}$	200–500		15–30	15–20	5–10
Aerosols at the tailings dump sites and nearby contaminated buildings	$10^{-6}\,Bq\,m^{-3}$	50–300		100–150	600–1,100	150–400
Aerosols in the inhabited area near the industrial sites	$10^{-6}\,Bq\,m^{-3}$	15–20		10–30	50–200	15–40

The highest exposures were found in some indoor working areas where highly contaminated facilities are still in place (uranium extraction facilities, transporting tubes, etc.). In such areas, indoor Rn concentrations were found to be in the range 10^3–$10^5\,Bq\,m^{-3}$.

Preliminary dose and risk assessments carried out recently have shown that the current levels of alpha-activity in surface waters are rather low and cannot lead to doses exceeding the permissible levels in Ukraine (IAEA, 2002). However, according to Skalsky and Riazantsev (2008), the pore water in aquifers around tailings dumps is highly contaminated (the highest concentrations of alpha-emitting radionuclides have reached $10^5\,Bq\,m^{-3}$) and can pose a potential risk in the event of the protective dyke being damaged, thereby spilling the highly polluted pore water into the drainage canal and further into the Dnieper river. Under natural conditions, the pore water moves very slowly in the aquifer towards the Dnieper river. At present, the water in the drainage canal (Konoplyanka river) has gross alpha

Figure 3 Tailings 'D' (left) covered by phosphogypsum and Centralny Yar tailings (right) covered by coniferous forests to be removed in the event of this tailing dump being relocated from the industrial site. Optionally, these could be moved to tailings 'D', tailings 'S' or to the rocky pit (see Figure 1).

activity levels between 0.3 and 0.6 Bq L^{-1}, this being 10–20 times higher than the background levels found in the Dnieper river upstream of the area of drainage water inlet to the reservoir. The most significant source of Dnieper river pollution is from tailings 'D', situated at the shortest distance to the river and drained by the Konoplyanka river.

According to recent studies, the groundwater flux from the legacy site may not significantly pollute the Dnipro water and thus may not be a main factor in the future potential radiological impact on the neighbouring residential areas, at least after 500 years, if the surface cover of the tailings is adequate and a drainage system is in place to protect the tailings body from precipitation and significant inundation. Therefore, highest priority at present is given to implementing actions to prevent workers in the territory of the former PChP site from becoming exposed to undue radiation (Figure 3).

4. Preliminary Dose Assessment and Conclusions Based on Monitoring Data Analyses

Screening models were selected for estimating radiation exposures through all pathways that may be relevant at uranium mining and processing sites. The 'Ecolego' tool was applied for preliminary dose assessments as a basis for further remediation strategies (Avila et al., 2000). A number of scenarios were chosen to demonstrate the applicability of these screening models to estimation of doses to workers in the PChP territory. These dose estimates provided a first indication of the dose range for population groups that receive the highest exposures at the site (ENSURE, 2008). The assessment calculates maximum individual doses under realistic assumptions.

Radiation exposures due to external exposure, as well as pathways due to inhalation of contaminated aerosols and radon, as well as soil ingestion were calculated. Results showed that external exposure and radon inhalation provide the highest contribution to the total doses. Depending on scenarios (worst cases), annual dose rates may exceed both dose limits for worker categories A and B, that is 20 and 5 mSv year^{-1}, respectively.

The doses for people living in the vicinity of the PChP site are estimated to be less than 1 mSv year^{-1} for most of calculated scenario of potentially possible environment impacts and safe management of the tailing dumps. However, in some accidental situation affecting safety barriers the tailings dams and removing the cover may lead to significant radiological consequences; hence, long-term surveillance and emergency preparedness are required at least for 100–1,000 years.

For most workers, whose work places are in non-contaminated buildings or who are mainly working in an uncontaminated area of the legacy site, the total annual dose values are estimated to be in the range 0.1–0.5 mSv year^{-1}. The annual doses (maximum) to workers whose work places are located near the tailings dumps or near the contaminated buildings (former uranium production workshops) or who are regularly inspecting the tailings dumps for monitoring purposes may vary from 1–2 to 8–12 mSv year^{-1} dependent upon their specific duties and time spent in the contaminated areas or contaminated premises. The highest doses (30–45 mSv year^{-1}) will be obtained by those who have regular access to the contaminated premises and will be involved in remediation activities, in removal and utilisation of the tailings materials and/or of the most contaminated equipment, thus representing a worst case scenario.

The preliminary assessment concluded that the remediation plan has to focus on clean-up of the former uranium extraction facilities, on proper surface coverage of the tailings or removal of the tailings to the specially prepared tailings sites containing engineering barriers. Predictions based on the radionuclide migration model incorporated into 'Ecolego' even for a worst case scenario (due to the large uncertainty of model parameters for uranium migration) revealed that proper soil coverage and removal of the tailings materials from the largest tailings site 'D' may decelerate radionuclide transfer into the Dnieper river for the next 500 years (Skalsky and Riazantsev, 2008). This fact allows justification for the development of a new concept for tailings conservation, taking into account the fact that all prior strategies and options such as re-treatment and removal cannot be economically justified when comparing risk reduction, other benefits and costs.

Still, the option to remove small tailings from the industrial site requires exhaustive long-term safety assessment studies in close collaboration with stakeholders, that is the regulatory body and the public. The assessment to be carried out should include hypothetical scenarios of landuse in this territory over a timescale of 100–1000 years, scenarios that illustrate the potential exploitation of the site which may comprise the following potential constraints in land use: prohibition of housing in waste areas because of extremely high levels of radon emission and also prohibition or restriction of housing near alternative sites for removal of the tailings materials from the industrial site that may demonstrate non-compliance. Scenarios of the future land use should include both realistic and even irrational scenarios for future developments at the site.

To carry out the long-term prediction with acceptable uncertainties for decision making, the relevant radioecological models should be well validated with the geochemical parameters to be studied (oxidation, acidification and other processes affecting radionuclide releases) and also with the baseline monitoring information such as hydrological modes and long-term possible development trends.

For optimisation of the engineering solution and additional data requirements, as well as prioritisation of the additional remedial investigations and measures, the use of a Conceptual Site Model (CSM) approach proved to be very helpful (Jakubick and Kahnt, 2002).

5. Compliance with Radiation Safety Requirements

As described above, the territory and facilities of the former PChP are still in use, for example in the industrial production of zirconium concentrate, ion-exchange resins, mineral fertilisers, concentrates of non-ferrous metals and other production. The government of Ukraine and local authorities plan to further develop these enterprises and their production within this territory. Such a concept requires an assurance that radiation and ecological impacts in this area will be in compliance with the national radiation safety standards and with environmental legislation.

The main sources of current and potential exposure to workers in this area are the uranium tailings and other facilities as a legacy of past uranium production. Therefore, in 2000, the Ministry of Fuel and Energy created the 'Barrier' special enterprise with the following tasks and functions:

- to develop a basis for radiation protection provisions in this territory;
- to make an inventory of the radionuclides and materials impounded at each tailings site;
- to develop an observational network and to establish ecological and radiation monitoring in this territory and
- to provide a surveillance service and safe management of the former facilities for uranium production in this territory.

Since 2003, the regular ecological monitoring and surveillance programmes at the industrial site have been established under the management and in cooperation with the 'Barrier' enterprise. Debris and industrial wastes were partly removed from the 'Yugo-Vostochnoe' uranium tailings surface (southeast). Proper cover with 0.5 m of clay materials and 1 m of local soils has been created at the 'Zapadnoe' tailings (Western). The drainage system around the 'Zapadnoe' tailings was also cleaned and partly restored. The slope of the tailings was adjusted. To prevent soil erosion, a special plastic fabric has been installed on the tailings slope. About 2 km of highly contaminated pipelines and tubes were removed from the territory; cut and packed into special containers and temporarily stored in a fenced storage area. During 2007, part of the ore residues were excavated from the former 'Base C' storage facility and relocated to the Zhevti Wody Mining Combine for reprocessing (see Figure 1).

These actions significantly reduced the gamma dose rate at many locations varying from 2–4 $\mu Sv\,h^{-1}$ to even 30 $\mu Sv\,h^{-1}$ compared to the average acceptable background level of 0.10–0.30 $\mu Sv\,h^{-1}$ in this territory. In fact, Rn exhalation and aerosol alpha-particle dispersion from this territory has also been significantly reduced due to such temporary remediation solutions.

However, these remediation actions, even when fully performed, do not guarantee long-term safety for radioactive residue management. They are considered as temporary options because most of the soil cover is installed without the multilayer usually recommended; some tailings dumps are situated on the unstable slopes of the Dnieper valley and may be significantly eroded during rainfall. The drainage system around some tailings in the PChP territory are not fully operational and do not allow full control of the transport of contaminated groundwater towards the riverside.

Therefore, the new concept for remediation (Voitsekhovych et al., 2008) suggests establishing the following pre-feasibility actions in further remediation planning:

- to re-consider some legislative and regulatory norms as a basis for safe management of the former uranium facilities (the new health-related rules to be improved according to BSS principles and recommended criteria (IAEA, 2003, 2005);
- to developed and establish the licensing procedures for remediation implementation for current practices;
- to extend tailings dump characterisation and their inspection programmes taking into account IAEA recommendations (IAEA, 2000, 2002, 2000b);
- to carry out long-term safety assessments taking into account the predictions of long-term groundwater transport of uranium and other toxic substances into the environment of the residential area and
- to consider re-treatment (re-processing) options for some tailings materials as a part of the remediation process.

The pre-feasibility studies are to be implemented by 2010 and will help to select the most appropriate and economically justified options for remediation at the PChP industrial site. The experiences gained from best uranium facility remediation practices, for example as described by Jakubick and Kahnt (2002), Hagen and Jakubik (2005) and others (UMTRA, 2005; UMREG, 2008), are to be considered.

6. New Approach to a Remediation Strategy

Among the most pressing remediation problems which are still awaiting an optimal strategy are: the fate of highly contaminated buildings at the industrial sites; the fate of phospogypsum cover, integration of the largest 'D' (Dnieprovskoe) tailings with other tailings materials and the fate of the wet uranium Sukhachevskoe tailings (tailings 'S'), which are still partly covered with water. One option to be considered is the deepening of the Konoplyanka creek, serving as a natural drainage canal for both areas of the industrial site and tailings dump 'D'.

The new approach (Voitsekhovych et al., 2008) considers the most preferable action to be the removal of relatively small tailings dumps over the territory and transportation of the tailings materials onto the surface of the largest tailings 'D' (about 1 km) with further conservation of the tailings with multilayer soil cover, while the State Programme (2003) suggested removal of all tailings dumps and contaminated scrubs to the tailings 'S' site (Sukhachevskoe, about 14 km away) dramatically increasing the project costs. However, both options are still to be finally evaluated taking into consideration the social and long-term ecological considerations using multi-attributive assessment procedures.

The following remediation plan with a two-phase approach is to be implemented in line with the concept of the New State Programme 2009 for remediation of the former PChP legacy site:

Phase 1 (2009–2010): Safety assessment; environment impact assessment; monitoring and surveillance programme implementation; pre-feasibility studies; conceptual planning and project designs; regulatory framework development and provisions; public consideration, agreement procedures; approval of the action plan and designs; preparatory administrative and engineering actions for further plan implementation (establishing sites for storage and waste sorting); start of first-priority actions.

Phase 2 (2011–2015): Implementation of the main set of engineering actions foreseen by the action plan and designed as a first-priority action at the territory of PChP; possible removal of the tailing materials from the industrial site to the tailing 'D'; preliminary cover of the relocated tailing materials at the surface of tailing 'D' and other engineering actions according to the approved action plan as first-priority actions at the industrial site and surrounding territories.

It is expected that, after completion of this plan, the former industrial site for uranium production at the PChP will be at least partly exempted from regulatory control and that enterprises will receive higher economic potential for development than they have now. Further remedial actions at a later phase (2016–2020) have to be focused on the neighbouring areas, which at the moment are covered by other industrial waste storage and landfills created by other metallurgical and geochemical plants of the town.

The complete remediation of this area is part of an integrated Environment Restoration Plan of Dnieprodzerzhinsk town. Complete restoration is foreseen by afforesting the area of the Dnieprovskoe tailings. The Sukhachevskoe wet tailings (situated 14 km from PChP) is also to be remediated, based on experiences in dry remediation options gained at the Helmsdorf tailings pond in Germany (Hagen and Jakubik, 2005). The stewardship programme and handover of the re-profiled facilities and neighbouring areas of the town will be accompanied by comprehensive compliance and environmental monitoring continuously carried out under the auspices of the regulatory bodies, also giving access to this information to the public and convincing them that the actions were effective for their safety.

REFERENCES

Avila, R., R. Broed, and A. Pereira. (2000). ECOLEGO – A toolbox for radioecological risk assessment. *Proceedings of the International Conference on the Protection from the Effects of Ionizing Radiation*. IAEA-CN-109/80. International Atomic Energy Agency, Stockholm.

Chernov, A. (1998). Uranium production plans and developments in the nuclear fuel industries of Ukraine/The Uranium Institute. In: *23rd Annual International Symposium*, 8–11 September. Uranium Institute, London, http://www.world-nuclear.org/sym/1998/chernov.htm

ENSURE. (2008). *ENSURE: Assessment of Risks to Human Health and the Environment from Uranium Tailings in Ukraine. FACILIA. Phase-1*, Final Report. Swedish Radiation Protection Institute, Stockholm, SIUS-under contract UA401A/2007-09-24.

Hagen, M., and A. T. Jakubik. (2005). Returning the WISMUT legacy to productive use. *Proceedings Of Uranium Mining and Hydrogeology IV, 11–16 September 2005*, Springer, Freiberg, Germany.

IAEA. (2000). *Regulatory Control of Radioactive Discharges to the Environment*. Safety Standards Series WS-G-2.3. International Atomic Energy Agency, Vienna.

IAEA. (2002). *Radiological Conditions in the Dnieper River Basin*. IAEA Radiological Assessment Reports. International Atomic Energy Agency, Vienna.

IAEA. (2003). *Remediation of Areas Contaminated by Past Activities and Accidents Safety Requirements*. IAEA Safety Standards Series WS-R-3. International Atomic Energy Agency, Vienna.

IAEA. (2005). Environmental contamination from uranium production facilities and their remediation. *Proceedings of an International Workshop*, Lisbon, February 2004; *IAEA Proceedings Series*. International Atomic Energy Agency, Vienna.

Jakubick, A. T., and R. Kahnt. (2002). Remediation-oriented use of conceptual site models at WISMUT GmbH: Rehabilitation of the Trünzig tailings management area. In: Uranium in the Aquatic Environment. *Proceedings of International Conference Uranium Mining and Hydrogeology III and the International Mine Water Association Symposium, September 2002*. Springer, Freiberg, Germany.

Korovin, V., Y. Korovin, G. Laszkiewicz, O. Lawrence, Y. Lee Koshik, G. Shmatkov, and G. Semenets. (2001). Problem of radioactive pollution as a result of uranium ores processing. In: *Scientific and Technical Aspects of International Cooperation in Chernobyl* (Collection of Scientific Articles). Kiev, pp. 461–469.

Skalsky, O., and V. Riazantsev. (2008). Problems of the hydrogeological monitoring at the Pridneprovsky Chemical Plant (Dneprodzerzhinsk, Ukraine). In: *Uranium Mining and Hydrogeology*. Springer, Freiberg, Germany.

SPLKP-91. (1991). *Sanitary Roles for Liquidation, Conservation and Re-profiling of the Enterprises on Exploration and Reprocessing of the Uranium Ores* (CII JIKII—91). Ministry of Health of USSR, Moscow, 44pp. (in Russian).

State Programme. (2003). State program for remediation of the former Pridneprovsky Chemical Plant, http://zakon.rada.gov.ua/cgi-bin/laws/main.cgi?nreg = 1846-2003-%EF (in Ukrainian).

STCU. (2007). *Substantiation of the Radionuclides Transfer Reduction to the Environment and Human Body in Uranium Sites*. STCU Project No. 3290, Interim Project Report. Ukrainian Hydrometeorological Institute.

UMREG. (2008). *Uranium Mine Remediation in Times of Revival of Production*. UMREG Monograph. Selected paper limited distribution IAEA, Vienna,http://www.geo.tu-freiberg.de/umh/Final%20Programme_UMHV.htm

UMTRA. (2005). *Uranium Mill Tailings Remedial Action. US Uranium Production Facilities: Operating History and Remediation Cost under Uranium Mill Tailings Remedial Action Project as of 2000*. Energy Information Administration, Official Energy Statistics from the US Government, http://www.eia.doe.gov/cneaf/nuclear/page/umtra/title1map.html

Voitsekhovych, O., Y. Soroka, I. Los, Y. Tkachenko, and T. Lavrova. (2008). *Conservation, Liquidation Re-profiling and Remediation of the Former Uranium facilities of the Pridneprovsky Chemical Plant in the Ecologically Safe Conditions. Status and Concept of the Remediation Strategy*. Report, CMSET, Ministry of Fuel and Energy of Ukraine, Kiev, Ukraine (in Ukrainian).

CHAPTER 8

Principles and Technologies for Remediation of Uranium-Contaminated Environments

Yong-Guan Zhu* *and* Bao-Dong Chen

Contents

Abstract

From the beginning of the 20th century, radioactive materials have accumulated on the earth's surface as a result of (1) the mining and processing of uranium (U) and thorium (Th) for the use and testing of nuclear weapons and for normal operations, (2) accidents in the civil nuclear power industry and, most recently, (3) the use of depleted uranium in conventional military weapons. As one of the most toxic radionuclides, uranium can disperse on soil surface by runoff, into the air by wind and to groundwater by leaching, subsequently endangering both human and animal health. Proper management of uranium-contaminated environments has therefore become an urgent need specifically in times of *Nuclear Renaissance*, which calls upon a holistic strategic approach from the exploitation of such natural resources,

*Corresponding author. Tel.: +86-106293 6940; Fax: +86-10-6292 3563
E-mail address: ygzhu@mail.rcees.ac.cn

Research Center for Eco-Environmental Sciences, Chinese Academy of Sciences, 18 Shuangqinglu, Haidian District, Beijing 100085, China

Radioactivity in the Environment, Volume 14
ISSN 1569-4860, DOI 10.1016/S1569-4860(08)00208-8

its processing in the nuclear fuel cycle to decommissioning with the appropriate considerations of environmental and radiation impacts.

1. Environmental Uranium Contamination

Even before its formal discovery by the German chemist, Martin Klaproth, in 1789, uranium (U) has been used for a variety of purposes starting from colouring glass and ceramics to its use in military and public industries. However, the centuries of mining and milling of uranium, and of other elements during their exploitation, have resulted in the production of considerable amounts of radioactive waste materials which are perceived to threaten the environment and public health. Generally, a mine capable of producing 100,000 t of uranium ores annually will simultaneously produce 100,000–600,000 t waste tailings.

At present, there are more than 10 countries that run uranium mines, from which 20 million tons of uranium tailings are produced annually. At numerous sites, the improper storage of these wastes has led to uranium contamination of the surrounding environment (Liator, 1995). For instance, in a uranium mining site in Germany, the tailings contain $275\,mg\,U\,kg^{-1}$, and in the groundwater $707\,mg\,U\,l^{-1}$ was detected (Junghans and Helling, 1998). In a 30-year-old uranium mining site in south-east Siberia, Russia, the uranium concentration in the top layer of soil of the vicinal grassland exceeded $1{,}000\,mg\,kg^{-1}$. Soil contamination by other radionuclides (e.g. thorium (Th)) and heavy metals (e.g. As) were also detected. In the pollution area, the abundance and diversity of soil arthropods were 3–37 times lower than at control sites, and the uranium and As concentrations in beetles in the polluted area were 2–41 and 2–26 fold higher than those at the control sites (Gongalsky, 2003).

In addition to mining activities, some radioactive contamination of the environment has also resulted from the extractive industries, such as those for phosphorus, oil, iron, coal and mineral sands. Direct application of uranium-rich phosphate rock as an alternative to commercially processed phosphorous fertilizers over years may have led to large-area contamination of arable soil, specifically in developing countries (Williams and David, 1976; Anderson and Siman, 1991; Van Kauwenbergh, 1997).

Naturally, uranium has three isotopes: ^{234}U, ^{235}U and ^{238}U, with relative abundance of 0.0055%, 0.720% and 99.27%, respectively (Allard et al., 1984). Among the naturally occurring actinides, uranium is actually the most abundant. Its concentration in the earth's crust may range from 1 to $4\,mg\,kg^{-1}$ in sedimentary rocks, to ten or even hundreds of milligrams per kilogram in phosphate-rich deposits (Langmuir, 1997; Qureshi et al., 2001) and uranium-ore deposits (Plant et al., 2003). It is considered to be more plentiful than mercury, antimony, silver or cadmium, and is about as abundant as molybdenum or arsenic. Uranium is naturally present in both aquatic and terrestrial environments (Cowart and Burnett, 1994). Anthropogenic uranium contamination of the environment, especially by mining and milling operations, however, can lead to ecosystem degradation.

Uranium speciation is closely related to soil properties (especially pH) (Ebbs et al., 1998; Langmuir, 1978; Mortvedt, 1994). It is most mobile as uranium(VI) (Campbell and Biddle, 1977), which exists in solutions predominantly as UO^{2+} and as soluble carbonate complexes (Langmuir, 1978; Ciavatta et al., 1981; Grenthe et al., 1992; Duff and Amrhein, 1996). Between pH 4.0 and 7.5, the pH range of most soils, uranium(VI) exists primarily in hydrolysed forms (Meinrath et al., 1996) and is readily taken up by plants from the exchangeable and soluble fractions of the soil. However, according to published work, negligible amounts of uranium(VI) remain in the soluble and exchangeable forms over any significant time, thereby limiting the amount available for plant uptake (Sheppard and Thibault, 1992).

Although uranium is not necessary for any biological function, a wide range of organisms, including plants (Ebbs et al., 1998; Huang et al., 1998a; Shahandeh et al., 2001), bacteria, algae and fungi (Abdelouas et al., 1999; Suzuki and Banfield, 1999), were shown to accumulate uranium in both terrestrial and aquatic environments. On the other hand, biological activity from bacteria, algae, fungi to plants can affect uranium speciation and, hence, uranium mobility by modifying the pH, extra-cellular binding, transformation and formation of complexes or precipitates (Kalin et al., 2004; Sar et al., 2004; Anderson et al., 2003; Dushenkov, 2003). Thus, these organisms can influence uranium transfer along the food chains and thus can also be used to develop bioremediation technologies to decontaminate uranium-contaminated environments.

Human beings and animals are exposed to uranium mainly by direct contact (e.g. contaminated drinking water). Besides ionic radiation, uranium behaves similarly to heavy metals, especially for Pb. Uranium is chemically toxic to kidneys and the insoluble uranium compounds are carcinogenic. However, there are conflicting results reported on the phytotoxicity of uranium; Gulati et al. (1980) reported that dry-weight yields of tomato plants were decreased by the presence of uranium higher than 1.0 mg kg^{-1} in soil. In contrast, other studies showed that 100-, even 1,000-, fold higher concentrations did not show any toxic effect. For example, *Brassica rapa* indicated normal growth, even produced seeds, in soil containing 10,000 mg U kg^{-1} (Sheppard et al., 1992). In order to confirm the critical uranium concentration that produces phytotoxicity, Sheppard et al. (1992) investigated the growth response of 5 plant species in 11 different soils, but no significant effects were recorded for concentrations below 300 mg U kg^{-1}.

2. Remedial Technologies for Uranium-Contaminated Environments

To reduce the detrimental effects of uranium on ecosystems and local communities, various strategies have been proposed for remediation of contaminated environments. These strategies consist of physical, chemical and biological technologies (Abdelouas et al., 1999; Suzuki and Banfield, 1999; Dushenkov, 2003). Obviously, there are both advantages and disadvantages for any remediation technology, and each may be applicable in certain circumstances only.

2.1. Physical and chemical technologies

2.1.1. Traditional methods

Traditionally, physical methods are the most widely practised methods for waste management in the mining industry. When tailings are generated in the uranium production process, terrestrial deposition is the preferred method of disposal utilising geomorphological depressions or valleys.

The principle of tailings dams (or ponds) is to dispose of the tailings in an accessible condition that provides for their future reprocessing (once improved technology or a significant increase in price make it profitable). Other disposal methods such as underground backfilling or deep-water disposal (lakes and sea) are also considered. Obviously, such kinds of management are, strictly speaking, not remediation but are more likely to store the wastes for future processing or remediation.

Because of the presence of large areas of uranium-contaminated soils around the world, engineering-based remediation methods, such as excavation, require millions of tons of soils which have to be disposed of as low-level radioactive waste. This process is expensive, fills up scarce landfill space and requires additional site restoration. Therefore, remediation of uranium-contaminated soils imposes significant extra expenses on industries and governmental agencies. The development of a cost-effective method to remove uranium from contaminated soils could accelerate the clean-up process and reduce remediation costs.

2.1.2. Novel chemo-physical technologies

Groundwater contamination with uranium occurs as a result of uranium mining and processing, which turned out to be one of the most difficult and expensive environmental problems at many locations worldwide. Removal of uranium from groundwater can be performed using active methods (pump-and-treat) or passive *in situ* methods (permeable reactive barriers (PRB)) (Simon et al., 2002). The processes applied for removal are the same in both technologies and imply precipitation, ion exchange, sorption and chemical reduction.

The most common technology used to remediate groundwater is the 'pump-and-treat' technology (pump the water and treat it at the surface). PRB have been developed for degradation or immobilisation of various pollutants, including halogenated hydrocarbons, chromium, nitrite and radionuclides. A permeable reactive wall is constructed from appropriate treatment media mixed with sand (to improve permeability) and installed down-gradient of a pollutant source. The pollutant has to be kept immobile in the wall. The most commonly used reactive material is granular zero-valent iron (ZVI). ZVI walls are assumed to be active for several decades, although the long-term reactivity of ZVI materials is currently under investigation. Even though a considerable amount of work has become available in the field of ZVI application to groundwater remediation, fundamental questions regarding the reaction mechanism remain. For example, field data did not confirm quantitative uranium(VI) reduction in ZVI reactive walls.

From a study of the effects of pyrite (FeS_2) and manganese nodules (MnO_2) on the uranium-removal potential of a selected ZVI material, a test methodology (FeS_2–MnO_2 method) is suggested to follow the pathway of contaminant removal by ZVI materials (Noubactep et al., 2005).

Simon et al. (2003) investigated the long-term performance of hydroxyapatite (HAP) as a reactive material for the removal of uranium in passive groundwater remediation systems. The stoichiometric ratio between uranium and HAP was found to be 1:(487 ± 19). Uranium removal by HAP is of pseudo first-order kinetics and the rate constant was measured to be $(1.1 \pm 0.1) \times 10^{-3}\,s^{-1}$. HAP can absorb more than $2{,}900\,mg\,kg^{-1}$ uranium. Another study by Simon et al. (2004) showed that the stoichiometric ratio between uranium and iron was 1:(1390 ± 62). The reaction between iron and uranium is of pseudo first-order kinetics and the rate constant was measured to be $(1.1 \pm 0.09) \times 10^{-3}\,s^{-1}$.

Three materials that are designed to treat uranium-contaminated water were investigated by Barton et al. (2004). These are a cation exchange resin, IRN 77; an anion exchange resin, Varion AP and a recently developed material called PANSIL (quartz sand coated with 2% amidoxime resin by weight). All three materials react rapidly in the pH range 5–7, reaching equilibrium in less than 4 h at 23°C. The anion exchange resin is very effective at removing anionic uranyl carbonate species from solutions with a pH above 5, with high specificity. Up to $50\,g\,kg^{-1}$ of uranium is removed from contaminated groundwater at neutral pH. PANSIL is effective at sequestering cationic and neutral uranyl species from solutions in the pH range 4.5–7.5, with very good specificity. In neutral groundwater containing carbonate, both the anion exchange resin and PANSIL exhibit conditional distribution coefficients exceeding $1{,}470\,ml\,g^{-1}$, which is about an order of magnitude higher than the comparable PRB materials reported in the literature.

Addition of an amendment or reagent to soil/sediment is a technique that can decrease mobility and reduce bioavailability of uranium and other heavy metals in the contaminated site. According to data from the literature and results obtained in field studies, the general mineral class of apatites was selected as the most promising amendment for *in situ* immobilisation/ remediation of uranium. Raicevic et al. (2006) presented a theoretical assessment of the stability of uranium(VI) in four apatite systems (HAP, North Carolina Apatite (NCA), Lisina Apatite (LA) and Apatite II) in order to determine an optimal apatite soil amendment. The results of this analysis indicate that all analysed apatite samples represent, from the point of view of stability, promising materials which could potentially be used in field remediation of uranium-contaminated sites.

3. Bioremediation of Uranium-Contaminated Environments

Bioremediation refers to the use of plants or microorganisms to remove or immobilise contaminants in soils or water bodies and to restore the ecological function of contaminated environments. Among the available remediation strategies for heavy metal and radionuclide-contaminated environments, bioremediation is one of the most promising measures in terms of both economy and efficiency (Zhu and Shaw, 2000; Dushenkov, 2003). Obviously, waste backfill or soil leaching is not only expensive and needs special instruments and trained staff, but also is not the ultimate solution to the contamination problem. Considering the costs, the traditional engineering for cleaning up 1 t of contaminated soil will cost 50–500 USD. If special technology is necessary, the cost will increase to 1,000 USD. On such a basis, remediation of 1 acre of soil (3 ft in depth and total weight ca. 4,500 t) will cost at least 250,000 USD (Cunningham et al., 1995). In contrast, the cost of phytoremediation of 1 acre of soil (50 cm in depth) ranges from 60,000 to 100,000 USD (Salt et al., 1995). Furthermore, bioremediation causes least disturbance to the ecosystem; it is generally well perceived by the public and can clean up the air and water in addition to remediation of the soil. During the course of phytoremediation of contaminated soil, the soil fertility could also be improved. Phytostabilisation can stabilise the soil surface in the long term with low cost for further maintenance. In general, bioremediation is an environment-friendly, economic and efficient technology. Thus, with the increasing global concerns of environmental sustainability, bioremediation or phytoremediation of any contaminated environments will certainly lead to an increase of research activities in this field.

3.1. Use of microbes

Microorganisms affect the solubility, bioavailability and mobility of uranium in radioactive wastes. Under appropriate conditions, uranium is solubilised or stabilised by the direct enzymatic or indirect non-enzymatic actions of microorganisms. Therefore, there are potential roles for microbes in remediation of uranium-contaminated environments.

3.1.1. Use of microbes for remediation of Uranium-contaminated soil

Laboratory experiments carried out on remediation of soil contaminated by heavy metals and radionuclides revealed that an efficient remediation of the soils can be achieved by an *in situ* treatment method based on the activity of the indigenous soil microflora (Groudev et al., 2001). The treatment was connected with the dissolution of the contaminants in the upper soil horizons and their transfer into the deeper soil horizons mainly to the horizon B2 where they were immobilised as different insoluble compounds.

The dissolution of contaminants was connected with the activity of both heterotrophic and chemolithotrophic aerobic microorganisms, and the immobilisation was mainly due to anaerobic sulphate-reducing bacteria (SRB).

The activity of these microorganisms was enhanced by suitable changes in the levels of some essential environmental factors such as water, oxygen and nutrient contents in the soil. On the basis of the above-mentioned laboratory results, the method was then applied under real field conditions in a heavily contaminated experimental plot of land located in the Vromos Bay area. Within eight months of treatment, the amount of radioactive elements and toxic heavy metals in the soil fell below the relevant permissible levels (Groudev et al., 2001).

3.1.2. Use of microbes for remediation of uranium-contaminated groundwater

Microbial reduction of soluble uranyl, uranium(VI), to insoluble uraninite by SRB is a promising remediation strategy for uranium-contaminated groundwaters. Certain strains of bacteria can combine the oxidation of an organic compound to the reduction of uranium(VI) to uranium(IV), which precipitates as uraninite. Abdelouas et al. (1999) tested the reduction of uranium(VI) in groundwaters of various origins and compositions. In all groundwater samples, uranium(VI) was reduced by SRB that had been activated by ethanol and trimetaphosphate. The reduction rate of uranium(VI) was dependent on the sulphate concentration in the water and the abundance of bacteria in the system. This study shows that bacteria capable of uranium(VI) reduction are ubiquitous in nature and suggests the potential for large-scale application of the enzymatic reduction of uranium(VI) for *in situ* clean-up of groundwater contaminated with uranium.

A *Pseudomonas* strain, characterised as part of a project to develop a biosorbent for removal of toxic radionuclides from nuclear-waste streams, is a potent accumulator of uranium(VI) and Th(IV), with the metal sequestration process being unaffected by culture age, presence of carbon/energy source and metabolic inhibitor but sensitive to the composition of the growth medium (Sar et al., 2004). Further characterisation of radionuclide biosorption using lyophilised biomass revealed rapid cation binding of >90% within 1–10 min of contact, and complete removal of uranium and Th was observed at initial concentrations up to 100 mg L^{-1}. Initial solution pH strongly affected radionuclide biosorption, with an optimum at pH 4.0–5.0. High-affinity, efficient and high-capacity uranium and Th binding was indicated, with a maximum loading of 541 mg U g^{-1} dry mass or 430 mg Th g^{-1} dry mass. More than 90% of biomass-bound radionuclide was recovered using sodium or calcium carbonate. For continuous process applications, an immobilised biomass system was developed and tested with multiple cycles of sorption–desorption. Overall, the biosorbent appeared suitable for realistic bioremediation.

Kalin et al. (2004) describe a three-step process for the removal of uranium from dilute wastewater. Step 1 involves the sequestration of uranium on, in, and around aquatic plants such as algae. Cell-wall ligands efficiently remove uranium(VI) from wastewater. Growing algae continuously renew the cellular surface area. Step 2 is the removal of uranium-algal particulates from the water column. Step 3 involves reducing uranium(VI) to uranium(IV) and transforming the ions into stable precipitates in the sediments. The algal cells provide organic carbon and other nutrients to heterotrophic microbial consortia to maintain the low Eh within which the uranium is transformed. Among the microorganisms, algae are of predominant interest for the ecological engineer because of their ability to sequester uranium and because some algae can live under extreme ecological conditions, often in abundance. Algae grow in a wide spectrum of water qualities, from alkaline environments (Chara, Nitella) to acidic mine drainage wastewaters (Mougeotia, Ulothrix). If they could be induced to grow in wastewater, they would provide a simple, long-term means to remove uranium and other radionuclides from uranium mining effluents.

A conventional technique such as 'pump-and-treat' may not be adequate for uranium removal because pumping the water may change the uranium speciation followed by sorption of uranium on the host rock (Abdelouas et al., 1998b). With *in situ* bioremediation, both soluble and sorbed uranium(VI) can be reduced and immobilised by bacteria. In natural aquifers, mixed cultures of nitrate-, metal- and sulfate-reducing bacteria are likely to be present (Hodgkinson, 1987; Ghiorse, 1997; Nealson and Stahl, 1997; Bachofen et al., 1998). However, to date, *in situ* biological remediation of uranium has not been clearly demonstrated as a success in the field.

Several laboratory studies have been devoted to the enzymatic reduction of uranium under a variety of conditions relevant to *ex situ* treatments of waste streams from radionuclide processing facilities (e.g. Macaskie, 1991; Ganesh et al., 1997). These studies used pure strains of bacteria, such as desulfovibrio species, to elucidate the impact of inorganic (e.g. nitrate, sulphate, bicarbonate) and organic (e.g. acetate, malonate, oxalate, citrate) ions on uranium removal from wastewaters. Only a few studies focused on uranium reduction with mixed cultures of bacteria in groundwater (Barton et al., 1996; Ganesh et al., 1997; Abdelouas et al., 1998a). Thus, in the case of *in situ* bioremediation, the presence of a mixed culture of bacteria is a prerequisite for uranium reduction.

Yi et al. (2007) investigated the effects of environmental factors, including pH and coexisting ions, on uranium(VI)-bioreduction processes (UBP) in anaerobic batch experiments. Kinetic investigations with varied pH demonstrated that uranium(VI) was reduced mostly within 48 h. The bioprecipitation yields depended strongly on pH, increasing from 12.9% to 99.4% at pH 2.0 and 6.0, respectively. A sulphate concentration of 4,000 mg L^{-1} did not affect UBP; however, a sulphate concentration of 5,000 mg L^{-1} significantly slowed UBP. At 20 mg L^{-1} Zn or 10 mg L^{-1} Cu, no UBP inhibition was observed and uranium was detected in metal sulphide precipitate. However, 25 mg L^{-1} Zn or 15 mg l^{L1} Cu stopped UBP completely. No uraninite could be detected before nitrate removal, suggesting nitrate strongly inhibited UBP, which may possibly be related to denitrification intermediates controlling the solution redox potential. However, as Finneran et al. (2002) suggested, anaerobic oxidation of uranium(IV) to uranium(VI) with nitrate serving as the electron acceptor may provide a novel strategy for solubilising and extracting microbial uranium(IV) precipitates from the subsurface.

3.2. Phytoremediation

Strictly speaking, phytoremediation is one of the kinds of bioremediation that utilises plants to remediate contaminated environments compared to bioremediation strategies using only microorganisms (Chaney et al., 1997). Based on the remediation principles, phytoremediation has been further classified into phytoextraction, phytostabilisation, rhizofiltration, phytovolatilisation, phytodegredation, etc. As for remediation of uranium-contaminated environments, the former three sub-categaries are most practicable.

3.2.1. Phytoextraction

Uranium phytoextraction, the use of plants to extract uranium from contaminated soils, is an emerging technology. The use of phytoextraction to remove heavy metals and radionuclides from contaminated soils provides a low-cost alternative to currently employed remediation procedures. First, plant cultivation and harvesting are relatively inexpensive processes as compared to traditional engineering approaches that involve intensive soil manipulation. Second, in phytoextraction, a minimum amount of secondary waste is generated as compared to heap leaching or soil washing that produces large amounts of heavy-metal-laden leachate. Furthermore, phytoextraction allows *in situ* treatment and greatly decreases the burden on existing hazardous or radioactive waste landfills. Multiple crops of metal-accumulating plants can be grown, harvested and dried during a single growing season. The dried plant materials may then be combusted or composted to reduce mass, and the residue may be vitrified or otherwise stabilised and deposited in a radioactive waste disposal facility, or even used for energy production. Commercial application of phytoextraction provides the opportunity to develop a new agriculture-based industry and a new application of crop plants that markedly reduces the cost of treating and reclaiming heavy-metal-contaminated soils at many of the world's hazardous waste sites.

Huang et al. (1998b) reported on the development of phytoextraction for the clean-up of uranium-contaminated soils. They investigated the effects of various soil amendments on uranium desorption from soil to soil solution, studied the physiological characteristics of uranium uptake and accumulation in plants, and developed techniques to trigger uranium hyperaccumulation in plants. A key to the success of uranium phytoextraction is to increase soil uranium availability to plants. They found that some organic acids can be added to soils to increase uranium desorption from soil to soil solution and to trigger a rapid uranium accumulation in plants. Of the organic acids (acetic acid, citric acid and malic acid) tested, citric acid was the most effective in enhancing uranium accumulation in plants. Shoot uranium concentrations of *B. juncea* and *B. chinensis* grown in a uranium-contaminated soil (total soil uranium, 750 mg kg^{-1}) increased from less than 5 mg kg^{-1} to more than 5,000 mg kg^{-1} in citric-acid-treated soils. Using this technique, uranium accumulation in shoots of selected plant species grown in two uranium-contaminated soil samples (total soil uranium of 280 and 750 mg kg^{-1}, respectively) was increased by more than 1,000-fold within a few days. The results suggested that uranium phytoextraction may provide an environment-friendly alternative for the clean-up of uranium-contaminated soils.

The experimental results of Huang et al. (1998b) suggested that the residual concentration of uranium-citrate complexes left in the treated soils can be reduced rapidly if the soil water pH is held between 8 and 9 after the extraction processes. The environmental risks of citrate-induced potential migration of residual uranium after the extraction process could thus be controlled.

3.2.2. Rhizafiltration

Despite the significant differences in uranium uptake and accumulation by different plants, uranium transfer from roots to shoots is very limited for most plant species (Campbell and Rechel, 1979; Zafrir et al., 1992; Saric et al., 1995; Shahandeh and Hossner, 2002). Therefore, extensive accumulation of uranium by roots could possibly be used for the clean-up of uranium-contaminated water bodies.

The elimination of natural uranium and ^{226}Ra from contaminated water by rhizofiltration was tested using *Helianthus annuus* L. (sunflower) seedlings growing in a hydroponic medium (Tomé et al., 2008). In every trial, a precipitate appeared which contained a major fraction of the natural uranium and ^{226}Ra. The results indicated that the seedlings themselves induced the formation of this precipitate. When four-week-old seedlings were exposed to contaminated water, a period of only two days was sufficient to remove the natural uranium and ^{226}Ra from the solution; about 50% of the natural uranium and 70% of the ^{226}Ra were fixed in the roots, and essentially the rest was found in the precipitate, with only very small percentages fixed in the shoots and left in the solution.

3.2.3. Phytostabilisation

Phytoextraction of uranium was induced by amending soil with organic acids (Huang et al., 1998a), but ultimate remediation or cleaning up of contaminants is in most cases impractical for economic and periodic reasons, especially on heavily contaminated sites or mine tailings. Phytostabilisation is an alternative strategy, which could prevent uranium dispersion in environments. The feasibility of uranium phytostabilisation is supported by the fact that only a few hyperaccumulating plant species have been identified (Whiting et al., 2004), whereas many plants can accumulate uranium in their roots despite large variation in uranium-uptake capability (Sheppard and Thibault, 1984; Saric et al., 1995).

Furthermore, the physiochemical techniques used for soil remediation may also remove other soil components that are essential for plant growth, such as microbes and nutrients (Khan et al., 2000).

Plant species differ in uranium accumulation, which predominantly occurs in the roots (Sheppard and Thibault, 1984; Saric et al., 1995), and it is therefore relatively simple to find plants suitable for phytostabilisation purposes. However, similar to other metal tailings, the uranium tailings are usually hostile environments for plants due to the presence of many growth-limiting factors, particularly high levels of residual metals, nutrient deficiencies and poor substrate structures (Tordoff et al., 2000). While tolerant plants should be selected from dominant species colonising the tailings, effective measures should also be taken to support plant survival in adverse environments. One commonly accepted strategy for the establishment of vegetation is to cover the mine tailings with topsoil from an unmined site (Wong, 2003). However, the establishment of plant cover is only part of the reclamation process; creation of a self-sustaining ecosystem after re-vegetation practices being of equal importance (Pichtel et al., 1994). The symbiotic microorganisms, especially for mycorrhizal fungi, may serve as multifunctional partners to assist plant survival on the contaminated sites, and play an important role in phytostabilisation of uranium.

3.2.4. Role of fungi in phytostabilisation of uranium-contaminated soils

It is known that arbuscular mycorrhizae (AM) are ubiquitous symbiotic associations between higher plants and soil fungi (Smith and Read, 1997) and their extraradical mycelium forms bridges between plant roots and soil and mediates the transfer of various elements into plants. The role of AM fungi in plant uptake of heavy metals has been extensively studied (Leyval et al., 1997). It is generally agreed that AM fungi often protect plants against high concentrations of non-essential metals in their shoots by enhancing metal retention in the roots (Leyval and Joner, 2001). Consequently, AM could possibly help in re-vegetation or phytostabilisation of metal-contaminated sites (Leyval et al., 2002). The symbiotic associations have been shown to stimulate re-vegetation of mine spoil sites by supplementing the nutrient-absorption capacity of the plant root systems, resulting in increased seedling survival and growth (Perry and Amaranthus, 1990). Huang et al. (1998a) and Shahandeh et al. (2001) therefore proposed that long-term rehabilitation of uranium-contaminated sites could be performed using phytoremediation techniques and that associated micro-biota should be considered. Among these microorganisms, mycorrhizal fungi are of particular interest (Declerck et al., 2002).

It was recently shown that the extraradical AM fungal mycelium took up and translocated uranium towards root in the *in vitro* culture system (Rufyikiri et al., 2002), and hyphae were more efficient in uranium translocation compared with roots (Rufyikiri et al., 2004). Soluble uranyl cations or uranyl-sulphate species that are stable under acidic conditions were translocated to a higher extent to roots through fungal tissues, while phosphate and hydroxyl species dominating under acidic to near-neutral conditions or carbonate species dominating under alkaline conditions were rather immobilised by hyphal structures (Rufyikiri et al., 2002).

Chen et al. (2005a, 2005b, 2005c) observed an increased uranium immobilisation in roots, a lower uranium accumulation in shoots and, consequently, a lower translocation of uranium from root to shoot when plants were colonised by AM fungi. Chen et al. (2005b) also noticed, using a compartmented pot system, that more uranium was partitioned to the shoots of *M. truncatula* when roots and hyphae could take up uranium than when only hyphae could develop in the contaminated soil compartment. This tended to indicate that uptake through the root pathway could be more easily translocated to the plant shoots than when uranium was taken up by the fungal hyphae. This further showed that uranium taken up by AM fungi is, at least partly, immobilised within the intraradical structures of the fungal symbionts. The mycorrhiza-mediated uranium uptake was most likely immobilised in root-internal fungal structures as suggested by the observed accumulation of uranium in fungal vesicles of mycorrhizal *Cynodon dactylon* roots (Weiersbye et al., 1999).

Although these results are encouraging to develop phytostabilisation strategies in which AM fungi could be used, further research would nonetheless be needed. Firstly, site trials are necessary to confirm the potential of AM fungi in uranium phytostabilisation strategies. Secondly, it would be most interesting to study AM fungal strains from uranium-polluted sites. Indeed, they might have adapted to the toxicity of uranium and of other contaminants often associated with uranium, and could thus be more resistant. This potential higher resistance would be highly valuable as it could allow better protection of AM plant hosts and eventually increase the stabilisation of uranium and other pollutants (Dupré de et al., 2008).

4. Conclusion

The mining and milling of uranium minerals have resulted in large quantities of waste materials which contain radionuclides in quantities causing environmental health concerns. Improper storage of uranium wastes can result in uranium dispersion into the environment. In addition to aesthetic impacts on the local environment and on public perception, the contaminants may also pose health risks to both humans and animals due to chemical toxicity and radiological effects. Therefore, there is a strong demand for proper management of the uranium-contaminated environments.

Currently available technologies include both chemico-physical and biological methods. Chemico-physical technologies, such as PRB techniques, provide the possibility of efficient clean-up of the contaminated groundwater. There is still a need to find novel materials favourable in different circumstances. Microorganisms could also be used to remediate contaminated water by means of bioreduction of uranium(VI) to uranium(IV). Similarly, more efficient microbes should be isolated for wider application of this technique. For remediation of uranium-contaminated soils, or ecological restoration of uranium mine tailings, phytoremediation turns out to be one of the best choices, although further research is still necessary to search for uranium hyperaccumulators for more efficient phytoextraction, and tolerant plant species for phytostabilisation.

Because most of the above technologies are still under laboratory research, there is a need to demonstrate their efficiency under field and large-scale conditions.

REFERENCES

Abdelouas, A., Y. Lu, W. Lutze, and H. E. Nuttall. (1998a). Reduction of U(IV) to U(IV) by indigenous bacteria in contaminated ground water. *Journal of Contaminant Hydrology*, 35, 217–233.

Abdelouas, A., W. Lutze, and H. E. Nuttall. (1998b). Chemical reactions of uranium in ground water at a mill tailings site. *Journal of Contaminant Hydrology*, 34, 343–361.

Abdelouas, A., W. Lutze, and H. E. Nuttall. (1999). Uranium contamination in the subsurface: Characterization and remediation. In: *Uranium: Mineralogy, Geochemistry and the Environment* (Eds P. C. Burns and R. Finch). *Reviews in Mineralogy*, Vol. 33. Mineralogy Society of America, Washington DC, pp. 433–473.

Allard, B., U. Olofsson, and B. Torstenfelt. (1984). Environmental actinide chemistry. *Inorganica Chimica Acta*, 94, 205–221.

Anderson, A., and G. Siman. (1991). Levels of Cd and some other trace elements in soils and crops as influenced by lime and fertiliser level. *Acta Agriculturae Scandinavica*, 41, 3–11.

Anderson, R. T., H. A. Vrionis, I. Ortiz-Bernad, C. T. Resch, P. E. Long, R. Dayvault, K. Karp, S. Marutzky, D. R. Metzler, A. Peacock, D. C. White, M. Lowe, and D. R. Lovley. (2003). Stimulating the *in situ* activity of Geobacter species to remove uranium from the groundwater of a uranium-contaminated aquifer. *Applied and Environmental Microbiology*, 69, 5884–5891.

Bachofen, R., P. Ferloni, and I. Flynn. (1998). Microorganisms in the subsurface. *Microbiological Research*, 153, 1–22.

Barton, C. S., D. I. Stewart, K. Morris, and D. E. Bryant. (2004). Performance of three resin-based materials for treating uranium-contaminated groundwater within a PRB. *Journal of Hazardous Materials*, 116, 191–204.

Barton, L. L., K. Choudhury, B. M. Thomson, K. Steenhoudt, and A. R. Groffman. (1996). Bacterial reduction of soluble uranium: The first step of *in situ* immobilization of uranium. *Radioactive Waste Management and Environmental Restoration*, 20, 141–151.

Campbell, M. D., and K. T. Biddle. (1977). Frontier areas and exploration techniques. Frontier uranium exploration in the south-central United States. In: *Geology of Alternate Energy Resources* (Ed. M. D. Campbell). Houston Geological Society, Houston, TX, pp. 3–44.

Campbell, W. F., and E. A. Rechel. (1979). Tradescantia a 'super-snooper' of radioactivity from uranium mill wastes. *Utah Science*, 3, 101–103.

Chaney, R. L., M. Malik, Y. M. Li, S. L. Brown, E. P. Brewer, J. S. Angle, and A. J. M. Baker. (1997). Phytoremediation of soil metals. *Current Opinion in Biotechnology*, 8, 279–284.

Chen, B. D., I. Jakobsen, P. Roos, O. K. Borggaard, and Y. G. Zhu. (2005a). Mycorrhiza and root hairs enhance acquisition of phosphorus and uranium from phosphate rock but mycorrhiza decreases root to shoot uranium transfer. *New Phytologist*, 165, 591–598.

Chen, B. D., I. Jakobsen, P. Roos, and Y. G. Zhu. (2005b). Effects of the mycorrhizal fungus *Glomus intraradices* on uranium uptake and accumulation by *Medicago truncatula* L. from uranium-contaminated soil. *Plant and Soil*, 275, 349–359.

Chen, B. D., Y. G. Zhu, X. H. Zhang, and I. Jakobsen. (2005c). The influence of mycorrhiza on uranium and phosphorus uptake by barley plants from a field-contaminated soil. *Environmental Science and Pollution Research*, 12, 325–331.

Ciavatta, L., D. Ferri, I. Grenthe, and F. Salvatore. (1981). The first acidification step of the tris(carbonato) dioxoourantantate(VI) ion, $UO_2(CO_3)_3^{4-}$. *European Journal of Inorganic Chemistry*, 20, 463–467.

Cowart, J. B., and W. C. Burnett. (1994). The distribution of uranium and thorium decay-series radionuclides in the environment – A review. *Journal of Environmental Quality*, 23, 651–662.

Cunningham, S. D., W. R. Berti, and J. W. Huang. (1995). Phytoremediation of contaminated soils. *Trends in Biotechnology*, 13, 393–397.

Declerck, S., C. Leyval, I. Jakobsen, Y. Thiry, C. Heine, and B. Delvaux. (2002). The European project MYRRH: Use of mycorrhizal fungi for the phytostabilisation of radio-contaminated environments. *Radioprotection*, 37, 337–339.

Duff, M. C., and C. Amrhein. (1996). Uranium(VI) adsorption on goethite and soil in soil carbonate solutions. *Soil Science Society of America Journal*, 60, 1393–1400.

Dupré de, B. H., E. J. Joner, C. Leyval, I. Jakobsen, B. D. Chen, P. Roos, I. Thiry, G. Rufyikiri, B. Delvaux, and S. Declerck. (2008). Impact of arbuscular mycorrhizal fungi on uranium accumulation by plants. *Journal of Environmental Radioactivity*, 99, 775–784.

Dushenkov, S. (2003). Trends in phytoremediation of radionuclides. *Plant and Soil*, 249, 167–175.

Ebbs, S. D., D. J. Brady, and L. V. Kochian. (1998). Role of uranium speciation in the uptake and translocation of uranium by plants. *Journal of Experimental Botany*, 49, 1183–1190.

Finneran, K. T., M. E. Housewright, and D. R. Lovley. (2002). Multiple influence of nitrate on uranium solubility during bioremediation of uranium-contaminated subsurface sediments. *Environmental Microbiology*, 4, 510–516.

Ganesh, R., K. G. Robinson, G. Reed, and G. S. Sayler. (1997). Reduction of hexavalent uranium from organic complexes by sulfate- and iron-reducing bacteria. *Applied and Environmental Microbiology*, 3, 4385–4391.

Ghiorse, W. C. (1997). Subterranean life. *Science*, 275, 789–790.

Gongalsky, K. B. (2003). Impact of pollution caused by uranium production on soil macrofauna. *Environmental Monitoring and Assessment*, 89, 197–219.

Grenthe, I., J. Fuger, R. J. M. Konings, R. J. Lemire, A. B. Miller, C. Nguyen-Trung, and H. Wanner. (1992). *Chemical Thermodynamics of Uranium*. Elsevier, Amsterdam.

Groudev, S. N., P. S. Georgiev, I. I. Spasova, and K. Komnitsas. (2001). Bioremediation of a soil contaminated with radioactive elements. *Hydrometallurgy*, 59, 311–318.

Gulati, K. L., M. C. Oswal, and K. K. Nagpaul. (1980). Assimilation of uranium by wheat and tomato plants. *Plant and Soil*, 55, 55–59.

Hodgkinson, D. P. (1987). The NIREX safety assessment research program on near-field effects in cementitious repositories. *Radioactive Waste Management and the Nuclear Fuel Cycle*, 9, 272–291.

Huang, F. Y. C., P. V. Brady, E. R. Lindgren, and P. Guerra. (1998a). Biodegradation of uranium–citrate complexes: Implications for extraction of uranium from soils. *Environmental Science and Technology*, 32, 379–382.

Huang, J. W., M. J. Blaylock, Y. Kapulnik, and B. D. Ensley. (1998b). Phytoremediation of uranium-contaminated soils: role of organic acids in triggering uranium hyperaccumulation in plants. *Environmental Science and Technology*, 32, 2004–2008.

Junghans, M., and C. Helling. (1998). Historical mining, uranium tailings and waste disposal at one site-can it be managed? A hydrogeological analysis. *Proceedings of the 5th International Conference on Tailings and Mine Waste '98*, Fort Collins, Colorado, January 26–29. Balkema, Rotterdam, pp. 117–126.

Kalin, M., W. N. Wheeler, and G. Meinrath. (2004). The removal of uranium from mining wastewater using algal/microbial biomass. *Journal of Environmental Radioactivity*, 78, 151–177.

Khan, A. G., C. Kuek, T. M. Chaudhry, C. S. Khoo, and W. J. Hayes. (2000). Role of plants, mycorrhizae and phytochelators in heavy metal contaminated land remediation. *Chemosphere*, 41, 197–207.

Langmuir, D. (1978). Uranium solution-mineral equilibria at low temperatures with applications to sedimentary ore deposits. *Geochimica et Cosmochimica Acta*, 42, 547–569.

Langmuir, D. (1997). *Aqueous Environmental Geochemistry*. Prentice Hall, NJ.

Leyval, C., and E. J. Joner. (2001). Bioavailability of heavy metals in the mycorrhizosphere. In: *Trace Elements in the Rhizosphere* (Eds G. R. Gobran, W. W. Wenzel, and E. Lombi). CRC Press LLC, London, pp. 165–185.

Leyval, C., E. J. Joner, C. del Val, and K. Haselwandter. (2002). Potential of arbuscular mycorrhizal fungi for bioremediation. In: *Mycorrhizal Technology in Agriculture* (Eds S. Gianinazzi, H. Schüepp, J. M. Barea, and K. Haselwandter). Birkhäuser Verlag, Basel, pp. 175–186.

Leyval, C., K. Turnau, and K. Haselwandter. (1997). Effect of heavy metal pollution on mycorrhizal colonization and function: Physiological, ecological and applied aspects. *Mycorrhiza*, 7, 139–153.

Liator, M. I. (1995). Uranium isotopes distribution in soils at the Rocky Flats Plant, Colorado. *Journal of Environmental Quality*, 24, 314–323.

Macaskie, L. E. (1991). The application of biotechnology to the treatment of wastes produced from the nuclear fuel cycle: Biodegradation and bioaccumulation as a means of treating radionuclide-contaminated streams. *Critical Reviews in Biotechnology*, 11, 41–112.

Meinrath, G., Y. Kato, T. Kimura, and Z. Yoshida. (1996). Solid-aqueous phase equilibria of uranium(VI) under ambient conditions. *Radiochimica Acta*, 75, 159–167.

Mortvedt, J. J. (1994). Plant and soil relationships of uranium and thorium decay series radionuclides – A review. *Journal of Environmental Quality*, 23, 643–650.

Nealson, K.H., and D. A. Stahl. (1997). Microorganisms and biogeochemical cycles: what can we learn from layered microbial communities? In: *Geomicrobiology: Interactions Between Microbes and Minerals* (Eds J. F. Banfield and K. H. Nealson). *Reviews in Mineralogy*, Vol. 35. Mineralogy Society of America, Washington DC, pp. 5–34.

Noubactep, C., G. Meinrath, and B. J. Merkel. (2005). Investigating the mechanism of uranium removal by zerovalent iron. *Environmental Chemistry*, 2, 235–242.

Perry, D. A., and M. P. Amaranthus. (1990). The plant-soil bootstrap: Microorganisms and reclamation of degraded ecosystems. In: *Environmental Restoration: Science and Strategies for Restoring the Earth* (Ed. J. J. Berger). Island Press, Washington DC, pp. 94–102.

Pichtel, J. R., W. A. Dick, and P. Sutton. (1994). Comparison of amendments and management practices for long-term reclamation of abandoned mine lands. *Journal of Environmental Quality*, 23, 766–772.

Plant, J. A., S. Reeder, R. Salminen, D. B. Smith, T. Tarvainen, B. De Vivo, and M. G. Peterson. (2003). The distribution of uranium over Europe: Geological and environmental significance. *Applied Earth Science, IMM Transactions Section B*, 112(3), B221–B238.

Qureshi, A. A., N. U. Khattak, M. Sardar, M. Tufail, M. Akram, T. Iqbal, and H. A. Khan. (2001). Determination of uranium contents in rock samples from Kakul phosphate deposit, Abbotabad (Pakistan), using fission-track technique. *Radiation Measurements*, 34, 355–359.

Raicevic, S., J. V. Wright, V. Veljkovic, and J. L. Conca. (2006). Theoretical stability assessment of uranyl phosphates and apatites: Selection of amendments for in situ remediation of uranium. *Science of the Total Environment*, 355, 13–24.

Rufyikiri, G., S. Declerck, and Y. Thiry. (2004). Comparison of ^{233}U and ^{33}P uptake and translocation by the arbuscular mycorrhizal fungus *Glomus intraradices* in root organ culture conditions. *Mycorrhiza*, 14, 203–207.

Rufyikiri, G., Y. Thiry, L. Wang, B. Delvaux, and S. Declerck. (2002). Uranium uptake and translocation by the arbuscular mycorrhizal fungus, *Glomus intraradices*, under root-organ culture conditions. *New Phytologist*, 156, 275–281.

Salt, D. E., M. Blaylock, P. B. A. Nanda-Kumar, V. Dushenkov, B. D. Ensley, I. Chet, and I. Raskin. (1995). Phytoremediation: A novel strategy for the removal of toxic metals from the environment using plants. *Biotechnology Journal*, 13, 468–474.

Sar, P., S. K. Kazy, and S. F. D'Souza. (2004). Radionuclide remediation using a bacterial biosorbent. *International Biodeterioration and Biodegradation*, 54, 193–202.

Saric, M. R., M. Stojanovic, and M. Babic. (1995). Uranium in plant species grown on natural barren soil. *Journal of Plant Nutrition*, 18, 1509–1518.

Shahandeh, H., and L. R. Hossner. (2002). Role of soil properties in phytoaccumulation of uranium. *Water, Air, and Soil Pollution*, 141, 165–180.

Shahandeh, H., J. H. Lee, L. R. Hossner, and R. H. Loeppert. (2001). Bioavailability of uranium and plutonium to plants in soil–water systems and the potential of phytoremediation. In: *Trace Elements in the Rhizosphere* (Eds G. R. Gobran, W. W. Wenzel, and E. Lombi). CRC Press LLC, London, pp. 93–124.

Sheppard, M. I., and D. H. Thibault. (1984). Natural uranium concentrations of native plants over a low-grade ore body. *Canadian Journal of Botany*, 62, 1069–1075.

Sheppard, M. I., and D. H. Thibault. (1992). Desorption and extraction of selected heavy metals from soils. *Soil Science Society of America Journal*, 56, 415–423.

Sheppard, S. C., W. G. Evenden, and A. J. Anderson. (1992). Multiple assays of uranium in soil. *Environmental Toxicology and Water Quality*, 7, 275–294.

Simon, F.-G., V. Biermann, C. Segebade, and M. Hedrich. (2004). Behaviour of uranium in hydroxyapatite-bearing permeable reactive barriers: investigation using ^{237}U as a radioindicator. *Science of the Total Environment*, 326, 249–256.

Simon, F.-G., T. Meggyes, and C. McDonald. (2002). *Advanced Groundwater Remediation – Active and Passive Technologies*. Thomas Telford, London.

Simon, F.-G., C. Segebade, and M. Hedrich. (2003). Behaviour of uranium in iron-bearing permeable reactive barriers: investigation with ^{237}U as a radioindicator. *Science of the Total Environment*, 307, 231–238.

Smith, S. E., and D. J. Read. (1997). *Mycorrhizal Symbiosis*. 2nd edition. Academic Press, London.

Suzuki, Y., and J. F. Banfield. (1999). Geomicrobiology of uranium. *Reviews in Mineralogy Geochemistry*, 38, 393–432.

Tomé, F. V., P. B. Rodríguez, and J. C. Lozano. (2008). Elimination of natural uranium and ^{226}Ra from contaminated waters by rhizofiltration using *Helianthus annuus* L. *Science of the Total Environment*, 393, 351–357.

Tordoff, G. M., A. J. M. Baker, and A. J. Willis. (2000). Current approaches to the revegetation and reclamation of metalliferous mine wastes. *Chemosphere*, 41, 219–228.

Van Kauwenbergh, S.J. (1997). Cadmium and other minor elements in world resources of phosphate rock. *The International Fertiliser Society Proceeding 400*, The International Fertiliser Society, New York.

Weiersbye, I. M., C. J. Straker, and W. J. Przybylowicz. (1999). Micro-PIXE mapping of elemental distribution in arbuscular mycorrhizal roots of the grass, *Cynodon dactylon*, from gold and uranium mine tailings. *Nuclear Instruments and Methods in Physics Research Section B: Beam Interactions with Materials and Atoms*, 158, 335–343.

Whiting, S. N., R. D. Reeves, D. Richards, M. S. Johnson, J. A. Cooke, F. Malaisse, A. Paton, J. A. C. Smith, J. S. Angle, R. L. Chaney, R. Ginocchio, T. Jaffré, R. Johns, T. McIntyre, O. W. Purvis, D. E. Salt, H. Schat, F. J. Zhao, and A. J. M. Baker.

(2004). Research priorities for conservation of metallophyte biodiversity and their potential for restoration and site remediation. *Restoration Ecology*, 12, 106–116.

Williams, C. H., and D. J. David. (1976). The accumulation in soil of cadmium residues from phosphate fertilizers and their effect on the cadmium content of plants. *Soil Science*, 121, 86–93.

Wong, M. H. (2003). Ecological restoration of mine degraded soils, with emphasis on metal contaminated soils. *Chemosphere*, 50, 775–780.

Yi, Z. J., K. X. Tan, A. L. Tan, Z. X. Yu, and S. Q. Wang. (2007). Influence of environmental factors on reductive bioprecipitation of uranium by sulfate reducing bacteria. *International Biodeterioration and Biodegradation*, 60, 258–266.

Zafrir, H., Y. Waisel, M. Agami, J. Kronfeld, and E. Mazor. (1992). Uranium in plants of southern Sinai. *Journal of Arid Environments*, 22, 363–368.

Zhu, Y.-G., and G. Shaw. (2000). Soil contamination with radionuclides and potential remediation. *Chemosphere*, 41, 121–128.

CHAPTER 9

Effectiveness of Remedial Techniques for Environments Contaminated by Artificial Radionuclides

Kasper G. Andersson* *and* Per Roos

Contents

*Corresponding author. Tel.: +45 4677 4173; Fax: +45 4677 5330
E-mail address: kasper.andersson@risoe.dk

Risø-DTU National Laboratory for Sustainable Energy, P.O. Box 49, NUK-204, DK-4000 Roskilde, Denmark

Radioactivity in the Environment, Volume 14
ISSN 1569-4860, DOI 10.1016/S1569-4860(08)00209-X

1. Introduction

The overall objective of introducing remedial measures in a contaminated ecosystem is to improve the radiological situation for the affected populations. In achieving this, the main goal is to reduce the adverse radiological impact imposed by the contamination. However, in compliance with the latest recommendations of the International Commission on Radiological Protection (ICRP, 2000, 2007), other factors need to be taken into account in the justification and optimisation of intervention. Specifically, it is stated that 'the immediate advantage of intervening in a prolonged exposure situation is the expectation of obtaining averted (individual and collective) doses', but 'other advantages are the consequent reassurance gained by the population and the decrease in anxiety created by the situation. Disadvantages introduced by the intervention include costs, harm and social disruption associated with it. If the advantages of intervening offset the disadvantages, the net benefit of intervening will be positive and the intervention is said to be justified'. Concerning optimisation of implementation, it is stated that 'the optimum protection option is not necessarily the option that results in the lowest residual annual doses, either individual or collective dose. Some options could result in a lower residual annual dose but give a smaller net benefit than the optimum option' (ICRP, 2000).

Optimisation of remediation strategies that may consist of combinations of a variety of remedial measures is treated Chapter 10. As demonstrated in Chapter 10 on the environmental, social, ethical and economic aspects of remediation, an optimised framework for decision making should take into account a host of factors, and a remediation strategy may include measures which do not reduce dose in any way.

The focus in this chapter is on the individual remedial measures designed to reduce doses received by humans due to the contamination in urban (inhabited), agricultural, aquatic and forest ecosystems following an airborne contaminant release. It should be stressed that a contaminated area may well consist of a combination of these types of ecosystems and, thus, may require a combination of very different types of remedial measures.

2. Influence of Release Scenario on Effectiveness of Remedial Measures

Numerous different types of scenarios might be envisaged, which could potentially lead to contamination of ecosystems that could subsequently be subject to cleanup. The 'classical' example, which has attracted by far the highest attention in remediation investigations and decision support developments over the latest two decades and more, is a large-scale nuclear power plant accident (the Chernobyl accident), where not only noble gases but also radionuclides originating from the reactor core as well as irradiated materials were released into the atmosphere. The Chernobyl accident showed that, contrary to the previous belief, very large areas could become significantly contaminated by such an accident, and these could comprise living areas, as well as food production and other natural areas.

Over the years that followed this accident, a wide range of remedial measures were identified for different types of contaminated areas, which were tested on different scales under more or less controlled conditions, either in the contaminated areas of the former Soviet Union or in areas in Western Europe where Chernobyl contamination levels were sufficiently high to allow measurements to be made with adequate accuracy (e.g. Nisbet, 1993; Alexakhin, 1993; Guillite and Willrodt, 1993; Roed and Andersson, 1996; Fesenko et al., 2007). Experimental work was also conducted in smaller scale using 'tracers' ranging from actual radionuclides on different physico-chemical forms sprayed onto the surfaces to, for instance, metal cylinders, which when retrieved with a metal detector yielded valuable information on the mixing of contaminated and uncontaminated soil layers when testing ploughing procedures (Roed et al., 1996). The efforts actually implemented by the authorities in attempts to solve the contamination problems in the highly affected areas in the former Soviet Union are summarised in Chapter 4 of this book.

The vast experience gained from these investigations forms the core of the state-of-the-art knowledge of the effect, limitations, constraints and other direct or indirect implications of remedial measures for radioactively contaminated areas. However, it needs to be stressed that the weakness and limitations in incorporating this knowledge into preparedness for future emergencies are that it all links to a specific single release incident and that the vast majority of the experimental work, particularly for inhabited areas, relates to radiocaesium in more or less readily soluble form. The physicochemical characteristics of contaminants will, in general, greatly influence the effectiveness of remedial measures. Indeed some methods, such as application of ammonium iron hexacyanoferrate (AFCF) boli to reduce gut uptake to animals grazing in contaminated areas or washing of building exteriors with solutions containing ammonium ions exploiting cationic similarities, are directly targeted for a specific contaminant cation – radiocaesium. It is well known that the caesium cation, particularly due to its low tendency for hydration and high polarisability compared with that of other cations, will have the potential to be fixed strongly in highly selective sites present in soil minerals and natural sediments (Cremers et al., 1988;

De Preter, 1990; Tamura and Jacobs, 1960; Jackson, 1963). Such sites are also present in minerals occurring in most common construction materials (e.g. concrete, roof clay tiles and asbestos roof sheets) as well as in street dust (Andersson, 1991; De Preter, 1990). They are, however, not present in common clay bricks, which are typically fired at such high temperatures that the clay structure becomes amorphous and only large macropores remain (Andersson, 1991; Andersson et al., 2002). This means that, after a period of months to years, when a strong, selective fixation of the caesium cation has occurred in minerals in many types of surfaces, it becomes exceedingly difficult to remove unless an entire outer layer of the surface is removed (even prolonged cationic exchange with ammonium will, in practice, have very limited effect).

In contrast, other contaminants, which are not selectively fixed in mineral structures, have been found to be much more easily removed from surfaces – particularly if they are in the form of large and insoluble particles (Salbu, 2000). It should also be noted that, due to these fixation mechanisms, caesium cations are generally fixed in soil in the upper few centimetres of the layer and remain there for many years without remedial action.

Other potentially important contaminants like strontium and ruthenium released during the Chernobyl accident have been reported to migrate much faster to deeper soil (Salbu, 2000; Salbu et al., 1998; Andersson and Roed, 1994). Nevertheless, an exception was observed in the immediate vicinity of the wrecked Chernobyl power plant. Here, the strontium contamination occurred in the form of large fuel particles with very low solubility, which after deposition on soil remained on the soil surface. Over time, they were subjected to a slow natural weathering process, leading to a slow release of radionuclides in the following years. This increased their mobility in the environment with time and thus led to a deeper vertical distribution in the soil. In turn, this meant that, if decontamination had not been carried out at an earlier stage, thicker topsoil layers would rather need to have been removed or placed deeper into the soil to reduce external doses or plant uptake. The dissolution of deposited particles with high chemical stability depends on the environmental conditions as well as the contaminant matrix. A comprehensive study in the heavily contaminated areas near the Chernobyl nuclear power plant (Kashparov et al., 2004) demonstrated a clear correlation between the fraction of non-dissolved fuel particles and soil acidity. It has been suggested that the particle dissolution process in soil can be described by a simple first-order kinetics equation (Kashparov et al., 2004; Petryaev et al., 1991; Konoplev et al., 1992). On the basis of measurements, the dissolution half-life was estimated to be of the order of 14 years at pH 7. However, as demonstrated by the model results of Kashparov et al. (2004) (see Figure 1), which are verified by measurements made by different authors (e.g. Krouglov et al., 1998;

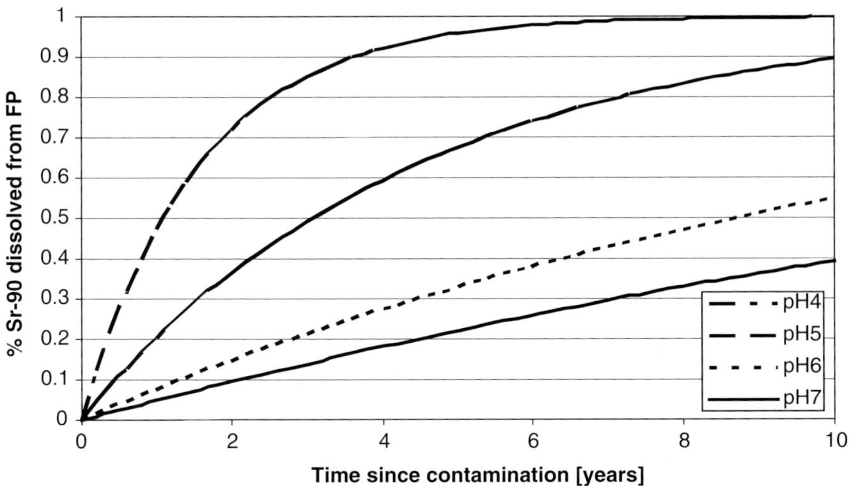

Figure 1 Time-dependence of the fraction of ^{90}Sr dissolved from fuel particles in the areas contaminated by the Chernobyl accident, according to an empirical Ukrainian model (Kashparov et al., 2004). Model curves are shown for different soil pH values.

Petryaev et al., 1991), the dissolution half-life is considerably shorter at lower pH.

Another important historical example of an accident that resulted in rather widespread atmospheric dispersion of radioactive contaminants was the Kyshtym accident in 1957, where a thermal explosion at a weapons-grade plutonium production facility led to contamination of large rural areas in the Southern Urals. However, the remediation experience gained here was largely limited to the context of agricultural production and natural areas (see Chapter 4), and the entire accident was kept secret until the late 1980s.

It should be noted that much smaller accidents can also lead to significant contamination that may require remedial measures, although on a smaller scale. One example is the accident that occurred in September 1987 in Goiânia, Brazil, where significant fractions of a 50.9 TBq ^{137}CsCl teletherapy source (in readily soluble powder form) were transported to various parts of Goiânia city (see Chapter 4). This caused contamination of a considerable area and of many people (Leao and Oberhofer, 1988). Although a multitude of transportation and migration pathways contributed to the spreading of the contamination, it has been reported that the primary cause of contamination of houses in the area was atmospheric dispersion (Da Silva et al., 1991). Remedial measures were employed to decontaminate some 50 houses, mostly very near to the location where the source had been recovered. An effort was made to decontaminate roofs with ^{137}Cs contamination levels ranging up to some 700 kBq m^{-2} with pressurised water jets (Da Silva et al., 1991), but the effect was a mere removal of about 20% of the contamination, and some of the roofs were eventually replaced. An extensive programme to decontaminate housing interiors was launched, and although good effect was often obtained using simple techniques, the affected population refused to accept any measurable residual contamination, and for instance clothes and cooking utensils were discarded if they showed even the lowest detectable level of contamination (here, in general, 0.4 Bq cm^{-2}) (Da Silva et al., 1991). The effort to monitor and decontaminate large numbers of people, who were concerned that they might have become contaminated, required a major and rapid effort in the Goiânia case. The lessons that could be drawn from the application of decontamination methods in Goiânia were insignificant, as only few techniques were applied, and it would seem from other works that implementation was far from optimal. However, importantly, the fact that the concept of acceptable levels of contamination was in this case rejected by the population demonstrated that in some cases where the anxiety is very great, as in this case reflected in symptoms like elevated blood pressure, gastrointestinal problems and cardiovascular symptoms (Steinhäusler, 2005), methods that have been proven sufficiently effective in eliminating radiation hazards may still not have the potential to solve the societal problems satisfactorily.

The Goiânia accident was also interesting in the sense that the affected area was probably similar in size to that which might become affected by the explosion of a 'dirty bomb' (a terror attack involving a conventional bomb dispersing radioactive matter, most likely in a densely populated area) (Andersson et al., 2008a). In the Goiânia case, the psychological and social impact was immense, and on public demand, practically all efforts possible were made to clean the affected areas as thoroughly as possible, in spite of the poor economic situation of the area. This would not have been possible if vast areas had been contaminated, as was the case with the Chernobyl accident. In the event of a malicious act, which is by definition aimed at provoking severe psychological stress in the public, a very high degree of anxiety and disruption can be expected, and authorities would be likely to be driven to go very far to ensure removal of even small traces of any substances that were intentionally placed in the environment to cause harm. This means that eventually effective remedial measure variants which were deemed too expensive and thus not sufficiently cost-effective might possibly become attractive means of restoring sites contaminated by a malicious dispersion event.

The idea of a radiological terror attack has raised increasing concern in recent years, particularly following the different terror attacks on the US World Trade Center, subway systems in Spain and Japan, embassies, nightclubs and hotels. Also, Al-Qaeda leader Osama Bin Laden has expressed interest in radiological and nuclear terror devices (Time, 2002).

In principle, a host of different sources might be envisaged for use in connection with a 'dirty bomb'. However, considering factors like availability (assuming that terrorists are unlikely to gain access to facilities for production of highly radioactive sources), sufficient source strength, transportability, physical half-life, types of radiation/energies and physicochemical form, the list of realistically potentially applicable sources can be reduced. Certainly, many different sources have been 'orphaned' and are not kept track of. In fact, thousands of potentially harmful sources are lost on an annual basis over the world, although probably only few of these would be strong enough to cause real harm if dispersed over an inhabited complex. According to the IAEA (Yusko, 2001), by 2001, 'orphaned' sources had been involved in 60 severe radiological accidents, due to which 266 persons had been overexposed and 39 had died. The Goiânia accident is a classic example of such an accident, and caesium sources for teletherapy are amongst those types of sources that cause most concern.

However, considerably stronger sources than that involved in the Goiânia case might also be envisaged in connection with a 'dirty bomb', depending on the ability of terrorists to transport and detonate the bomb from a shielding arrangement and their wish to survive the attack. An example of a very strong source turning up completely unexpectedly and out of context occurred in December 2001, when three residents of Tsalenjikha in Georgia suffered severe radiation sickness and skin burns after having found a 1,300 TBq ^{90}Sr source in a forest (Wedekind, 2002). This source in the Soviet days had been in use to power a radioisotope thermoelectric generator (RTG) for a communication tower in a remote natural area. Hundreds of RTGs with ^{90}Sr sources with strengths typically ranging between 1,000 and 10,000 TBq were distributed over the Soviet Union to provide power for lighthouses, beacons and other unmanned facilities, and many of these are no longer kept track of (Ferguson et al., 2003). In comparison, the total ^{90}Sr release from the Chernobyl accident has been estimated to amount to about 8,000 TBq (Sohier, 2002). Table 1 shows a selection of strong sources that might be of particular concern (Harper et al., 2007; Ferguson et al., 2003; Argonne, 2005).

It is evident from recent experimentation (Harper et al., 2007) that the aerosolisation process in connection with a 'dirty bomb' explosion is strongly dependent on the initial physicochemical characteristics of the contaminating material, as well as the geometrical construction of the device. For instance, if the source is a metal (^{60}Co and ^{192}Ir would here be among the primary radionuclides of concern), nearly all the contamination would be dispersed as large fragments, and atmospherically dispersible particles would only occur if sufficient pressure is generated to create phase transition.

The condensation particles created after evaporation of the metals would be likely to be of the same size range as the caesium condensation particles observed after the Chernobyl accident (i.e. slightly

Table 1 A selection of abundant radioactive sources. Typical physico-chemical forms of large existing sources and maximum activity estimates are also given.

Radionuclide	Typical physicochemical form	Existing sources and their strengths
^{60}Co	Metal-soluble in acid	Sterilisation irradiator (maximum 400,000 TBq). Teletherapy source (maximum 1,000 TBq).
^{90}Sr	Ceramic ($SrTiO_3$) – insoluble, brittle, soft	Radioisotope thermoelectric generator (1,000–10,000 TBq).
^{137}Cs	Salt (CsCl) – readily soluble	Sterilisation irradiator (maximum 400,000 TBq). Teletherapy source (maximum 1,000 TBq).
^{192}Ir	Metal – soft, insoluble in water	Industrial radiography source (maximum 50 TBq).
^{226}Ra	Salt ($RaSO_4$) – very low solubility	Old therapy source (maximum 5 TBq).
^{238}Pu	Ceramic (PuO_2) – insoluble	Radioisotope thermoelectric generator (maximum 5,000 TBq).
^{241}Am	Pressed ceramic powder (AmO_2) – insoluble	Well logging source (maximum 1 TBq).
^{252}Cf	Ceramic (Cf_2O_3) – insoluble	Well logging source (maximum 0.1 TBq).

submicroneous), but slightly larger liquid particles could also be formed. Such particles would be readily soluble. If the source were initially in ceramic form (e.g. a ^{90}Sr source from an old Soviet RTG), no phase transition would occur according to Harper et al. (2007), and a particle-size spectrum with much of the mass associated with large, insoluble, but still widely dispersible particles would be expected. This is illustrated by the removal of as much as 98% of the contamination, as observed by Clark and Cobbin (1964), when hosing water on a street contaminated by insoluble 44–100 μm particles. In case contamination occurs in the form of submicroneous soluble Cs particles, as demonstrated after the Chernobyl accident, hosing at the same pressure would be expected to remove only about half of the contamination (Brown et al., 2008; Andersson et al., 2003). All such particles would not only deposit very differently on different surfaces, but also behave differently in environmental media after deposition and thus influence the remediation regimes to be applied.

Another example illustrating the importance of taking into account the fact that large insoluble particles are considerably easier to remove from surfaces by both natural and forced processes, compared to small particles and contaminants in solution, was found in different inhabited areas contaminated by the Chernobyl accident. The countermeasure applied comprised rinsing with hosing water at the same pressure and over the same duration on similar sandstone walls of houses that had been contaminated in Pripyat, only about 3 km from the power plant, and in Vladimirovka, some 65 km away. In Pripyat, where much of the contamination had been in the form of large and insoluble particles, the treatment removed some two-thirds of the caesium, but as far away as Vladimirovka, where the contaminants were primarily in the form of condensation particles, only about one-fifth of the caesium could be removed (Roed and Andersson, 1996). This means that the parameters traditionally applied to describe the effectiveness of remedial measures in connection with atmospherically long-range transported small particles from nuclear power plant accidents cannot be applied for all scenarios involving for instance 'dirty bomb' contaminants. Moreover, since the deposition distribution and post-deposition contaminant behaviour on the different surfaces are highly dependent on particle characteristics, remediation priorities to achieve optimal dose reducing effect would differ widely between for instance different types of 'dirty bomb' scenarios.

It should be noted that, in connection with different types of airborne releases, contaminants with significantly different physicochemical properties, and thus different potentials for remediation, are generated dependent on, for example, radionuclide composition, temperature and pressure at release. For instance, the shape of the uranium fuel particles originating from the low temperature prefire Windscale releases was reported to be flake-like, indicating different releasing processes (Salbu et al., 1998). It has been reported that these particles could not be dissolved in 1 M HCl (Salbu, 2000). Also, plutonium particles spread over the ice after the crash of a B52 aeroplane carrying nuclear weapons in 1968 near Thule air base in Greenland were found to be virtually insoluble (Eriksson, 2002). These particles were found to be very similar to those produced by another plane crash also leading to dispersion of radioactive material from nuclear weapons – the Palomares accident in 1966 (Garcia-Olivares and Iranzo, 1997; Pöllänen et al., 2006). According to autoradiographic assessments made on the contaminated ice shortly after the Thule accident, particle size distributions were lognormally distributed here with a mean of 2 μm and a log standard deviation of 1 (Danish Atomic Energy Commission, 1970). This means that only 1.3% of the particles were larger than ca. 18 μm, but these carried nearly 80% of the activity (Eriksson, 2002). This size distribution pattern is in line with observations made by Harper et al. (2007) for 'dirty bombs' (also contaminant dispersion through a conventional bomb explosion) with a radioactive material like a ceramic strontium matrix, with low solubility and high melting point. Such large insoluble particles are comparatively easy to remove from impermeable surfaces in the environment.

Livens and Baxter (1988) and Bunzl et al. (1989) found that radiocaesium originating from Chernobyl was in a form that was at the time more labile in mineral soil than radiocaesium from the early weapons tests. This was attributed to either differences in physico-chemical form or the time dependence of the 'locking up' of caesium in soil. In support of this, Hilton et al. (1992) reported that, in aerosol samples collected in the United Kingdom, the water-soluble fraction of the radiocaesium released from Chernobyl was about 70%, whereas in the fallout particles from weapons testing, it was less than 10%.

Dorrian and Bailey (1995) compiled results from 52 different publications covering a wide range of industries and workplaces, where releases of radioactive aerosols have been measured. It was found that the overall particle size distribution could be adequately described by a lognormal distribution with a median value of 4.4 μm. Many of the reported cases were related to largely insoluble plutonium particles, but a wide range of other, much more readily soluble, radioactive substances were also recorded.

3. Different Remediation Needs: Options and Implications

3.1. Reducing contributions from the most significant exposure pathways

Depending on the type of affected area, as well as on the characteristics of the released contaminants, numerous different components may contribute to the total dose after an airborne release, and at least some of these different dose contributions may be reduced through the application of different targeted remedial measures. The different exposure pathways may include:

- internal dose from inhalation during plume passage;
- internal dose from inhalation of resuspended contaminants;
- external dose from the passing contaminated plume;
- external dose from contamination on outdoor surfaces;
- external dose from contamination on indoor surfaces;
- external and internal doses from contamination deposited on humans;
- external dose from contaminants transferred onto humans by contact; and
- internal dose from ingestion.

A lesson learned from the Chernobyl accident as an example of a severe nuclear power plant accident is that internal doses from ingestion and external doses from contamination on outdoor surfaces can provide a higher contribution compared to doses from inhalation of resuspended contaminants (particularly in the indoor environment) and external doses from radionuclides of the contaminated plume (Andersson and Roed, 2006).

This is however not the case, if the scenario involves malicious dispersion. A terrorist might well, if possible, strive towards producing a particularly unexpected scenario, and the released contaminants might be either a single radionuclide or a series of different radionuclides. As demonstrated by the list in Table 1 of some possible sources of concern, the contaminant could be a single, pure alpha emitter, in which case all external dose components can be excluded, or the contaminant could be in the form of very large particles, in which case inhalation might play a minor role. Therefore, an operational preparedness needs to build in remedial measure options to reduce the different components of the final dose. Some remedial measures, like different means of burying contamination (Andersson et al., 2003; Nisbet et al., 2004), can reduce both external and internal exposure pathways, whereas others are targeted very specifically to a single dose component.

For instance, the relative importance of cleaning different surfaces in a contaminated inhabited complex would strongly depend on factors like the weather during deposition, the characteristics of the contaminants and the specific constructional characteristics of the environment such as building sizes, vegetation, street widths and construction materials (Andersson et al., 2008b). Therefore, complex dose assessment models are often useful in pinpointing where remediation would have the greatest dose-reducing effect in a given contaminated environment, and such models are being implemented in the operational standard decision support systems in Europe – ARGOS (2008) and RODOS (2008).

3.2. Considerations relating to timing and size of the affected area

Frequently, remedial measures are distinguished into two categories, according to time phases: (i) an initial emergency phase where urgent measures are required to protect individuals from short-term, relatively high risks, for example sheltering or evacuation from an area, and (ii) a subsequent recovery phase. Here the overall ultimate aim of introducing remedial measures in a contaminated inhabited area is the return to 'normal lifestyles'; that is, people can live and work in an area without the radiological emergency and its consequences being at the forefront of their minds (Brown et al., 2008).

Different types of contaminating airborne release scenarios would be likely to lead to very different time periods between the actual occurrence of the incident (e.g. an accident or a terror attack) and the time at which a given area becomes contaminated. For instance, if contaminants are contained in a nuclear power plant for some time after an accident, it may be possible to issue a very early warning that could facilitate the implementation of early measures like sheltering or evacuating populations living at certain distances from the power plant. Naturally, since it takes time for the contaminant plume to reach more remote areas, this may give more time to implement early remedial measures, possibly including sheltering of animals and covering of precious objects. In contrast, a terror attack would not be likely to be preceded by any warning, and due to the comparatively low effective contaminant release altitude, a rather small area would be likely to become affected. This means that the dispersion is likely to occur over a very short time, and it is questionable whether it would be possible to effectively and systematically implement even simple remedial measures like indoor sheltering with closed air ducts before much of the contaminant inhalation would have occurred.

A number of simple remedial measures may be advantageous to reduce doses, if they can be implemented over a limited time. For instance, if the contamination occurs in the absence of precipitation, a mature crop stand will receive high levels of dry deposition, whereas the underlying soil will receive comparatively little deposition. If these crops can be harvested and removed prior to the first heavy rain shower following the contamination, which would typically transfer much of the crop contamination to the soil, a long-term and severe soil contamination problem may largely be avoided. In urban areas, under such circumstances it is equally advantageous to rapidly mow contaminated lawns and remove and dispose of the contaminated cut-off grass. Extra doses to remedial workers would have to be considered. If in the early phase the dose rate is dominated by contributions from short-lived radionuclides, it may even be advisable to postpone remediation for some time.

Remedial measures can be implemented in different stages following an emergency. Often a good effect will gradually become more and more difficult to achieve as time passes, although some remedial measures are aimed at reducing very long-term contamination problems and can be carried out after a few years (Roed et al., 1996; Brown et al., 2008; Jacob et al., 2001). The covering of limited areas used for growing particularly valuable crops prior to the arrival of the contaminated plume may prove to be advantageous, but it would also place workers at risk, as no-one can precisely predict when the contamination will occur in a given area. Early harvesting – as mentioned earlier – can in some cases prevent soil contamination. If this is not done or does not have adequate effect, and long-lived radionuclides still pose a problem, more long-termed (and often more expensive) measures may need to be implemented, to limit the transfer of contaminants from soil to edible crop parts. If this does not solve the problem, it may be considered whether it is necessary to change the production completely in the area, possibly to other items than food. Often, the economic repercussions and social acceptance implications of the various remedial measures that can be implemented will become more severe the later the measures are implemented.

It should also be noted that some remedial measures will take considerable time to implement, for instance if the equipment or sufficiently skilled workers are not readily available, if weather conditions (e.g. frost) make it impossible to start remediation immediately or if the measures generate much radioactive waste, for which at least temporary repositories of adequate capacity and safety must first be made available.

Finally, as illustrated earlier in connection with the Goiânia accident, the resources that could be applied to reduce a contamination problem in a given area would greatly depend on the size of that area, as well as on the contamination level, the societal implications of the contamination and the economic situation in the area in general. Different priorities may need to be given to the remediation of parts of large contaminated areas, according to the estimated severity in different zones.

3.3. To remove or not to remove contamination

In many situations, there will be a choice between different remediation solutions for essentially the same contamination problem, and in some cases, the choice will be between applying a remedial measure that entirely removes the contamination from the area (decontamination) and one that reduces dose through placing the contamination under a layer of a shielding material (e.g. uncontaminated soil), whereby both internal (by reducing ingestion dose if edible crops are subsequently grown and by limiting contaminant resuspension that can cause inhalation doses) and external dose components may be reduced.

One advantage of decontamination methods is that much of the contamination is *removed* from the area, although no remediation technique is in practice capable of cleaning an area 100% (Brown et al., 2008). If the contamination is instead placed under a shielding material, the public perception may be that the problem still exists. However, decontamination methods also require that the removed contamination is stored or buried somewhere, and a solution in the local area might be preferred, since it would often be the least expensive, and it could also be argued that it would be most fair not to export this problem to a different area.

An important concern in connection with shielding methods like ploughing or deep digging is that it needs to be ensured very carefully in advance that the method is satisfactory and sufficiently effective, so that it will not be necessary at a later date to remove the contamination from the area. Once the contamination is buried at some depth, a very large amount of waste would need to be removed to decontaminate the area. Also, if contamination is buried, it would be necessary to instruct inhabitants not to dig in the future beyond the burial depth, which might be seen as a restriction on their ability to return to 'normal life'.

3.4. Management of wastes generated by remedial measures

Some remedial measures will generate waste (either liquid or solid), which must be managed safely and should be regarded as an inherent part of the remediation strategy, as it is a direct consequence of the choice of remediation methods. This means that the various costs of managing the contaminated waste should enter the decision matrix in selecting the optimal remediation strategy. Current legal demands may in some countries restrict the applicability of cost-effective strategies for waste disposal.

Waste disposal schemes for solid contaminated waste must be selected with care. For coping with an emergency and its implications, it must be possible to plan for waste management, including creation of the needed repositories, over a short period of time. This means that it must be secured in advance that the required materials, transport vehicles, skilled workers, infrastructure and so on are available to enable the construction of the type(s) of repositories that would be considered most suitable. It also means that concrete trenches and other complicated and time-consuming engineering projects, which might be considered for wastes originating from controlled practices, could not realistically be finalised within the required time frame. In any case, there would be limits as to how contaminated the wastes generated by remedial measures could be, and generally the more expensive and complicated repository options would not be likely to be called for. It should be kept in mind that the waste disposal strategy should be adapted for the task. In selecting the disposal site, the following factors should be considered (Brown et al., 2008):

- Local repositories would generally be less expensive and give less total worker dose than centralised/distant repositories.
- The hydrogeological characteristics of the site should be considered. From the bottom of the constructed repository to the groundwater level, there should be a distance of at least 3 m. If trench repositories cannot be constructed to meet this requirement, it may be possible to construct elevated surface mound repositories on the ground. Old gravel pits should not be exploited for this purpose, as they are likely to provide too little distance to the groundwater. It should also be ensured that the area is not prone to flooding, for example, by a nearby river.
- The area should not be prone to earthquakes.

- The future land use needs to be considered. Edible crops, for instance, should not be grown on top of a waste repository, both to completely rule out the possibility of uptake by plant roots and to reassure the local inhabitants.

In constructing the repository, a number of other factors need to be taken into account:

- The shielding effect of the repository construction should be sufficient, such that the contamination in the repository does not contribute significantly to the external dose rate in the area. As the shielding of the radiation provided by the contaminated soil itself (self-attenuation) is great, this problem can largely be overcome even with very simple repository designs. An example of this is the formation of simple, uncovered waste pile 'hills' in connection with a decontamination exercise in Chernobyl – ^{137}Cs-contaminated Novozybkov area in Russia in 1995 (Roed et al., 1996). It was found that the dose rate to a person standing on top of one of these hills containing contaminated topsoil removed from a vast area was only 15% higher than that in the surrounding contaminated area. By covering the contamination with, for example, a layer of uncontaminated soil excavated from deeper soil layers of the same area, this dose rate can be greatly reduced. Further, the formation of a pile of earth in the area will shield well against radiation from contamination far away.
- The ecological characteristics in the area should be assessed. The construction of the repository should as far as possible reasonably prevent against intrusion by humans, animals and the roots of plants. In practice, this means that the waste should be stored under suitable protective layers (e.g. soil and plastic membranes).
- Migration of radionuclides from the waste with rainwater must be prevented. This can be done by draining the rainwater off in a gravel layer on top of the waste layer. Also flow barriers made of clay, which effectively retain many contaminant ions (especially caesium), and thick plastic are useful. A ditch should be dug around the repository to collect the drained-off rainwater. If the repository is arched at the top, the rainwater will conveniently run off the repository side and end up in the ditch.
- The repository should be protected against erosion. This can be accomplished, for instance, by growing grass in the topsoil layer of the repository.
- The acceptability of constructing repositories, possibly on private property, needs to be assessed. Communication issues are of great importance, in carefully explaining the implications and possible alternatives. Information leaflets may be produced in advance of an incident.

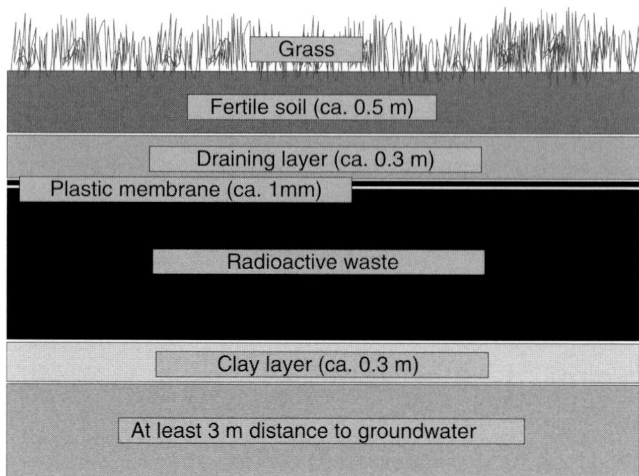

Figure 2 The principles of a simple ground repository for radioactively contaminated remediation waste (e.g. soil), as recommended by Junker et al. (1998).

- The repository may be equipped with a drainpipe at the bottom for inspection of the radionuclide contents in any water passing through the construction.
- The design should be simple, inexpensive and possible to construct in a short time. It has been estimated that the construction of a simple surface mound or trench repository (based on the principles shown in Figure 2) for ca. 500 m^3 solid waste would cost some 3,000 euros.

Figure 3 shows a schematic representation of suggested management strategies for wastes generated by different remedial measures targeted for decontamination of surfaces in an inhabited environment.

3.5. Self-help measures

A certain category of the more simple remedial measures that do not require any specific skills or experience to implement (e.g. lawn mowing, digging) are considered potentially suitable for implementation by affected inhabitants. This could, for instance, be beneficial in introducing an extra labour force in situations where large areas need to be treated over a short time.

The involvement of the affected persons in actions to improve their own situation can be psychologically important and can give a better feeling of control of the situation, which also prevents undue anxiety. The ethical aspects of 'self-help' measures are dealt with in Chapter 10 of this book. It should always be considered that people carrying out such 'self-help'

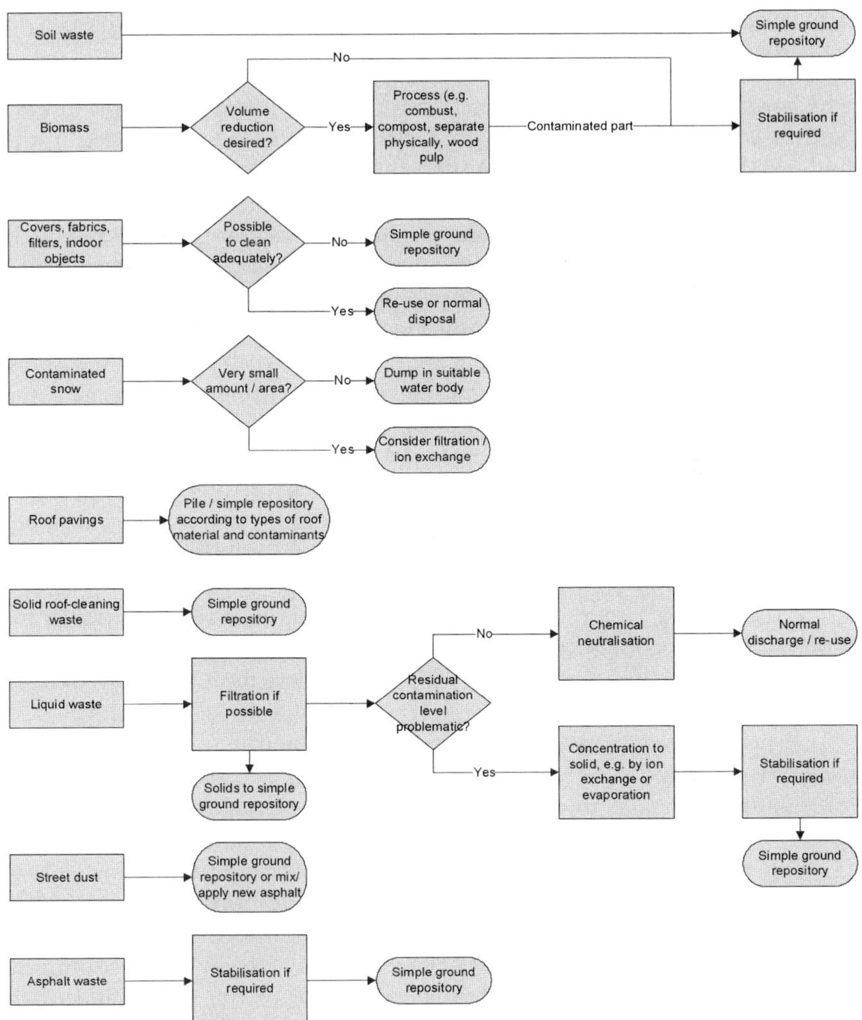

Figure 3 Suggested management schemes for different remediation wastes (adapted from Brown et al., 2008).

options may be unfamiliar with the type of work, and all possible arrangements should be made to ensure that implementation does not lead to accidents or other harm. Also, people may not be physically fit for the scale of work required.

'Self-help' remedial measures need to be conducted on a voluntary basis, and careful communication with/supervision of these individuals would be required, as some remedial measures could have irreversible negative outcomes if implemented wrongly (e.g. soil digging to a wrong depth).

3.6. Use of templates for recording and presenting information on remedial measures

The concept of systematically describing remedial measure features, on the basis of 'hands-on' experience, in a datasheet format was developed in the mid-1990s (Roed et al., 1995; Andersson, 1996; Voigt et al., 2000).

Standardised datasheets capturing key information required to evaluate options can be useful in providing a clear and consistent overview of a wide range of options, aiding the decision-making process and ensuring that potentially important issues are not overlooked.

They can also outline the requirements (e.g. equipment, consumables and number of skilled workers) that are necessary to carry out the measures. On this basis, for the most practicable options within a given area, the required resources can be secured prior to any contaminating incident.

It should however be noted that the information in the datasheets developed so far is to a very great extent based on practical experience from remediation investigations relating to nuclear power plant accident scenarios and does not adequately reflect what would be expected in connection with other scenarios.

As mentioned earlier, effectiveness of remedial measures in particular for some malicious contaminant dispersion scenarios may be very different from that which has, for instance, been recorded for Chernobyl-related contamination.

Over the years, various 'generations' of remedial measure datasheets have emerged, and the level of detail in descriptions has increased steadily, for example, to take into account the need for data representing more diverse conditions and to look at a wider range of radiological hazards.

Non-radiological considerations have been incorporated into the later generations of remedial measure datasheet descriptions (Howard et al., 2005; Andersson et al., 2003; Eged et al., 2003; Brown et al., 2008), introducing viewpoints from experts on social, ethical and economical sciences, and also considering observed reactions of the public and other participants in connection with remedial measure trials in the former Soviet Union.

The latest suite of remedial measure descriptions are integrated parts of European decision-aiding handbooks for contaminated food production areas and inhabited areas, respectively (Brown et al., 2008; Nisbet et al., 2007). These handbooks contain advice on development of optimised strategies for remediation under very specific conditions. This is facilitated through the inclusion of a series of diagrams, tables and decision charts, aimed at guiding the user through a series of eliminations of unsuitable options.

Naturally, local factors need to be taken into account in this process, and it is strongly recommended in the handbooks that they be customised and tailored for the particular audience. Also the importance of adequately involving different groups or persons in decisions that somehow affect them is stressed. The applied template includes the following general criteria:

- the objectives of the measure;
- a short description of the measure;
- constraints on its implementation;
- effectiveness;
- requirements;
- waste generated;
- safety precautions;
- costs;
- side-effects; and
- practical experience.

As an example to illustrate the format developed in the European STRATEGY project for the remedial measure database, the pages taken from Andersson et al. (2003) show the datasheets for the remedial measure 'Shallow ploughing' (to 25 cm depth) (Table 2). The 'remedial measures' in these datasheets are termed 'countermeasures'.

4. Remediation in Contaminated Urban Ecosystems

4.1. Specific needs for remediation in urban ecosystems

The urban ecosystem is the one where most people live. It is characterised by a multitude of different types of elements (e.g. grassed areas, paved areas, exterior walls, roof pavings, trees, shrubs, interior surfaces) that can all become contaminated and contribute to dose. Many of these elements are also found in suburban or rural living environments, and the contamination problems encountered in the urban ecosystem thus apply to inhabited areas in general. As mentioned earlier, the relative magnitude of the dose contributions from the contamination, and thus the specific needs for remediation, depend strongly on the release scenario.

One issue that should be noted particularly carefully when implementing remedial measures for contaminated surfaces is the need for a sufficiently thorough identification of the problem, including the type and spatial distribution of the contamination. This can be illustrated by the results of the efforts of the Soviet army in 1989 to decontaminate 93 rural settlements in the Bryansk region that had been affected by the Chernobyl accident. The army units removed a layer of topsoil from contaminated garden areas, and this method was expected to be efficient in reducing dose,

Table 2 Example of template (Shallow ploughing) suggested by the STRATEGY project for evaluation of information on remedial action effectiveness.

	Shallow ploughing
Objective	To reduce the external dose rate in the area.
Other benefits	Limited reduction of ingestion dose if food is produced.
Countermeasure description	Without intervention, it is generally expected that much of an airborne Cs deposition to soil will remain distributed throughout several years in the upper few centimetres of the soil profile. By shallow ploughing with an ordinary mouldboard plough to a depth of some 25 cm, the contamination is buried in the soil. The contamination is thus shielded against, and it may be brought out of the uptake zone of some plants.
Target	Large open urban areas of soil (e.g. parks), which have not been tilled since contamination.
Targeted radionuclides	Caesium (plus other radionuclides if edible products are grown).
Scale of application	Achievable on a large scale.
Contamination pathway	None (possibly root uptake if food is produced).
Exposure pathway	Mainly external exposure from contaminated land. Possibly also dose from consumption of food products.
Time of application	Should generally be carried out as early as possible, when the radiological situation is clear, but worker doses must be considered. After a decade, can still save a significant fraction of the 70-year dose. As the procedure would often have nearly same effect on dose rate after one week as after two years, one set of equipment can treat a large area.
Constraints	
Legal constraints	Cultural heritage protection, especially in conservation areas or equivalent. Liabilities for possible damage to property. Requirement for radiation protection training of workers.
Social constraints	Acceptability of smothering flora and fauna and destruction of planting. Aesthetic consequences of landscape/amenity changes.
Environmental constraints	Soil texture (big rocks), snow covers and frost may be restrictions. Application of fertilisers may be called for. Soil depth >0.3 m required for shallow ploughing.
Communication constraints	Need for public explanation of countermeasure and dialogue regarding selection of areas for treatment.
Effectiveness	
Countermeasure effectiveness	Reduction in dose rate contribution by ca. 50%–75%. Internal dose reduction: all contamination in upper 10 cm can be reduced by 80%–90%. Consumption dose reduction depends on, for example, root system of crops.

Table 2. (*Continued*)

	Shallow ploughing
Factors influencing effectiveness of procedure (technical)	Soil type and conditions ('loose' soil will be more difficult to treat optimally). Uniformity of vertical distribution of Cs. Time (downward Cs migration in soil). Contaminant resuspension could possibly have an impact on effectiveness if the method is carried out very early.
Factors influencing effectiveness of procedure (social)	Compliance with appropriate process of application of countermeasure.
Feasibility	
Required specific equipment	Plough (readily available in European areas, where ploughing is possible).
Required ancillary equipment	Tractor.
Required utilities and infrastructure	Roads for plough transport.
Required consumables	Petrol.
Required skills	Can be carried out by agricultural workers, who are familiar with ploughing, but must be instructed carefully about the objective.
Required safety precautions	Particularly, early after accident and under very dusty conditions, respiratory protection and protective clothes may be recommended.
Other limitations	Contaminants will be brought a bit closer to the groundwater level.
Waste	
Amount and type	None
Possible transport, treatment and storage routes	–
Factors influencing waste issues	–
Doses	
Averted dose	Highly dependent on environment type. See separate chapter.
Factors influencing averted dose	Consistency in carrying out the procedure over a large area. Population density and behaviour pattern. Age of persons exposed. If edible crops are grown, the method may reduce consumption dose, depending on crop root system.
Additional dose	Depends on short-lived radionuclides (time). The dose over a day to an operator may be 2–3 times higher than that to an individual living in the contaminated area. Influenced by measures taken to protect operators against inhalation of contaminants and contamination of skin/clothes, where required.

Table 2. (*Continued*)

	Shallow ploughing
Intervention costs	
Equipment	Plough: ca. 2,000 euros. Tractor: ca. 50,000 euros.
Consumables	Petrol: ca. 7 l ha^{-1}.
Operator time	ca. 1.2 h ha^{-1} (one operator).
Factors influencing costs	Operator skills. Soil type and conditions (e.g. moisture, season), vegetation, topography, labour costs.
Communication costs	Provision of information for public on rationale for countermeasure. Information for operators on correct application of countermeasure. Dialogue costs reselection of areas for application of countermeasure.
Compensation costs	–
Waste cost	None.
Assumptions	
Side-effect evaluation	
Ethical considerations	Need for informed consent from amenity users. Potential redistribution of dose from amenity users to operators and others via groundwater. Free informed consent of workers to risks of radiation exposure. Compensation for increased radiation dose (workers). Compensation for amenity damage/change. Liability cover for unforeseen health or environmental effects.
Environmental impact	The procedure brings contamination closer to the groundwater. Cs will however normally be very strongly bound. Future restriction on land use: must not be tilled.
Agricultural impact	May require seeding/replanting.
Social impact	Maintenance of use of urban spaces, although partial change in usage likely (temporary loss of amenity area). Acceptability and potential for dispute regarding selection of areas to be treated.
Other side effects, positive or negative	Adverse aesthetic effect of treatment. Severely complicates subsequent *removal* of the contamination. Public reassurance issues.
Practical experience	Tested widely in CIS and on limited scale in Denmark.
Key references	Hubert et al. (1996), Andersson et al. (2000), Roed et al. (1995), Vovk et al. (1993) and Andersson and Roed (1994).
Comments	–

since practically all the radiocaesium would usually be present in a thin topsoil layer.

However, the dose-reducing effect was found to be disappointingly low; the dose rate was reduced by a factor of only 1.1–1.5, and the operation was thus not deemed worthwhile. Therefore, decontamination of inhabited areas was not considered to be a realistic option over the years that followed, and relocation of populations continued.

The operation was clearly flawed by at least two factors. One was that not all landowners allowed intervention on their ground (Anisimova et al., 1994). Therefore, the coherent-treated areas may not have been very large. Roed et al. (1999) estimated that the application of such a method in a very large area could in some cases result in five times as great a reduction in the external dose rate from the ground as the application of the method would do in a 10 × 10 m area. The other factor is that, judging from the sparsely reported results of the Soviet army's effort, the thickness of the removed soil layer was probably insufficient, and the work was not done consistently, even over small garden areas. Figure 4 shows an example of a vertical radiocaesium distribution in a soil sample collected in the same area only a few years later (Roed et al., 1996).

In this case, if the contamination distribution had not been assessed, one might remove a 5–10 cm topsoil layer, only to discover that the dose rate had risen, since effectively no contamination, but only a shielding soil layer, had been removed. This is consistent with the registration of increased dose rates at some measurement points after the Soviet army had finalised their operation. In 1997, however, scientists demonstrated in the same area that if the removal depth is optimised in relation to the distribution of

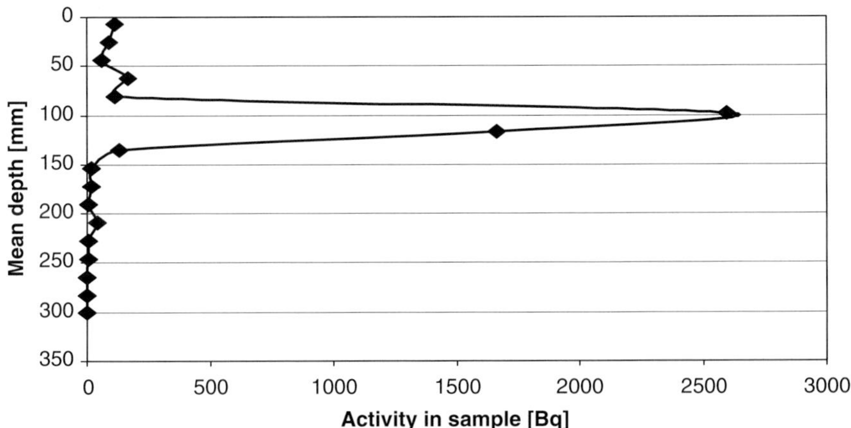

Figure 4 Example of a vertical distribution of ^{137}Cs in undisturbed soil recorded in an area in the Bryansk region, Russia, in 1993.

Figure 5 Removal of contaminated topsoil with a 'Bobcat' mini-bulldozer in the Bryansk region in 1997.

the contaminants, topsoil removal (e.g. using a mini-bulldozer) can reduce the external dose rate from the ground by at least a factor of 6 (Roed et al., 2006) (Figure 5).

Although most remediation techniques for urban ecosystems are designed for reduction of external dose (Andersson et al., 2003, 2008b), inhalation doses both received during the passage of the contaminating plume and from resuspended contamination can also have importance, as outlined earlier. Thus, measures like sheltering and closing windows and air ducts while controlling air exchange can be effective means of reducing dose in an initial emergency phase, whereas techniques to fix contamination to the surface may be useful for dust suppression in a later recovery phase (Brown et al., 2008). The relationship between indoor and outdoor air contaminant concentrations, determining the reduction in inhalation dose by sheltering, is at equilibrium given by:

$$\frac{C_{i,a}}{C_{o,a}} = \frac{f \bullet \lambda_v}{\lambda_v + \lambda_d}$$

(Roed and Cannell, 1987; Andersson et al., 1995), where f is the filtering factor (the fraction of aerosols, which are not retained in cracks and fissures of the building structure, as air enters the building), λ_d is the rate coefficient of indoor deposition (the fraction of aerosols in the building depositing per unit time) and λ_v is the rate coefficient of ventilation (the fraction of air in the building that is exchanged per unit time). It should be noted that both f and λ_d are highly dependent on particle size. For instance, f will generally be close to unity for the particles smaller than

2 μm, but only to about 0.1 for particles in the 10 μm range (Roed, 1990; Long et al., 2001). Over the same particle size interval, λ_d increases by more than one order of magnitude (Andersson et al., 2004; Long et al., 2001). A representative value of the rate coefficient of ventilation for modern, naturally ventilated European dwellings is 0.4 h^{-1} (Hiemstra et al., 1997; Janssen et al., 1998; Malanca, 1993; Roed, 1990).

Direct deposition of contaminants on humans can give significant doses (Andersson and Roed, 2006). Efficiencies of forced removal measures (washing, scrubbing) strongly depend on contaminant characteristics. Large particles in the 10 μm range have typically been found to have a natural clearance half-life on human skin of only some 3 h, whereas slightly submicroneous condensation particles would tend to lodge in such cavities as hair follicles, where they may remain until the shedding of the stratum corneum after a few weeks (Hession et al., 2006). Even particles as large as several microns are not easily removed from the skin surface by force (Andersson et al., 2004).

4.2. Brief descriptions of remedial measures for contaminated urban ecosystems

The following sections outline a number of techniques designed to be used in the recovery phase following a contaminating incident to reduce doses in an urban ecosystem (based on the results of the European STRATEGY project; www.strategy-ec.org.uk). Full descriptions of these methods in the standardised datasheet format discussed earlier can be found on the STRATEGY web site. Quoted effectiveness estimates are derived from the STRATEGY templates (Howard et al., 2005).

4.2.1. Road planning

A thin contaminated top layer of asphalt surfaces is grounded off with a road planer (a rotating grinding equipment used by contractors). In some cases, subsequent re-paving is necessary. Reduction in contamination level by 80%–90% is achievable, but the method is relatively expensive, and waste must be managed appropriately.

4.2.2. Vacuum sweeping roads and walkways

Roads and walkways are vacuum swept with a type of machine used in many areas of Europe for routine street cleaning. It often has three rotating brushes and a vacuuming attachment. The road dust is accumulated in a vessel behind the operator. If applied early, the method can remove 50%–70% of the contamination. It is rapid and relatively inexpensive, but generates waste, which can have rather high specific activity.

4.2.3. Firehosing roads and walkways

Hosing with water can remove loosely bound contamination from surfaces of roads or other horizontal pavings. This is particularly efficient if applied early after contaminant deposition, when 50%–75% of the contamination may be removed. It is generally not possible to collect the loosened radioactive material, which will end in drains.

4.2.4. Topsoil removal

A thin topsoil layer containing most of the contamination from a soil surface is removed using a 'bobcat' mini-bulldozer or similar equipment. This can remove more than 90% of the contamination from the surface if it is optimised according to the vertical contaminant distribution in the soil. The procedure takes some 50–100 h ha^{-1}. It can also be carried out manually with spades, in which case it typically takes at least 10 times longer.

4.2.5. Application of clean sand/soil around dwellings and in frequently occupied areas

Sand or soil from a radiologically clean area can be applied around dwellings, to shield against radiation. This is typically considered for reduction of residual radiation after removal of a topsoil layer. Application of 10 cm soil over a large area may reduce dose rate by 80%. Time consumption is comparable to that of topsoil removal if the work is done with similar equipment.

4.2.6. Snow removal

If contaminant deposition occurs to a snow surface, mechanical removal of a snow layer before the first thaw can prevent contamination of underlying (soil or asphalt) surfaces. If carried out optimally, more than 90% of the contamination can be removed from the surface. Time consumption is a factor of 2–3 less than that of topsoil removal if the work is done with similar equipment.

4.2.7. Triple digging

By manually triple digging, the order of three vertical soil layers is changed; the thin contaminated top layer is buried at the bottom, with the turf facing down, the bottom layer is placed on top of this and the intermediate layer (not inverted) is placed at the top. Thus, shielding is achieved with a minimised impact on soil fertility. Optimised application over a large area can reduce dose rate by 80%–90%. However, the method is time consuming (20–30 min m^{-2}).

4.2.8. Ploughing (park areas)

By ploughing, much of the contamination on a soil surface will be buried deep in the vertical profile (depending on ploughing depth), so that radiation from the contaminants is substantially reduced. A special 'skim-and-burial' plough ideally skims off a thin top layer, which is placed at the bottom of the vertical soil profile. A deeper soil layer (up to about 45 cm) is placed on top of this (this layer is not inverted). Thus, shielding is achieved with a minimised impact on soil fertility. A reduction in dose rate by a factor of 10 is here achievable at a rate of 3 h ha^{-1}.

4.2.9. Lawn mowing

If deposition occurs without precipitation, much of the contamination on a lawn, in the earliest phase (days to weeks), will be present on the grass rather than on the soil. Cutting the lawn (and removing the grass), therefore, to some extent, can prevent soil contamination. Up to some 90% of the contamination on a lawn can be removed in this way. It is a rapid and relatively inexpensive method, but generates waste, which can have rather high specific activity.

4.2.10. Pruning or removal of trees and shrubs

If deposition occurs without precipitation, shrubs and trees (particularly if in leaf) may receive and retain very high levels of contamination compared with other vegetation. Their pruning or removal from areas such as gardens may therefore significantly reduce dose to inhabitants. The fraction of the initial deposit on the vegetation that can be removed is great, although fallen leaves and needles may pose problems. Time consumption depends greatly on the amount and type of vegetation.

4.2.11. High-pressure water hosing of walls

Water hosing with high-pressure nozzles can remove part (typically 35%–80%) of the contamination deposited on walls of buildings, especially if applied early after deposition. The waste is practically impossible to collect and is led to drains. The hosing of 1 m^2 of wall surface is estimated to take about 1 min.

4.2.12. Sandblasting of walls

By sandblasting, using high-pressure air with sand injected, a thin layer of the surface of a wall is removed, taking with it much (typically 75%–85%) of the contamination. Wet sandblasting is recommended, as both the efficiency and the control of the generated dust are better. The method generates rather large amounts of waste (ca. 3 kg m^{-2}), which is virtually impossible to collect and is relatively time consuming (ca. 4 min m^{-2}).

4.2.13. Ammonium treatment of walls

The ammonium ion has similarities to the caesium ion and can (on non-specific sorption sites) exchange with caesium. A solution of ammonium nitrate is sprayed onto the wall at low pressure, ensuring a continuous flow over the wall.

The method is generally most efficient soon after contamination has occurred, when up to half of the caesium may be removed. It generates 6 l of liquid waste per square metre, which is virtually impossible to collect. It is relatively time consuming (ca. 5 min m^{-2}).

4.2.14. High-pressure water hosing of roofs

Water hosing with a high-pressure nozzle can remove part (typically 35%–80%) of the contamination deposited on the roofs of buildings, if applied early after deposition. Time consumption is some 1–2 min m^{-2}. The generated waste (an estimated 20 l m^{-2} liquid and 0.2 kg m^{-2} solid) can be collected through drainpipes.

4.2.15. Roof cleaning by cleaning device

Various designs exist and are commercially available, involving rotating brushes or oscillating high-pressure jet systems. They are typically mounted on an extendible rod and operated from the ground or from the rooftop. They are about equally effective in reducing the roof contamination as ordinary high-pressure hosing. They often have the advantage that the cleaning takes place under an enclosure, thus minimising spreading of loosened contamination and, for example, asbestos.

4.2.16. Intensive indoor surface cleaning

Dose contributions from indoor contamination may be significant, especially over the first year. Also over longer periods, contamination may be brought into dwellings, for instance, attached to the soles of shoes.

The remedial measure is self-help advice to carefully and thoroughly clean floor surfaces, particularly carpets, in a short time after contamination occurs. For instance, the contamination level on a hard floor material can typically be reduced by 35%–65% by washing.

5. Remediation in Contaminated Agricultural Ecosystems

5.1. Specific needs for remediation in agricultural ecosystems

The agricultural ecosystem is the area where food that we eat is produced. Naturally, agricultural workers and rural populations may experience the contamination problems described in Section 4.1 in their living environment, but the specific problems pertaining to the agricultural ecosystem relate to production of food and other items. Agricultural production lines that may become contaminated and require remediation can be distinguished broadly in categories comprising the following (Nisbet et al., 2007; Fesenko et al., 2007):

Crop food production

- cereals (wheat, rye, barley, oats, etc.);
- industrial crops (oil seeds, pulses, sugar beets, etc.);
- vegetables and horticultural crops (roots, tubers, leafy vegetables, brassicas, legumes, salad, etc.);
- fodder plants (silage, hay, grass, root vegetables, etc.); and
- fruits (grapes, berries, citrus, olives, apples, etc.).

Animal food production

- meat of cattle, sheep, goats, pigs, poultry and so on and
- milk and other dairy products, eggs and so on.

Non-food production

- wool, cotton, flax, crop oil, leather and so on. Generally, contamination of such products is less problematic.

One group of remedial measures is directed at soil and crops. These include methods for removing contamination from the contaminated area, either together with a topsoil layer or by early removal of contaminated crops from the field. An alternative involves methods that reduce soil-to-plant transfer by physical or chemical means. These comprise various ploughing procedures and application of different soil nutrients to limit contaminant uptake. Another alternative is to process harvested crops on an industrial scale, so that the final product contains less contamination.

A second group of remedial measures is directed at contaminated animal products. Here, an objective may be to remove contamination from the human food chain by, for example, slaughtering dairy cows or suppressing cattles' lactation before slaughter. Another objective may be to reduce livestocks' ingestion of contaminated fodder, either through feeding with uncontaminated fodder over a certain period or by moving farm animals to less-contaminated pastures.

Another series of remedial measures is designed to reduce gut uptake of radionuclides, for example, by feeding with chemicals or clay minerals that can bind specific contaminants that are likely to cause problems. Also, animal slaughter times may be changed to reduce meat contamination levels. Further, food processing at industrial scale may be considered, through decontamination techniques for milk or meat.

Some remedial measures can be applied generally to any food chain. These include measures to restrict the entry of contaminated food into the food chain by bans and measures taking land areas out of food production systems and seeking alternative use to maintain an income for the local community. Such measures will often have a rather drastic impact on the local populace and thus low acceptability (Lauritzen, 2001).

It should be noted that a fraction of the consumed food is, to a varying extent in different areas of the world, produced in small single-household ground lots located in inhabited areas (commonly termed 'kitchen gardens'). For instance, in Sweden, nearly half of the potatoes and some 10% of the berries, carrots and onions consumed are home grown (Andersson et al., 2008c).

In large village areas of the former Soviet Union, the diet is primarily made up of the yield of kitchen gardens in inhabited areas and food collected in natural areas (Tsalko, 2004). Food production in kitchen gardens in inhabited areas would be subject to very different conditions than that in agricultural areas. For instance, fertilising, irrigation and tilling practices as well as environmental impact would be different in such small lots in inhabited areas. This distinctive feature of the farming practice in kitchen gardens, together with the generally small size of kitchen garden areas, requires different remediation practices compared with those traditionally described, which are aimed at the agricultural industry. For instance, ploughing procedures would generally require more space than is available in a kitchen garden area, and here digging procedures, as described for external dose reduction in the urban ecosystem, which can essentially change the order of vertical soil layers in the same way, may be advantageous.

Many of the remedial actions that have been suggested for both agricultural and other types of ecosystems are very simple and are basically often variants of procedures that are conducted on a routine basis in the area. This gives strength in that requirements, such as equipment, consumables and skilled personnel, are often in practice readily available in the affected area, so that remedial measures can be implemented quickly and consistently over large areas. However, similarity to routinely conducted procedures can also carry an inherent risk of failure. For instance, for ploughing procedures, a number of skilled operators can usually be found in the locality, but it can be difficult to ensure that these carry out the ploughing in the specific way that gives the optimal dose-reducing effect, rather than simply the way that they are used to doing it. For instance, when the so-called skim-and-burial plough was tested as a remedial measure in a contaminated area in the Bryansk region in Russia, the local plough operator was carefully instructed on how to operate the plough to achieve the best dose-reducing effect. However, he continuously and rapidly slipped into conducting 'business as usual', as can be seen from Figure 6, which shows an instructor rather desperately running behind the plough, trying to optimise the ploughing depth, while the operator is more concerned about keeping a straight line and doing the job quickly.

This type of finding from 'hands-on' experience, also in connection with methodological constraints and public reactions, is important to document in reference materials like the above-mentioned datasheets targeted for the user community and affected population groups, so that errors need not be repeated and the remedial measures can be carried out optimally and correctly. Also, the effectiveness of a procedure like 'skim-and-burial' ploughing needs to be described with caution, as the reduction in crop root uptake of contaminants depends on the rooting depth of the specific crop and the fertilisation status of the different soil layers.

Further, it is important to note that the 'burial' effect is not equally good in all soils, and for instance, in very dry, loose or sandy soils, the contamination is likely to be mixed over a topsoil layer rather than buried.

Figure 6 A test of the 'skim-and-burial' plough in the Bryansk region, Russia.

5.2. Brief descriptions of some remedial measures for contaminated agricultural ecosystems

The following sections outline a number of techniques designed to be used in the recovery phase following a contaminating incident to reduce doses from food production in an agricultural ecosystem (based on the results of the European STRATEGY project; www.strategy-ec.org.uk). Full descriptions of these methods in the above-mentioned standardised datasheet format can be found on the STRATEGY web site. Quoted effectiveness estimates are derived from the STRATEGY templates, unless specifically stated otherwise. It should be noted that since, by far, most of the investigations that have been made of remedial measures are linked to caesium contamination, the effectiveness is often only described with respect to caesium. Many techniques (e.g. bans, dilution, change of land use) are obviously equally effective for all radionuclides, but some others (e.g. involving the use of AFCF) are specifically designed for caesium contamination, whereas yet others require adjusting to be effective for a given type of contaminant (e.g. by changing the thickness of the topsoil layer to be removed or buried, according to the extent of migration of the contaminant in question).

5.2.1. Food ban

If food products have activity concentrations that exceed intervention limits, they may be banned. This will completely remove the contaminated food from the consumers' market. However, large quantities of waste can result from this, which must be managed.

5.2.2. Dilution

By mixing a contaminated food product with an uncontaminated one, the contaminant concentration in the mixed product may be lowered sufficiently so that it no longer exceeds council food intervention levels (CFILs).

5.2.3. Early crop removal

Standing crops may become highly contaminated, particularly if the contamination occurs in the absence of precipitation. The transfer of this contamination to the underlying soil can be minimised if the crops are removed from the field prior to the first heavy rainfall. Typically 50%–70% of the contamination in a field can be removed by rapid action. However, the removed crops constitute waste that must be managed safely.

5.2.4. Processing of crops

Blanching or boiling of crops can generally remove more than 50% of caesium contamination. Canning can over time reduce the crop contamination level by a further ca. 50% through transfer to the canning solution. However, the products may still not be sellable on any market (IAEA, 2009).

5.2.5. Processing of milk

Processing procedures that convert milk into butter and cheese can significantly reduce the concentration of radionuclides like caesium and strontium in foodstuffs, by highly variable factors. For short-lived radionuclide contamination, processing into products that can be stored more easily may be advantageous. However, the products may still not be sellable on any market (IAEA, 2009).

5.2.6. Salting of meat

Salting of meat is a method that can be applied to achieve a reduction (of up to about 80% for caesium and strontium) in the concentration of radiocaesium in the final meat product. However, the products may still not be sellable on any market (IAEA, 2009). Salting does not reduce the meat contaminant concentration of a number of other potentially important radionuclides, including ^{60}Co, ^{95}Nb, ^{95}Zr, ^{140}Ba, ^{140}La, ^{141}Ce/^{144}Ce, ^{192}Ir, ^{138}Pu, ^{241}Am and ^{252}Cf.

5.2.7. Feeding animals with foodstuffs exceeding intervention levels

Food exceeding intervention level for human consumption (e.g. contaminated grains) may still be used as animal fodder, as only a fraction of the contamination would be transferred to animal products, which could still be produced with acceptably low contamination levels.

5.2.8. Ploughing

As mentioned earlier, different ploughing procedures exist which can all place contamination deeper in the vertical soil profile, whereby root uptake may be reduced. At the same time, the external dose rate will also be reduced. The effectiveness depends on the ploughing design, depth, soil type, fertilisation status and crop root depth.

5.2.9. Topsoil removal

A thin topsoil layer, which contains only the contaminated layer, is removed. The effectiveness depends on, amongst other things, how well the procedure is optimised in relation to the vertical contaminant profile. It can generate substantial amounts of waste that must be managed safely.

5.2.10. Application of chemicals to soil

In soils with low potassium content, addition of potassium can reduce crop uptake of radiocaesium. Similarly, in soils with low pH or calcium status, addition of lime can reduce crop uptake of radiostrontium. For both techniques, uptake reduction by a factor of up to about 5 has been recorded.

5.2.11. Change in land use

This drastic method may be deemed necessary to generate an income for the population in strongly contaminated areas, where food production is no longer practicable. For instance, non-food items like cotton, flax or bio-ethanol could be produced.

5.2.12. Decontamination of milk

Different techniques may be applied for decontaminating milk, including magnetic separation, ion exchange, electrodialysis and ultrafiltration. Such techniques are generally rather expensive, but can reduce the contaminant concentration in milk by more than a factor of 10. However, the products may still not be sellable on any market. Applicability is generally only known for caesium and strontium contaminants.

5.2.13. Change in slaughtering time

One option is to slaughter animals shortly after the area becomes contaminated, so that there is no time for the animal meat to become contaminated. Fattening periods for animals could instead be extended until meat contaminant concentrations have declined to an acceptable value. Also the seasonal changes in animal diets may be exploited, so that animals are slaughtered during the season where their body content of contaminants is lowest.

5.2.14. Clean feeding

If animals are fed with clean fodder over a period prior to slaughtering, the contaminant concentrations in their meat may be reduced to an acceptable level. The animals may be prevented from eating pasture over a few weeks and fed with uncontaminated fodder (e.g. silage) over this period. Animals may also be effectively prevented from grazing in contaminated fields by fencing them in over a period of time. In principle, the technique can be applied to all contaminants, but due to comparatively long physical and biological half-lives, the effect of applying the technique to reduce doses from strontium and actinides will be limited (Nisbet et al., 2007).

5.2.15. Feeding with concentrates containing AFCF

By adding AFCF to cows' diets, their uptake of radiocaesium from the gut can be reduced. This in turn reduces the content of radiocaesium in meat and milk (typically by a factor of 5–10).

6. Remediation in Contaminated Forest Ecosystems

6.1. Specific needs for remediation in forest ecosystems

Forested areas may, in general, be used by humans both for industrial production and as recreational resorts. Therefore, contamination in forests can affect humans in a number of different ways. Relevant dose pathways specific to this type of areas may include (Junker et al., 1998):

- external exposure from contamination on the forest floor;
- external exposure from contamination in trees;
- external exposure from handling of forestry material;
- external exposure from industrial production based on contaminated wood;
- exposure from consumption of forest foods (e.g. mushrooms, nuts, berries and game);
- exposure from inhalation following forest fires; and
- exposure from inhalation following combustion of contaminated wood.

A number of these dose contributions may require reduction in connection with an atmospheric contamination scenario. A perhaps dramatic alternative may be to restrict access to the contaminated forested areas. Selection of effective remedial actions is highly problematic in that nearly all the specific technical knowledge that has been accumulated over the years relates to contamination of forest with ^{137}Cs.

6.1.1. Reduction in external exposure from contamination on the forest floor

The organic top layer in the forest floor will after some time following the contamination be expected to contain a very large fraction of the total contaminant inventory in the forest. The existing forest models (IAEA, 2002) indicate that, within the first year, 80%–95% of the caesium contamination in a coniferous forest may occur in this layer. After some 10–20 years, a significant part of this will have reached the underlying, more-mineral-rich soil. On the basis of investigations of the radiocaesium distribution in soil in the contaminated Belarusian forests, it was estimated that some 80% still lay in a combustible top layer 12 years after the Chernobyl accident (Junker et al., 1998).

In comparison, investigations by Grebenkov (1996) indicate that more than 90% of the radionuclides in most forested regions of Belarus were at that point located in biomass. This agrees reasonably well with the ecological half-lives reported by Rühm et al. (1996) for the transfer of radiocaesium between the different soil layers in a coniferous Bavarian forest. It should be stressed that ecological half-lives are highly site specific, since radiocaesium dynamics may be influenced, for instance, by pH, layer thickness, clay mineral content, climatic conditions and type of forest (e.g. deciduous or coniferous) (Rühm et al., 1998). On this basis, it is clear that the effect of harvesting the top, say 5–7 cm, organic layer of the forest floor will vary with time, and the process should be carried out as early as possible following the first leaf fall.

The effect would be expected to be substantial, provided that a large fraction of the organic layer is removed. Efficient mechanised techniques for removal of forest litter/organic topsoil may include rotating brushes powered by hydraulic engines and mounted on a tractor. This type of device was successfully tested in contaminated areas of Belarus (Hubert et al., 1996; Roed et al., 1995). In 1995 the cost of the brushing device was of the order of 5,000 euros, and the method operates at a speed of about $500\,m^2\,h^{-1}$. Alternative techniques for this operation are described by Hubbard et al. (2002). Removal of the organic top layer of the forest floor would reduce not only the external dose rate in the forest, but also the contamination level in subsequently growing forest fruits (mushrooms, berries, etc.). Alternatives that might well not be publicly acceptable involve restrictions on access to the area by forest workers and/or the general public (Shaw et al., 2001; Fesenko et al., 2005).

6.1.2. Reduction in external exposure from contamination in trees

A very large part of the total dry deposition of small particulate contaminants to a forest in leaf occurs on tree leaves or needles acting as a very efficient 'aerosol filter'. However, modelling of the fate of radiocaesium in coniferous forests (Tikhomirov and Shcheglov, 1994; Linkov et al., 1997) indicates that, without action, the contamination in the trees will rather rapidly (within the first year) decline to a vanishingly small fraction of the total contaminant inventory in the forest. The level will then again build up to a maximum of about 5%–15% of the total inventory after some 15–20 years, if the forest floor is not treated. After this point, the physical half-life of the contaminant will largely govern the wood contamination level.

The effect of felling trees in relation to reduction in the contaminant inventory and radiation field in the forest will thus depend on the time at which the operation is carried out. If the felling (and possibly treatment of the forest floor) is conducted in 'chessboard' fields, thus leaving parts of the forest for some time, while other parts can re-establish, the forest remains and the adverse effect on biodiversity will be limited (Junker et al., 1998).

More or less sophisticated tree-felling machines are quite generally available over the world. Such a machine was tested in a French forest area in connection with the CEC-supported ECP/4 project (Hubert et al., 1996). The machine is designed for felling coniferous trees with diameters between 30 and 50 cm. It is mobile and mounted with an arm equipped with a cutting head. This head cuts the tree with a circular chain saw. The arm can then move the tree to another area (within a radius of 10 m), where the same cutting head, in combination with two wheels 'feeding' the tree (at a maximum speed of $4\,m\,s^{-1}$), can be used to remove branches and bark. It was estimated that such a machine can perform the work of about 12 manual felling workers with chainsaws. Similar techniques for this action have been described by Theilby and Bøllehuus (1999) and Bøllehuus (1990).

An alternative technique is to reduce the uptake to the tree rather than felling it. Here fertilisation of forest soil (lime for strontium and potassium for caesium) has been reported to be useful (Moberg et al., 1999; Guillitte et al., 1994; Levula et al., 2002). The effect is lasting, particularly if fertilisation is repeated over the following years (Kaunisto et al., 2002). Also wood ash may be used as fertiliser (Hubbard et al., 2002; Levula et al., 2002). A further beneficial effect can be obtained in areas where trees have been felled and it is possible to harrow the soil. Also this will improve subsequent tree growth and thus increase growth dilution of contaminants in trees (Shaw et al., 2001). Another alternative might be to leave the forest for many decades until attenuation in the thick trunks and radioactive decay reduce the external dose rate adequately.

6.1.3. Reduction in external exposure from handling of forestry material

According to calculations made with the RESRAD BUILD model (Junker et al., 1998), the closer contact to the contamination due to handling wood rather than just staying in the forest would increase the dose rate contribution from the trees by a factor of about 1.4. However, as the dose rate contribution from the trees, in many cases, is likely to be relatively small compared to that from the forest floor, the governing factor in connection with forestry worker doses is often simply the amount of time spent in the forest, which may be reduced through shorter work shifts or, if needed, access bans to the most contaminated areas (Fesenko et al., 2005).

6.1.4. Reduction in external exposure from industrial production based on contaminated wood

To comply with the recommendations of ICRP Publication 82 (2000), the maximum permissible contamination levels in forest products (e.g. timber,

Table 3 Conversion factors from ^{137}Cs content (Bq kg^{-1}) in wood to annual individual dose (mSv) from different types of exposure in connection with the use of contaminated wood or wood products (IAEA, 2003).

Type of exposure	Conversion factor (mSv $Bq^{-1} kg^{-1}$ in wood)
External dose from stored wood at sawmill	1×10^{-4}
External dose from wooden floor	3×10^{-6}
External dose from wooden bed	3×10^{-5}
External dose from wooden furniture set	3×10^{-5}

pulp wood and firewood) should be those which lead to an annual individual dose of 1 mSv. Table 3 gives a number of examples of calculated dose conversion coefficients for different types of applications of wood products, as suggested by IAEA (2003). The values should be regarded as examples of possible conversion factors, based on a series of assumptions regarding, for example, types and amounts of wood applied and geometries. Such values may be applied to determine whether the contamination in a particular product may be problematic and whether the product needs to be replaced or discarded. Early felling may prevent contamination of the core wood, and also fertilisation may reduce the wood contamination (particularly with strontium and caesium) through dilution and increased growth.

Alternatively, delaying felling for long periods may also reduce wood contamination with long-lived radionuclides, through radioactive decay and reduced root uptake due to increased fixation of some contaminants in soil (Shaw et al., 2001).

6.1.5. Reduction in exposure from consumption of forest foods (e.g. mushrooms, nuts, berries and game)

In contaminated areas, people may be advised against gathering forest foods for consumption. However, it may still be safe to spend time in the contaminated forest, so that restrictions on entry may be unnecessary. This is because external dose components are often much lower than the ingestion dose from consuming forest foods gathered in the same forest (Hubbard et al., 2002; Levula et al., 2002).

By pinpointing differences in contaminant uptake by forest products, for instance by different mushroom species, ingestion doses can be reduced (e.g. mycorrhizal fungi accumulate more caesium than saprophytic fungi; Nisbet et al., 2007). However, in areas contaminated by the Chernobyl accident, the reduction in the total annual effective dose achievable in this way has been found to be only 3%–4%, whereas restricting all mushroom collection gave a corresponding dose reduction by 9%–19% (Fesenko et al., 2005).

Fertilisation has been used as a countermeasure to reduce the contaminant contents of berries, but in the former Soviet Union, the overall resultant reduction in the total annual effective dose after the Chernobyl accident was insignificant (Fesenko et al., 2005).

Distribution of salt licks to reduce caesium uptake in animals is a method that can significantly reduce doses received from consumption of game products. Alternatively, changing or restricting hunting seasons may be necessitated (Nisbet et al., 2007).

In some areas of the former Soviet Union that were contaminated by the Chernobyl accident, even domestic animals graze in forested areas. By feeding these animals with uncontaminated grass, the total annual effective dose to a critical population group could be reduced by 54%–58% (Fesenko et al., 2005).

Various food preparation procedures can reduce the contaminant content in forest foods (Rantavaara, 1990; Fesenko et al., 2005); however, the effectiveness of such actions was highly variable.

6.1.6. Reduction in exposure from inhalation following forest fires

A forest fire is the type of scenario that could generate in the air the greatest resuspension of contaminated material from a forest. This type of scenario could affect populations living at some distance from the forest. The particles created by a fire would often be comparatively large (Junker et al., 1998; Roed et al., 2000), and sheltering indoors with air ducts closed off (as described under Section 5.4.1) would therefore be likely to be highly effective in reducing such inhalation dose problems.

6.1.7. Reduction in exposure from inhalation following combustion of contaminated wood

Huge forested land areas of Belarus were contaminated by the Chernobyl accident. At the same time, this country has a deficient energy production, and forest wood is frequently combusted in domestic furnaces. This combination, which could also occur in other places, inevitably leads to combustion of contaminated wood, whereby contaminants are released to the atmosphere. A consequence analysis, however, demonstrated that this exposure pathway might not be of major concern in comparison with other pathways (Junker et al., 1998). Nevertheless, combustion of the contaminated wood in specially designed power plants would be a solution that could practically eliminate entirely any atmospheric release problem. With a bag filter, it has been demonstrated in the Rechitza power plant in Belarus that some 99.5% of the caesium contamination could be removed from the flue gas prior to release from the stack (Roed et al., 2000). Although quite high concentrations of contaminants were observed in ash produced at the plant, an analysis showed that both internal and external doses to workers at a combustion plant fired with Belarusian forest biomass from areas contaminated by about 1 MBq m^{-2} of ^{137}Cs could be maintained at a level that is unproblematic (Andersson et al., 1999).

6.2. Remedial measures for contaminated forests considered in the STRATEGY project

In the previous section, a number of possible remediation options for dealing with various radiological problems originating from contamination in forest ecosystems were described. This section provides descriptions of a number of remedial measures that were specifically suggested for such areas in connection with the European STRATEGY project (Howard et al., 2005). Full descriptions of the pros and cons of these methods in the above-mentioned standardised datasheet format can be found on the STRATEGY web site (www.strategy-ec.org.uk).

6.2.1. Modification of tree-felling time

As mentioned earlier, forest wood can over time become increasingly contaminated through root uptake, and if felling can be carried out in advance of the original plan, radionuclide root uptake to stem wood can be minimised, so that it may not exceed intervention levels. Conversely, felling may be postponed if it is deemed that soil contaminant immobilisation, migration and/or physical decay of radionuclides will reduce the wood contamination over time.

6.2.2. Forest soil treatment with fertiliser

The addition of fertilisers (NPK, PK or potash) to forest soil will, as already mentioned, reduce the uptake of radiocaesium by trees and other vegetation (including forest foods like berries, mushrooms and nuts). Growth dilution will also occur due to soil improvement with fertilisers, which also reduces concentrations of other radionuclides in vegetation.

6.2.3. Restrictions on the use of wood

Restrictions may be implemented on the use of different parts of trees for specific industrial purposes. It is then important to consider the radionuclide concentrations both in the raw wood material and in wood that has undergone various processing procedures. Wood that is not deemed suitable, for example, for furniture production may be applicable to paper manufacturing, where chemical pulping with bleaching can reduce contaminant concentrations in the product.

A problem in application of contaminated wood in a pulp mill recycling system is, however, that the radiocaesium is accumulated in the digesting liquor. Ravila and Holm (2000) reported that typically 80%–95% of the radiocaesium in the recovery boiler is not supplied by the actual batch of wood, but from the recycled digesting agent. Ravila and Holm (1994) estimated that, even in Sweden, where contamination levels are not very high, projected 50-year doses to wood pulp factory workers may reach 50 mSv.

7. Remediation in Contaminated Aquatic Ecosystems

Aquatic reservoirs including lakes, rivers, groundwater and seas, all have characteristic and site-specific behaviour governed by combinations of hydrological and morphological parameters of the reservoir and its drainage area. Remediation of such reservoirs following a contamination of any type of substance is therefore largely dependent on site-specific parameters. This is an important fact since it severely limits the general preparedness that may be applied to aquatic reservoirs. Also, in general, remediation plans for contaminated waters may be expensive and may include large engineering costs. The effect of such remediation efforts therefore normally must be on a cost-benefit basis and chosen according to the well-known ALARA principles and be compared to risks from other toxic substances present in the water. Situations where other principles may be applied certainly may occur during the early stage of a contamination event where psychological, social and political factors must be taken into account.

In general, population doses from aquatic pathways are often less than those from terrestrial pathways, depending on our food habits. The main dose pathways from aquatic sources originate both from their use as drinking water supplies and from aquatic foodstuff. In some cases, transfer of radioactivity to land through water irrigation may become important, and similarly lakes used in power production may alter in volume, sometimes exposing contaminated bottom sediments, thereby increasing the risk of aeolian transport. Due to the self-shielding of water, external doses in connection with recreational use of lakes may be considered less important.

The experience of remediation of aquatic reservoirs following radioactive contamination largely comes from the Chernobyl accident and from work at weapons production sites in the United States (e.g. Hanford). For obvious reasons, remediation plans have never been thought of in connection with marine waters even though locally (e.g. Sellafield, UK) huge amounts of radioactive waste have been transferred to the sea. The combinations of volume, dispersion rate and chemistry make the health hazard from ingestion of seafood minimal in spite of the large amounts of artificial radionuclides dumped at sea. Remediation of marine waters is, due to the above-mentioned characteristics, a nearly impossible task, and any human effort is likely to be completely masked by the continuous variations in natural processes affecting seawater biogeochemistry. However, it may be worth mentioning in this context that due to the worries about global climate change, serious attempts have been made to fertilise (iron) parts of the world oceans (Southern Ocean) in order to increase bioproductivity and thereby the vertical flux of carbon from the surface ocean. Such an increased flux of particulate matter would also have an effect on particle-reactive radionuclides, and in fact, naturally occurring particle-reactive radionuclides (e.g. ^{234}Th) are commonly used to monitor this vertical flux. From a dose-forming point of view, such efforts have little impact since fishing today is done in already fertile waters for several reasons, and vertical fluxes of particle-reactive material and removal rates are determined by natural processes. The contribution to the collective dose originating from seafood in all the world oceans is by far dominated by naturally occurring radioisotopes and largely by ^{210}Po.

The situation may be completely different in the case of contamination of lakes, groundwater and streams where several factors may combine to create a real radiation risk following a release of radioactive material. Due to the often-limited volume of these reservoirs, they are relatively more dependent on their drainage areas than are marine waters. Remedial action must therefore be focused on both aquatic ecosystems and the drainage or source area. Even though site-specific measures are entirely necessary, some general actions can be applied.

Drainage area:

- preventing flooding events when the ground is contaminated in order to limit transport to the water reservoir;
- removing contaminated soil;
- increasing the vegetation in order to better bind soil and radionuclides; and
- Different types of reverse flow away from the reservoirs (channels, groundwater drainage, etc.).

Aquatic system:

- if used as a drinking water reservoir, a change of intake point to water depths with less-contaminated water or change of reservoir;
- in running water, creating sedimentation basins to remove suspended matter;

- adding chemicals like phosphorus and nitrogen to fertilise lake water and thereby increase removal rate of radionuclides through an increased amount of particulate matter and changing the partitioning coefficient of the radionuclides to suspended matter;
- adding chemicals like potash and suspended particles like clay to lake water in order to increase pH and change the partitioning coefficient of the radionuclides to suspended matter;
- building filtration dams made of mixture of sorbents (e.g. zeolites, charcoal, iron hydroxides, etc) and supplementary purification of drinking water;
- restrictions on consuming aquatic foodstuff; and
- alternative foodstuff preparation.

In general, the situation is very different if the reservoir is used for drinking water or if it is mainly used for foodstuffs, and actions should be grouped into those aimed for drinking water and those for reducing doses from aquatic foodstuff. The rapid transfer of radioactivity to the population from a drinking water reservoir drastically limits remedial measures other than moving intake points or changing drinking water supply. In practice, this is a matter of available resources and cost versus gain. Surface reservoirs like lakes that become contaminated by atmospheric deposition may still be used for drinking water, depending on season. Lakes in the temperate regions of the world are normally stratified during winter and summertime, while mixing of the water occurs in spring and autumn when the water column temperature becomes homogenous. Thus, if the surface water becomes contaminated, the water volume below the wind-mixed and thermally mixed layers is still unaffected as long as the lake is stratified. The effect of stratification is exemplified in the drinking water dams that supplied Kiev during the Chernobyl accident. At the time of the accident, the water was sufficiently stratified so that the contaminated water largely remained as surface water. However, it was believed that radionuclides were associated with the bottom sediments, and therefore, the surface water at an initial stage was used as drinking water. If the bottom dam gates had been opened instead, relatively clean water could still have been used.

It is worth learning from the efforts in remediation of the Chernobyl aquatic reservoirs. The large number of attempts to reduce the radioactivity in the waters follows a trial-and-error pattern, which may be expected to take place whenever a contamination event occurs due to the significant stress and demand among the population to take actions even if they are not always effective. The extremely complex interaction of a number of natural processes such as flooding, changes in water chemistry and suspended load, ice formation, wind-blown dust or soil erosion in a dynamic way makes most solutions effective only temporarily. The problems and solutions experienced during the Chernobyl accident have been summarised in several reports and can be found in IAEA (2000) and IAEA (2006), while the reports on remedial actions for contaminated groundwaters in the world's largest cleanup (US DOE Hanford) can be found at the web site http://www.hanford.gov/cp/gpp/library/programdocs.cfm.

Apart from the hydrological parameters, the effect of an aquatic contamination depends naturally on the type of radionuclide(s) and its(their) physicochemical properties. In general, elements that relatively strongly attach to surfaces (e.g. actinides, caesium) in freshwater systems will be removed by sedimentation, and the long-term water concentrations will depend on the partitioning between water and sediment, and the main risk from foodstuffs is where the food chain includes bottom-living species. If the radionuclide behaves conservatively in freshwater (e.g. strontium), removal is more difficult. Sorbents may be applied but will lead to a small-volume source that may need sophisticated handling and shielding, and the drinking water quality may change significantly. In addition to the radioisotope itself, the interaction between the element and water chemistry and microbial life causes a speciation of the radioelement which governs its behaviour. The partitioning between 'dissolved' and particulate state changes depending on the ligands present. In practice, this is governed by suspended load, pH, ionic strength and concentration and composition of natural dissolved organic matter (DOM). For lakes high in DOM, the fraction of reactive radionuclides in dissolved state is generally much higher than for clear water lakes, and a substantial portion of even reactive elements like the actinides are present in association with submicron-sized humic matter colloids. Colloidal transport has been of major concern in many aquatic reservoirs and has also been a major issue in long-term nuclear waste repositories where the presence of colloidal matter in groundwater may greatly enhance the transport of reactive elements. Examples of speciation studies in contaminated groundwater and wetland (Chalk River Laboratories) and how the speciation of some radionuclides may give insight into transport phenomena are given in Cooper and McHugh (1983) and Caron and Mankarios (2004). An example of a similar study of plutonium in the Hanford area is presented in Dai et al. (2005). Such studies may appear to be of academic nature but are absolutely necessary with respect to being able both to employ correct sorption techniques and to model groundwater transport and transfer contaminants to biota. For the US DOE Hanford site alone, more than 100 publications in scientific journals relating to its geology, biogeochemistry and hydrochemistry appeared during 2000–2004.

Keeping this in mind, direct remedial actions applicable to a drinking water reservoir in the early stage (days to weeks) are most likely only restrictions and redirected use of the water as well as change of location of the water supply. At a later stage (weeks to months), the water residence time in the reservoir as well as the biochemical nature of contaminant plus water determines what measures can be undertaken. For running water, the residence time is so short that remedial measures to the water itself are unpractical. Sedimentation dams may become of importance to save downstream use of the water. For lakes with long water residence time, increased partitioning of the radionuclides to suspended matter may become possible by increasing the suspended load through fertilisation (phosphorus, nitrogen), liming or dispersing clay. In combination with sand filters and selective filtration for drinking water (zeolites, charcoal, iron oxides, etc), a large fraction of most radionuclides may be removed. However, the specific design of such measures will, as mentioned above, be highly dependent on the type of reservoir and radionuclide. As an example, research on the $^{99}TcO_4^-$ ion in US DOE Hanford groundwater resulted in the development of a novel sorbent based on a strongly magnetic iron sulphide material produced by sulphate-reducing bacteria in a bioreactor (Watson and Ellwood, 2003).

In the long term, a contaminated lake will gradually transfer activity to the sediments, and at a certain point the water concentrations will depend on the release rate from sediments, which in turn depends on a number of diagenetic processes and on interstitial water exchange with the overlaying water mass. In this case, it may be worth either dredging the lake of contaminated sediments or covering the sediments with uncontaminated material. Both processes are complicated and associated with large uncertainties with respect to the gain achieved. Contamination of groundwater differs from events in lakes and streams mostly because the flow pattern and communication between groundwater reservoirs and surface water are usually less well known. In order to perform useful remedial measures, thorough studies first need to be conducted. At an early stage when the groundwater is used as a drinking water supply, restrictions of its use are likely to be the only solution.

With respect to foodstuffs from contaminated lakes, remedial measures are on a long-term timescale (years). Peak values in biota normally appear several months or even years after the contamination event, all depending on the residence time of the radionuclide in the lake system. Restrictions of foodstuffs are the most direct measure. For radionuclides having a long residence time in the lake, significant removal through remedial measures will probably be almost impossible. Most measures tried out after the Chernobyl accident in boreal lakes in Sweden showed that reduction in concentration of ^{137}Cs in fish by more than a factor of 2 over an extended period of several years was difficult (IAEA, 2000). In most cases, the reduction was insignificant. Alternative food preparation methods may be considered (e.g. excessive salting of meat to reduce radiocaesium), but often result in loss in food quality.

If, as in the Chernobyl case, long-term contamination of the water reservoirs are mainly dependent on transfer from land, it becomes important to limit soil erosion and run-off from land to water. Soil erosion may effectively be limited by planting vegetation (grass, bushes trees), and such actions may also limit run-off and transfer to groundwater by stabilising otherwise non-fixed radionuclides (phytostabilisation), while run-off often requires engineering solutions like dykes. The seasonal changes in precipitation and water level of the lake must be considered, and flooding of the surrounding areas must be prevented by dykes that lead water in other directions. Removal of soil is another possibility but is practically possible only when the contaminated area is limited.

In conclusion, it merits mentioning that the complex interaction between hydrological and morphological parameters and different contaminants results in strongly site-specific solutions, and in many cases, detailed knowledge of the site is a prerequisite for effective remedial measures. The need for long-term solutions requires knowledge of the dynamics of the system, and site-specific modelling will be essential. Most measures to reduce foodstuff contamination should be anticipated to be of limited effect.

The European STRATEGY project dealt also with remedial measures for the aquatic environment, and some method descriptions in the above-mentioned standardised datasheet format can be found on the STRATEGY web site (www.strategy-ec.org.uk).

REFERENCES

Alexakhin, R. M. (1993). Countermeasures in agricultural production as an effective means of mitigating the radiological consequences of the Chernobyl accident. *Science of the Total Environment*, 137, 9–20.

Andersson, K. G. (1991). Contamination and decontamination of urban areas. PhD Thesis, Risø National Laboratory.

Andersson, K. G. (1996). *Evaluation of Early Phase Nuclear Accident Clean-Up Procedures for Nordic Residential Areas.* NKS Report NKS/EKO-5(96)18, ISBN 87-550-2250-2.

Andersson, K. G., and J. Roed. (1994). The behaviour of Chernobyl Cs-137, Cs-134 and Ru-106 in undisturbed soil: implications for external radiation. *Journal of Environmental Radioactivity*, 22, 183–196.

Andersson, K. G., and J. Roed. (2006). Estimation of doses received in a dry-contaminated living area in the Bryansk Region, Russia, since the Chernobyl accident. *Journal of Environmental Radioactivity*, 85, 228–240.

Andersson, K. G., M. Ammann, S. Backe, and K. Rosén. (2008c). *URBHAND-Decision Support Handbook for Recovery of Contaminated Inhabited Areas.* NKS Report. Nordic Nuclear Safety Research, Roskilde, Denmark, 140pp.

Andersson, K. G., J. Brown, K. Mortimer, J. A. Jones, T. Charnock, S. Thykier-Nielsen, J. C. Kaiser, G. Proehl, and S. P. Nielsen. (2008b). New developments to support decision-making in contaminated inhabited areas following incidents involving a release of radioactivity to the environment. *Journal of Environmental Radioactivity*, 99, 439–454.

Andersson, K. G., C. L. Fogh, M. A. Byrne, J. Roed, A. J. H. Goddard, and S. A. M. Hotchkiss. (2002). Radiation dose implications of airborne contaminant deposition to humans. *Health Physics*, 82, 226–232.

Andersson, K. G., C. L. Fogh, and J. Roed. (1999). Occupational exposure at a contemplated Belarussian power plant fired with contaminated bio-mass. *Radiation Protection Dosimetry*, 83, 339–344.

Andersson, K. G., T. Mikkelsen, P. Astrup, S. Thykier-Nielsen, L. H. Jacobsen, L. Schou-Jensen, S. C. Hoe, and S. P. Nielsen. (2008a). Estimation of health hazards resulting from a radiological terrorist attack in a city. *Radiation Protection Dosimetry*, 131(3), 297–307.

Andersson, K. G., J. Roed, M. A. Byrne, H. Hession, P. Clark, E. Elahi, A. Byskov, X. L. Hou, H. Prip, S. K. Olsen, and T. Roed. (2004). *Airborne Contamination in the Indoor Environment and Its Implications for Dose*. Risoe-R-1462(EN), ISBN 87-550-3317-2. Risø National Laboratory, Roskilde, Denmark.

Andersson, K. G., J. Roed, K. Eged, Z. Kis, G. Voigt, R. Meckbach, D. H. Oughton, J. Hunt, R. Lee, N. A. Beresford, and F. J. Sandalls. (2003). *Physical Countermeasures to Sustain Acceptable Living and Working Conditions in Radioactively Contaminated Residential Areas*. Risoe-R-1396(EN), ISBN 87-550-3190-0. Risø National Laboratory, Roskilde, Denmark.

Andersson, K. G., J. Roed, H. G. Paretzke, and J. Tschiersch. (1995). Modelling of the radiological impact of a deposit of artificial radionuclides in inhabited areas. In: *Deposition of Radionuclides, Their Subsequent Relocation in the Environment and Resulting Implications* (Ed. J. Tschiersch). EUR 16604 EN, ISBN 92-827-4903-7, pp. 83–94.

Anisimova, L. I., A. V. Chesnokov, A. P. Govorun, O. P. Invanov, B. N. Potapov, S. B. Shcherbak, L. I. Urutskoev, V. I. Kovalenko, C. V. Krivonosov, A. V. Ponomarev, V. P. Ramzaev, and V. A. Trusov. (1994). *Analysis of Some Countermeasures Efficiency Based on Radioecological Data, Deposit Measurements and Models for External Dose Formation. Case Study of Two Russian Settlements Yalovka and Zaborie*. Annual Progress Report of the ECP4 Project Supported by the European Commission. EMERCOM, Moscow, Russia.

Argonne National Laboratory. (2005). Radiological Dispersal Device (RDD), Human Health Fact Sheet. Argonne National Laboratory, EVS, USA.

ARGOS. (2008). ARGOS CBRN Whitepaper. Decision Support for Emergency Management, www.pdc.dk/argos

Bøllehuus, E. (1990). *Test af Valmet 701 og Silvatec 454TH*. maskinrapport nr. 5 (in Danish), Miljøministeriet, Skov- og Naturstyrelsen, Denmark, ISBN 87-503-8617-4.

Brown, J., K. Mortimer, K. Andersson, T. Duranova, A. Mrskova, R. Hänninen, T. Ikäheimonen, G. Kirchner, V. Bertsch, F. Gallay, and N. Reales. (2008). *Generic Handbook for Assisting in the Management of Contaminated Inhabited Areas in Europe Following a Radiological Emergency, Parts I-V*. Final Report of the EC-EURANOS Project Activities CAT1RTD02 and CAT1RTD04, EURANOS(CAT1)-RP(07). Radiation Protection Division, Health Protection Agency, UK.

Bunzl, K., W. Schimmack, K. Kreutzer, and R. Schierl. (1989). The migration of fallout ^{134}Cs, ^{137}Cs and ^{106}Ru from Chernobyl and ^{137}Cs from weapons testing in forest soil. *Zeitschrift für Pflanzenernährung und Bodenkunde*, 152, 39–44.

Caron, F., and G. Mankarios. (2004). Pre-assessment of the speciation of ^{60}Co, ^{125}Sb, ^{137}Cs and ^{241}Am in a contaminated aquifier. *Journal of Environmental Radioactivity*, 77, 29–46.

Clark, D. E., and W. C. Cobbin. (1964). *Removal of Simulated Fallout from Pavements by Conventional Street Flushers*. Report. US Naval Radiological Defence Lab, USNRDL-TR-797.

Cooper, E. L., and J. O. McHugh. (1983). Migration of radiocontaminants in a forested wetland on the Canadian shield: nuclide speciation and arboreal uptake. *Science of the Total Environment*, 28, 215–230.

Cremers, A., A. Elsen, P. de Preter, and A. Maes. (1988). Quantitative analysis of radiocaesium retention in soils. *Nature*, 335, 247–249.

Da Silva, C. J., J. U. Delgado, M. T. B. Luiz, P. G. Cunha, and P. D. de Barros. (1991). Considerations related to the decontamination of houses in Goiânia: limitations and implications. *Health Physics*, 60, 87–90.

Dai, M., K. O. Buesseler, and S. M. Pike. (2005). Plutonium in groundwater at the 100K-Area of the U.S. DOE Hanford Site. *Journal of Contaminant Hydrology*, 76, 167–189.

Danish Atomic Energy Commission. (1970). *Project Crested Ice – A Joint Danish American Report on the Crash near Thule Air Base on 21 January 1968 of a B-52 Bomber Carrying Nuclear Weapons*. Riso Report No. 213, ISBN 87-550-0006-1. Danish Atomic Energy Commission, Risø, DK-4000 Roskilde, Denmark.

De Preter, P. (1990). Radiocaesium retention in the aquatic, terrestrial and urban environment: a quantitative and unifying analysis, Doctoraatsproefschrift Nr. 190 aan de Fakulteit der Landbouwwetenschappen van de Katholieke Universiteit te Leuven.

Dorrian, M. D., and M. R. Bailey. (1995). Particle size distributions of radioactive aerosols measured in workplaces. *Radiation Protection Dosimetry*, 60, 119–133.

Eged, K., Z. Kis, G. Voigt, K. G. Andersson, J. Roed, and K. Varga. (2003). Guidelines for planning interventions against external exposure in industrial area after a nuclear accident – part I: a holistic approach of countermeasure application, GSF-Bericht 01/03, GSF-Forschungszentrum für Umwelt und Gesundheit GmbH, Neuherberg, Germany, ISSN 0721-1694, 67pp.

Eriksson, M. (2002). *On Weapons Plutonium in the Arctic Environment (Thule, Greenland)*. Riso Report Riso-R-1321, ISBN 87-550-3006-8. Riso National Laboratory, Denmark.

Ferguson, C. D., T. Kazi, and J. Perera. (2003). *Commercial Radioactive Sources*. Occasional Paper No. 11, Center for Nonproliferation Studies, Monterey Institute of International Studies, California 93940, USA, ISBN 1-885350-06-6.

Fesenko, S. V., R. M. Alexakhin, M. I. Balonov, I. M. Bogdevitch, B. J. Howard, V. A. Kashparov, N. I. Sanzharova, A. V. Panov, G. Voigt, and Y. M. Zhuchenka. (2007). An extended critical review of twenty years of countermeasures used in agriculture after the Chernobyl accident. *Science of the Total Environment*, 383, 1–24.

Fesenko, S. V., G. Voigt, S. I. Spiridonov, and I. A. Gontarenko. (2005). Decision making framework for application of forest countermeasures in the long term after the Chernobyl accident. *Journal of Environmental Radioactivity*, 82, 143–166.

Garcia-Olivares, A., and C. E. Iranzo. (1997). Resuspension and transport of plutonium in the Palomares area. *Journal of Environmental Radioactivity*, 37, 101–114.

Grebenkov, A. J. (1996). *Preliminary Socioeconomic, Environmental, and Technical Evaluation of the Proposed Biomass Power Project*. Draft Report on the Chernobyl Bioenergy Project. Institute of Power Engineering Problems, Sosny, Minsk, Belarus.

Guillitte, O., G. Tikhomirov, G. Shaw, and V. Vetrov. (1994). Principles and practices of countermeasures to be carried out following radioactive contamination of forest areas. *Science of the Total Environment*, 157, 399–406.

Guillite, O., and C. Willrodt. (1993). An assessment of experimental and potential countermeasures to reduce radionuclide transfer in forest ecosystems. *Science of the Total Environment*, 137, 273–288.

Harper, F. T., S. V. Musolino, and W. B. Wente. (2007). Realistic radiological dispersal device hazard boundaries and ramifications for early consequence management decisions. *Health Physics*, 93, 1–16.

Hession, H., M. A. Byrne, S. Cleary, K. G. Andersson, and J. Roed. (2006). Measurement of contaminant removal from skin using a portable fluorescence scanning system. *Journal of Environmental Radioactivity*, 85, 196–204.

Hiemstra, Y., P. Stoop, and J. Lembrechts. (1997). *The Second Dutch National Survey on Radon in Dwellings: Set-Up of the Project.* Report No. 610058005. National Institute of Public Health and the Environment, Bilthoven, The Netherlands.

Hilton, J., R. S. Cambray, and N. Green. (1992). Chemical fractionation of radioactive caesium in airborne particles containing bomb fallout, Chernobyl fallout and atmospheric material from the Sellafield site. *Journal of Environmental Radioactivity*, 15, 103–111.

Howard, B. J., N. A. Beresford, A. Nisbet, G. Cox, D. H. Oughton, J. Hunt, B. Alvarez, K. G. Andersson, A. Liland, and G. Voigt. (2005). The STRATEGY project: decision tools to aid sustainable restoration and long-term management of contaminated agricultural ecosystems. *Journal of Environmental Radioactivity*, 83, 275–295.

Hubbard, L., A. Rantavaara, K. Andersson, and J. Roed. (2002). *Tools for Forming Strategies for Remediation of Forests and Park Areas in Northern Europe After Radioactive Contamination: Background and Techniques.* NKS Report NKS-52, ISBN 87-7893-105-3.

Hubert, P., L. Annisomova, G. Antsipov, V. Ramsaev, and V. Sobotovitch. (1996). Strategies of decontamination. Experimental Collaboration Project 4, European Commission, EUR 16530 EN, ISBN 92-827-5195-3.

IAEA. (2000). *Modelling the Transfer of Radiocesium from Deposition to Lake Ecosystems.* IAEA-TECDOC-1143. IAEA, Vienna, Austria.

IAEA. (2002). *Modelling the Migration and Accumulation of Radionuclides in Forest Ecosystems.* IAEA BIOMASS-1. IAEA, Vienna, Austria.

IAEA. (2003). *Assessing Radiation Doses to the Public from Radionuclides in Timber and Wood Products.* IAEA-TECDOC-1376. IAEA. Vienna.

IAEA. (2006). *Environmental Consequences of the Chernobyl Accident and Their Remediation: Twenty Years of Experience.* Report of the Chernobyl Forum Expert Group 'Environment'. *IAEA Radiological Assessment Report Series.* Vienna, Austria.

IAEA. (2009). *Quantification of Radionuclide Transfer in Terrestrial and Freshwater Environments for Radiological Assessments.* IAEA-TECDOC. IAEA, Vienna.

ICRP. (2000). *Protection of the Public in Situations of Prolonged Radiation Exposure.* ICRP Publication 82. *Annals of the ICRP*, 29 (1/2). Elsevier, Amsterdam.

ICRP. (2007). *The 2007 Recommendations of the International Commission on Radiological Protection.* ICRP Publication 103. *Annals of the ICRP*, 37 (2–4). Elsevier, Amsterdam.

Jackson, M. L. (1963). Interlayering of expansible layer silicates in soils by chemical weathering. *Clays and Clay Minerals*, 11, 29–46.

Jacob, P., S. Fesenko, S. K. Firsakova, I. A. Likhtarev, C. Schotola, R. M. Alexakhin, Y. M. Zhuchenko, L. Kovgan, N. I. Sanzharova, and V. Ageyets. (2001). Remediation strategies for rural territories contaminated by the Chernobyl accident. *Journal of Environmental Radioactivity*, 56, 51–76.

Janssen, M. P. M., L. de Vries, J. C. Phaff, E. R. van der Graaf, R. O. Blaauboer, P. Stoop, and J. Lembrechts. (1998). *Modelling Radon Transport in Dutch Dwellings.* Report No. 610050005. National Institute of Public Health and the Environment, Bilthoven, The Netherlands.

Junker, H., J. M. Jensen, H. B. Jørgensen, J. Roed, K. G. Andersson, P. D. Kofman, E. Bøllehuus, L. Baxter, and A. Grebenkov. (1998). *Chernobyl Bioenergy Project.* Final Report, Phase 1, ELSAMPROJEKT Report No. EP 8204-98-n600hju.

Kashparov, V. A., N. Ahamdach, S. I. Zvarich, V. I. Yoschenko, I. M. Maloshtan, and L. Dewiere. (2004). Kinetics of dissolution of Chernobyl fuel particles in soil in natural conditions. *Journal of Environmental Radioactivity*, 72, 335–353.

Kaunisto, S., L. Aro, and A. Rantavaara. (2002). Effect of fertilisation on the potassium and radiocaesium distribution in tree stands (*Pinus sylvestris*) and peat on a mire. *Environmental Pollution*, 117, 111–119.

Konoplev, A. V., A. A. Bulgakov, V. E. Popov, and T. I. Bobovnikova. (1992). Behaviour of long-lived Chernobyl radionuclides in a soil-water system. *Analyst*, 117, 1041–1047.

Krouglov, S. V., A. D. Kurinov, and R. M. Alexakhin. (1998). Chemical fractionation of ^{90}Sr, ^{106}Ru, ^{137}Cs and ^{144}Ce in Chernobyl contaminated soils: an evolution in a course of time. *Journal of Environmental Radioactivity*, 38, 59–76.

Lauritzen, B. (Ed.) (2001). *Huginn – A Late-Phase Nuclear Emergency Exercise*. NKS Report NKS-23, ISBN 87-7893-073-1. Nordic Nuclear Safety Research, Roskilde, Denmark.

Leao, J. L. B., and M. Oberhofer. (1988). The radiological accident at Goiânia. *Proceedings of the International Conference on Radiation Protection in Nuclear Energy*, International Atomic Energy Agency, Vienna, Austria, Vol. 2.

Levula, T., A. Saarsalmi, and A. Rantavaara. (2002). Effects of ash fertilisation and prescribed burning on macronutrient, heavy metal, sulphur and ^{137}Cs concentrations in lingonberries (*Vaccinium vitisidaea*). *Forest Ecology and Management*, 126, 267–277.

Linkov, I., B. Morel, and W. R. Schell. (1997). Remedial policies in radiologically-contaminated forests: environmental consequences and risk assessment. *Risk Analysis*, 17, 67–75.

Livens, F. R., and M. S. Baxter. (1988). Chemical associations of artificial radionuclides in Cumbrian soil. *Journal of Environmental Radioactivity*, 7, 75–86.

Long, C. M., H. H. Suh, P. J. Catalano, and P. Koutrakis. (2001). Using time- and size-resolved particulate data to quantify indoor penetration and deposition behaviour. *Environmental Science and Technology*, 35, 2089–2099.

Malanca, A. (1993). Indoor pollutants in a building block in Parma (Northern Italy). *Environment International*, 19, 313–318.

Moberg, L., L. Hubbard, R. Avila, L. Wallberg, E. Feoli, M. Scimone, C. Milesi, B. Mayes, G. Iason, A. Rantavaara, V. Vetikko, R. Bergman, T. Nylén, T. Palo, N. White, H. Raitio, L. Aro, S. Kaunisto, and O. Guillitte. (1999). *An Integrated Approach to Radionuclide Flow in Seminatural Ecosystems Underlying Exposure Pathways to Man (LANDSCAPE)*. Final Report SSI: 19, Research Contract FI4P-CT96-0039. European Commission Nuclear Fission Safety Programme, Swedish Radiation Protection Authority, Stockholm, Sweden.

Nisbet, A. F. (1993). Effect of soil-based countermeasures on solid–liquid equilibria in agricultural soils contaminated with radiocaesium and radiostrontium. *Science of the Total Environment*, 137, 99–118.

Nisbet, A. F., J. A. Mercer, N. Hesketh, A. Liland, H. Thørring, T. Bergan, N. A. Beresford, B. J. Howard, J. Hunt, and D. H. Oughton. (2004). *Datasheets on Countermeasures and Waste Disposal Options for the Management of Food Production Systems Contaminated Following a Nuclear Accident*. NRPB-W58, ISBN 0 85951 539 7. National Radiological Protection Board, Chilton, UK.

Nisbet, A. F., H. Rice, N. Hesketh, A. Jones, T. Jullien, V. Pupin, H. Ollagnon, F. Hardeman, B. Carlé, C. Turcanu, C. Papachristodoulou, K. Ioannides, R. Hänninen, A. Rantavaara, D. Solatie, E. Kostiainen, and D. Oughton. (2007). *Generic Handbook for Assisting in the Management of Contaminated Food Productions Systems in Europe Following a Radiological Emergency*. EURANOS (CAT1)-TN(06)-06, CEC-EURANOS Project.

Petryaev, E. P., S. V. Ovsyannikova, S. Y. Rubinchik, I. J. Lubinka, and G. A. Sokolik. (1991). State of radionuclides of the Chernobyl fallout in soils of Belarus. Proceedings of AS of BSSR1. *Physico-Energetical Sciences*, 4, 48–55.

Pöllänen, R., M. E. Ketterer, S. Lehto, M. Hokkanen, T. K. Ikäheimonen, T. Siiskonen, M. Moring, M. P. Rubio Montero, and A. Martín Sánchez. (2006). Multi-technique characterization of a nuclear bomb particle from the Palomares accident. *Journal of Environmental Radioactivity*, 90, 15–28.

Rantavaara, A. (1990). Transfer of radionuclides during processing and preparation of foods – Finnish studies since 1986. *Proceedings of the CEC Seminar*, 18–21 September 1989, Cadarache, France, Report CEC DG XI, 69-94.

Ravila, A., and E. Holm. (1994). Radioactive elements in the forest industry. *Science of the Total Environment*, 157, 339–356.

Ravila, A., and E. Holm. (2000). A survey of present levels of radiocaesium in Swedish pulp mill liquors and the implications for wood radiocaesium transfer factors: using Kraft mill liquors and an indicator of wood radiocaesium contamination. *Journal of Radioanalytical and Nuclear Chemistry*, 243, 573–577.

RODOS. (2008). RODOS homepage at Forschungszentrum Karlsruhe, Germany, www.rodos.fzk.de/Overview/Docs/RODOS_Brochure.pdf

Roed, J. (1990). *Deposition and Removal of Radioactive Substances in an Urban Area.* Final Report of the NKA Project AKTU-245, ISBN 87 7303 514 9. Nordic Liaison Committee for Atomic Energy at Risø, Denmark.

Roed, J., and K. G. Andersson. (1996). Clean-up of urban areas in the CIS countries contaminated by Chernobyl fallout. *Journal of Environmental Radioactivity*, 33, 107–116.

Roed, J., K. G. Andersson, A. N. Barkovsky, C. L. Fogh, A. S. Mishine, A. V. Ponamarjov, and V. P. Ramzaev. (2006). Reduction of external dose in a wet-contaminated housing area in the Bryansk Region, Russia. *Journal of Environmental Radioactivity*, 85, 265–279.

Roed, J., K. G. Andersson, C. L. Fogh, A. N. Barkovski, B. F. Vorobiev, V. N. Potapov, and A. V. Chesnokov. (1999). Triple digging – a simple method for restoration of radioactively contaminated urban soil areas. *Journal of Environmental Radioactivity*, 45, 173–183.

Roed, J., K. G. Andersson, C. L. Fogh, S. K. Olsen, H. Prip, H. Junker, N. Kirkegaard, J.-M. Jensen, A. J. Grebenkov, V. N. Solovjev, G. G. Kolchanov, L. A. Bida, P. M. Klepatzky, I. G. Pleshchenkov, A. A. Gvozdev, and L. Baxter. (2000). *Power Production from Radioactively Contaminated Biomass and Forest Litter in Belarus – Phase 1b.* Risoe-R-1146(EN), ISBN 87-550-2619-2. Risø National Laboratory, Roskilde, Denmark.

Roed, J., K. G. Andersson, and H. Prip. (Eds.) (1995). *Practical Means for Decontamination 9 Years after a Nuclear Accident.* Risoe-R-828(EN), ISBN 87-550-2080-1, ISSN 0106-2840. Risø National Laboratory, Roskilde, Denmark.

Roed, J., and R. J. Cannell. (1987). Relationship between indoor and outdoor aerosol concentration following the Chernobyl accident. *Radiation Protection Dosimetry*, 21, 107–110.

Roed, J., C. Lange, K. G. Andersson, H. Prip, S. Olsen, V. P. Ramzaev, A. V. Ponomarjov, A. N. Barkovsky, A. S. Mishine, B. F. Vorobiev, A. V. Chesnokov, V. N. Potapov, and S. B. Shcherbak. (1996). *Decontamination in a Russian Settlement.* Risoe-R-870, ISBN 87-550-2152-2. Risø National Laboratory, Roskilde, Denmark.

Rühm, W., L. Kammerer, L. Hiersche, and E. Wirth. (1996). Migration of ^{137}Cs and ^{134}Cs in different forest soil layers. *Journal of Environmental Radioactivity*, 33, 63–75.

Rühm, W., M. Steiner, L. Kammerer, L. Hiersche, and E. Wirth. (1998). Estimating future radiocaesium contamination of fungi on the basis of behaviour patterns derived from past instances of contamination. *Journal of Environmental Radioactivity*, 39, 129–147.

Salbu, B. (2000). Source-related characteristics of radioactive particles: a review. *Radiation Protection Dosimetry*, 92, 49–54.

Salbu, B., T. Krekling, and D. Oughton. (1998). Characterisation of radioactive particles in the environment. *Analyst*, 23, 843–849.

Shaw, G., C. Robinson, E. Holm, M. J. Frissel, and M. Crick. (2001). A cost-benefit analysis of long-term management options for forests following contamination with ^{137}Cs. *Journal of Environmental Radioactivity*, 56, 185–208.

Sohier, A. (Ed.) (2002). *A European Manual for 'Off-Site Emergency Planning and Response to Nuclear Accidents'*. SCK CEN Report R-3594. Belgian Nuclear Research Centre.

Steinhäusler, F. (2005). Chernobyl and Goiânia lessons for responding to radiological terrorism. *Health Physics*, 89, 566–574.

Tamura, T., and D. G. Jacobs. (1960). Improving cesium selectivity of bentonites by heat treatment. *Health Physics*, 5, 149–154.

Theilby, F., and E. Bøllehuus. (1999). Nye maskiner på Skogs-Elmia (in Danish), Skoven 8/99, Journal issued by Dansk Skovforening, Frederiksberg C, Denmark, ISSN 0106-8539.

Tikhomirov, F. A., and A. I. Shcheglov. (1994). Main investigation results on the forest radioecology in the Kyshtym and Chernobyl accident zones. *Science of the Total Environment*, 157, 45–57.

Time. (2002). Time online edition, 10 June, www.time.org

Tsalko, V. (2004). Statement by the Chairman of the Chernobyl Affairs Committee at the Council of Ministers of the Republic of Belarus at the International Conference 'Chernobyl Forum', www.belarusembassy.org/humanitarian/tsalko.htm

Voigt, G., K. Eged, J. Hilton, B. J. Howard, Z. Kis, A. F. Nisbet, D. H. Oughton, B. Rafferty, C. A. Salt, J. T. Smith, and H. Vandenhove. (2000). A wider perspective on the selection of countermeasures. *Radiation Protection Dosimetry*, 92, 45–48.

Watson, J. H. P., and D. C. Ellwood. (2003). The removal of the pertechnetate ion and actinides from radioactive waste streams at Hanford, Washington, USA, and Sellafield, Cumbria, UK: the role of iron-sulfide-containing adsorbent materials. *Nuclear Engineering and Design*, 226, 375–385.

Wedekind, L. (2002). Upgrading the Safety and Security of Radioactive Sources in the Republic of Georgia. IAEA News Centre, International Atomic Energy Agency, Vienna, http://www.iaea.org/NewsCenter/News/2002/georgia_radsources.shtml

Yusko, J. G. (2001). The IAEA action plan on the safety of radiation sources and security of radioactive material. In: *Radiation Safety and ALARA Considerations for the 21st Century*. Medical Physics Publishing, Madison, WI, pp. 137–142.

CHAPTER 10

Social, Ethical, Environmental and Economic Aspects of Remediation

Deborah Oughton[1,*], Ingrid Bay-Larsen[1] *and* Gabriele Voigt[2]

Contents

*Corresponding author. Tel.: +47 64965544; Fax: +47 64966007
E-mail address: deborah.oughton@umb.no

[1]Department of Plant and Environmental Sciences, P.O. Box 5003, Norwegian University of Life Sciences, 1432 Ås, Norway
[2]IAEA, Department of Nuclear Sciences and Applications, Agency's Laboratories Seibersdorf and Headquarters, 1400 Vienna, Austria

Radioactivity in the Environment, Volume 14
ISSN 1569-4860, DOI 10.1016/S1569-4860(08)00210-6

1. Introduction

Remediation measures can contribute much to reducing doses, alleviating anxiety and restoring the way of life in contaminated communities. However, these measures are not without side-effects; they can be expensive, socially disruptive or damaging to the environment. The Chernobyl accident showed that the consequences of a nuclear accident go far beyond health issues, and that there can be serious social and economic consequences, particularly with a breakdown in social infrastructure (UNDP, 2002; Bay and Oughton, 2005). Evaluation of remediation measures needs to address both the social and ethical costs, as well as the non-dose benefits of remedial action, such as increasing public understanding and control (Oughton et al., 2004; Fesenko et al., 2006). Remediation can also include measures to restore ecosystems, to secure the livelihood and social structure of affected populations or to stabilise the economic situation.

Although the primary objective of remediation is usually dose reduction, for an action to be justified, the benefits from dose reduction or averted dose should outweigh the costs of implementing the countermeasure (ICRP, 1989, 1991). It follows that a decision on how to reduce exposure to radiation will be an ethical judgement; we are making choices about which lives to save and at what cost. Until recently, the main criteria for remediation of contaminated areas concentrated on technical or economic constraints such as that the slope of the hill is too steep to allow ploughing, that there is insufficient labour or simply that the countermeasure is too expensive. However, in addition to these technical and economic arguments, an evaluation of remediation measures needs to take account of a wide range of social factors such as public perceptions of risk and dialogue with affected communities, as well as ethical aspects such as informed consent and the fair distribution of costs and benefits (Oughton, 1996; Oughton et al., 2003; Morrey and Allen, 1996; Hériard Dubreuil et al., 1999; EURANOS, 2006).

In most cases, remedial measures are applied after considerations of dose reduction to humans in the short, medium and long terms. Few thoughts have been given to the global effects of remediation, such as those affecting the environment and the society at large. In this chapter, therefore, the so-called side or secondary effects of remedial actions are addressed.

As a consequence, short-term and long-term remedial actions primarily address the reduction of internal and external doses to humans with the assumption that these will in a net benefit (IAEA, 1996). Any remedial measure, however, will comprise a number of factors, such as anxiety, reassurance and discomfort, and economic costs. The balance of these factors should be on the positive side to justify remediation. Therefore, a multi-attribute utility analysis which takes into account scores or weighting factors can be a useful tool to evaluate the impact of remediation. In the past, these analyses paid little attention to the environmental impacts of remediation, either to biota or to the ecosystem, nor were the effects of radiation exposure on biota taken into account. The latter has only recently attracted attention in radiation protection with the development of frameworks to assess the radiological impacts on animals and plants.

This chapter reviews the main environmental, social and ethical issues associated with remediation. Although environmental secondary effects and social and ethical aspects of remediation are strongly interlinked (e.g. economic and environmental effects heavily influence the wellbeing and comfort of an affected population), they are discussed in separate sections due to their specific nature. The chapter begins with an overview of environmental side-effects, continues with some of the consequences of remediation measures carried out after Chernobyl, including an estimate of economic costs, and then summarises the main ethical and social aspects of dose-reduction measures. Finally it addresses some of the remedial actions that are not primarily intended to reduce dose, such as compensation or information schemes.

2. Environmental Secondary Effects of Remediation

After any accidental or incidental release of radioactivity into the environment, immediate countermeasures and remedial actions are needed to varying degrees, depending on the exposure routes and the individuals or population groups that are most vulnerable. Assessment of the radiation exposure can identify the areas of most concern due to ecological function and is expressed as *radioecological sensitivity*. Radioecological sensitivity analysis is defined as the extent to which an ecosystem contributes to an enhanced radiation exposure to man and biota, integrating spatial and temporal current knowledge on pathways together with underlying parameters and attributes (Howard, 2000). Prior identification of radio-ecologically sensitive areas and exposed individuals should improve the focus for emergency planning and preparedness, and contribute to environmental impact assessment and improvement of remediation strategies.

In the past, after incidents such as the Chernobyl accident or East Ural Trail, measures were implemented in a rapid and in the most effective manner and mainly driven by dose reduction and cost efficiency (Alexakhin et al., 1996; Alexakhin et al., 2004; Roed and Andersson, 1996; Fesenko et al., 2007). However, depending on the nature of the contamination event and the radioecological sensitivity of an affected landscape, radioactive contamination of the ecosystem and food produced therein might be affected for decades. The physical and biological half-lives of radionuclides can be influenced by their ecological half-life which can often be greater than former two as it is driven by the ecosystem peculiarities. This is specifically true for semi-natural ecosystems where recycling and bio availability are major factors. Therefore, actions implemented only to reduce the dose to humans might be short-sighted and often not the best option if they do not consider also the whole society and living conditions in a radioactive contaminated land. The driving forces will be less concerned with radiological constraints and more with environmental, ecological and socio-psychological factors.

The environmental side-effects or secondary effects of remediation activities can be of positive or negative nature. These include the changes in soil properties, in air and water quality and also change of landscape and its resulting impact on the economic or perceived leisure value of an area. These effects are often difficult to quantify and therefore only qualitative judgments are applicable. An additional factor is the change of perceived damage or improvement in relation to time after a pollution or contamination event, as affected population groups adapt to the risks and change their behaviour or attitude towards their living conditions.

Decision matrices or decision-support systems which also introduce environmental secondary effects have been created. These matrices allow the assessment of the impact of remediation in a wider sense. They are mainly the result of studies involving geostatistical and geochemical modelling, directed and dedicated experiments as well as expert judgement related to contamination with radiocaesium and radiostrontium, for example in the consequences of the Chernobyl accident (Salt et al., 1999; Voigt et al., 2000; Salt and Rafferty, 2001; Fesenko et al., 2006).

Some of these have been included into decision-support systems for decision makers and are part of large research projects, such as CESER, RESTORE, TEMAS, STRATEGY, FORECO etc., supported during the last decade by the European Commission (Semioshkina et al., 2004). However, the environmental side-effects described in the following text are of generic nature and should be considered in any contamination scenario calling upon countermeasure applications as costs might become unexpectedly high.

The recommended remedial actions in agriculture or forestry are often a modified (extreme) form of agricultural/forestry practices such as changes in ploughing, fertilisation schemes, feeding and maintenance practices for animals. Therefore, knowledge of agriculture or forestry will assist the understanding on how remedial measures might impact agricultural or forest ecosystems. In the CESER project, a diverse range of potential side-effects for agricultural ecosystems were investigated and finally it focused on the 10 most significant side-effects which have the highest environmental and economic impacts. These have been assigned different scores to be selected by each individual stakeholder who wants to make use of the CESER decision-making system.

Soil erosion and sedimentation. The loss of soil via water or wind-induced transport is initiated by different ploughing and cultivation regimes. Sedimentation relates to deposition or accumulation of eroded soil in downhill land areas, in surface waters or in oceans having a detrimental impact on water quality, soil quality and nutritional value.

Soil organic matter loss. The humus content of the top soil can be affected by measures such as deep ploughing or liming or any other additives favouring mineralisation of soils. The total effect can be positive or negative.

Soil nutrient transport. Nutrient transport occurs mainly via water percolation and surface run-off. Loss of nutrients from the soil and their transfer into groundwater or surface water contributes to eutrophication of rivers and lakes, which can have a negative impact on water quality.

Soil contaminant transport. Soil pollutants such as heavy metals or organic pollutants such as herbicides and pesticides can be dissolved and transported via water to surface and groundwater reservoirs resulting in water contamination.

Ammonia emission. The volatilisation process releases ammonia from nitrogen contained in animal faeces and manure as well as in fertilisers. This should be specifically considered when changing over from crop production to animal husbandry.

Biodiversity. Biodiversity is defined as the variability among living organisms and the ecological complexes of which they are part. Normally, it is considered as the biological richness of species in an ecosystem, and the rarity and distinctiveness of species in habitats.

Landscape quality is the value of a landscape based on the perceived and predicted preferences of people and is dependent on their cultural background, educational level and economic dependence.

Food product quality is the quality of any food product in relation to saleability and consumer satisfaction and requirements (e.g. fat content, appearance, nutritional value, taste, etc.).

Water quality refers to the access to sufficient and safe drinking water resources, as well as uncontaminated water for irrigation and animal consumption.

Food product quantity is the amount of high value and high quality of food produced.

Animal welfare is the maintenance of the health and wellbeing of animals through humane handling, care and treatment (e.g. battery versus free-ranging).

Note that most remedial actions address the efficiency in reducing radioactivity in food products arising from agricultural systems, whereas semi-natural ecosystems, although widely contributing to the internal dose to humans, have not been as heavily studied. Research on agricultural systems have concentrated on soil-based actions as these are easily applicable and demonstrate immediate response in reducing transfer of radionuclides to vegetation. For semi-natural ecosystems, management-based actions seem to be the most effective. However, although secondary effects in semi-natural ecosystems seem to be more diverse and variable and far-reaching compared to agricultural systems are hardly reported, except for indirect reference to productivity or biomass composition. In forests these are more pronounced if the forest has extra functions such as an amenity, a grazing area or a source of wild game and other forest products. In addition, environmental side-effects in forests can have a knock-on impact on the habitat and amenity value, leading to leading to social and economic consequences. In contrast to agricultural ecosystems, forests are not simple monocultures but are living systems with natural balances sensitive to any disturbances.

Despite the known – although less investigated and quantified – secondary effects of remediation in ecological systems, some of the most pronounced and grave impacts lie in the behaviour and change of attitude of humans, as investigated in the field of social sciences. In the following sections, the social and ethical considerations related to countermeasure applications are outlined and practical examples are given.

3. The Chernobyl Experience

The Chernobyl accident has been the subject of much research in many fields, from medicine and natural sciences to sociology and anthropology (Bay and Oughton, 2005). One of the most comprehensive studies of the social and economic consequences of the Chernobyl accident was carried out during 2001, on behalf of the United Nations Development Programme (UNDP) and the United Nations Children's Fund (UNICEF), with the support of the UN Office for the Coordination of Humanitarian Affairs (UN-OCHA) and the World Health Organisation (WHO). The final report, *The Human Consequences of the Chernobyl Nuclear Accident*, represents one of the most comprehensive sources of information on the impacts of the accident on the former Soviet Union (fSU) (UNDP, 2002).

Table 1 Benefits and costs of remediation efforts (after UNDP/UNICEF, 2002).

Successes	Failures
Reducing collective dose by technical, economic and administrative measures	A significant number of rural people in high-risk groups are still exposed to substantial and, probably, increasing doses of radiation
Significantly improving scientific understanding of possible causes, scenarios and consequences of accidents in nuclear power plants	Environmental contamination still imposes significant economic constraints associated with a variety of protective measures, many of which are not effective in the new economic and political conditions
Improving preparedness to deal with the consequences of nuclear accidents, including understanding of the effectiveness of different protective measures	Economies and social structures in affected communities are deteriorating, alongside an apparent increase in poverty
Building the national capacity in Belarus, Russia and Ukraine to deal with contamination of the environment by radioactive material, including development of expertise, instrumentation and institutions	The activities undertaken so far have failed to increase trust and reduce anxiety
	Low local capacity to deal with health, economic and environmental challenges

The UNDP provided a summary of the successes and failures of remediation efforts in the FSU (Table 1). They and other workers (e.g. Fesenko et al., 2006) have noted that remediation strategies have significantly improved the scientific understanding, the authorities' preparedness, the national capacity and the effectiveness of different measures, in particular reducing the collective dose (except for rural high-risk groups). However, according to UNDP, it appears that, to date, the affected populations have not experienced much benefit from such gains in technical understanding. They suggest that protective measures have done little to increase living standards and to improve the local capacity to deal with health, economic and environmental challenges (UNDP, 2002). Increase in alcoholism, depression, anaemia and drug use has been reported, along with increased risks of associated disease. Indeed since for many people the risk of dying of a stress-related heart disease is greater than the reduction in radiation-induced cancer, one might question whether populations may have been better off without remediation (Bay and Oughton, 2005).

It should be realised that in many cases it is not the remedial measures that are to blame for the state of affairs, but the way these were implemented, with too little attention being paid to social and ethical side-effects. Based on previous work published by Bay and Oughton (2005), the following sections give an overview of the main social and economic impacts of remediation. These have been divided into (1) evacuation and resettlement; (2) agricultural measures; (3) compensation and (4) economic costs.

3.1. Evacuation and resettlement

Within 10 days of the accident on 26 April, 1986, a total of 175,000 inhabitants were evacuated from a 30-km radius around the reactor (Botsch, 2000). Then, following widespread monitoring of contamination in the early 1990s, more inhabitants were evacuated from 'hot spots' outside the exclusion zone. People in less-contaminated regions are permitted to stay in their homes for most of the year, with the exception of children who are sent out of the region for several months during summer (UNDP, 2002). This massive resettlement has resulted in a wide range of social and economic consequences. Many resettlers lost their jobs, social network and connection to places of particular community or historical value like graveyards or places where they played as children. In many cases, these losses were aggravated by the fact that people were forced to move rather than having moved by choice (Bay and Oughton, 2005).

Resettlement changes not only the lives of the evacuated people, but also the social structure of the villages or city districts that become their new home. The Gomel region lost about 43% of its population between 1986 and 2000, and demographic parameters, such as mortality and birth rates, have changed dramatically as elderly people in particular did not want to leave their villages while young people did. Despite the official prohibition on living in the exclusion zone, at least 800, mostly older, people have returned to their former villages (Botsch, 2000; Sahm, 1999).

The emigration of young people, combined with the (actual and perceived) health risks from radiation exposure, impeded the whole social and economic development of the region. A shortage of teachers and doctors is proving detrimental to the education and healthcare infrastructure, and factories and farms have had to close down, because of shortage of skilled workers. This, in turn, led to more families moving away (Botsch, 2000; Sahm, 1999; UNDP, 2002).

UNDP (2002) noted that, when evacuees were relocated to existing villages, tensions often arose between old and new inhabitants. The Chernobyl victims tended to socialise mainly with each other, and especially elderly resettlers found it difficult to adapt to their new environment. This was reinforced by the stigma of radiation contamination. A fear of congenital anomalies made it difficult for resettlers to find partners. Women who have moved out of the region try to keep their origin a secret, for fear that men will not want to marry them (UN-OCHA, 2000; Sahm, 1999; UNDP, 2002).

Resettlement and evacuation, especially immediately after the accident, undoubtedly played a major part in reducing the collective dose. However, the effectiveness of the measures declines after some time, as the negative impacts start to outweigh the benefits, especially if other potential uses of the resources are being considered. Resettlement appears to have been least successful when implemented inconsistently, for example when large number of people were left behind in villages designed for evacuation. This has contributed to a lack of trust between the authorities and the population. Research shows that the psycho-social welfare of people who stayed in their homes is actually better than that of those who were relocated (UNDP, 2002). Hence, recent studies propose that the present regime of residency restrictions might be relaxed to enable people who wish to return to make informed decisions about the risk to their health.

3.2. Agricultural measures

In 2001, 4 million people were living in contaminated areas which required some form of remedial actions, either active to reduce levels or passive such as monitoring. Agricultural measures include both a restriction on land use and measures undertaken at later stages in the food supply chain (Fesenko et al., 2007). These may include (a) actions to reduce contamination levels in food, such as improving pastures by adding fertilisers and liming or using radiocaesium binders to prevent uptake to animals; or (b) measures that remove contaminated food from the market, for example setting standards for radioactive contamination of foodstuffs, systematic monitoring and providing compensation to buy 'clean' food. Because of the economic aspects of agriculture and forestry, maintenance of production can be a primary reason for application of agricultural countermeasures. Hence, the main objective, and benefit, of remediation is both to significantly reduce the contamination level in the food product as well as to maintain trust in the market and economic activity in the region.

While effectively reducing the collective dose, experience showed that restrictions on land use in fact undermined the economic activity in the most affected areas (UNDP, 2002). Furthermore, studies on the social consequences of the remediation implementation in both the fSU and Europe have identified a number of factors that can have a knock-on effect on the economic activity of the contaminated areas (Hunt and Wynne, 2002). For example, many of the remedial measures were too expensive to allow production at competitive costs, and required large subsidies from the state to support the proper maintenance of abandoned land and forests. Money had to be spent on policing the countermeasures and averting the growth of a black market (Gould, 1990). In some cases, remedial actions resulted in stigma from having been identified as an area with significant contamination. This can have profound consequences both for a range of industries and for the local identities of people and groups (Hunt and Wynne, 2002; Flynn et al., 2001). Stigma associated with the perceived contamination of products where the countermeasure has been applied can generate mistrust of the farming industry in relation to food production. This phenomenon is not limited to the Chernobyl area, but has also been seen in the United Kingdom and Scandinavia (Gould, 1990).

3.3. Compensation

A policy for compensation was introduced in the aftermath of the accident. Belarusian and Russian legislations provide more than 70, and Ukrainian more than 50, different privileges for Chernobyl 'victims', depending on factors such as the degree of invalidity and the level of contamination. People with the right to direct compensation for damage to health or property include the liquidators (people involved in the clean-up), people who had been resettled and people who continued to live in the contaminated areas (UNDP, 2002). Other benefits include healthcare (e.g. free medicine and medical check-up); housing, including provision of heating and gas; travel (e.g. health recuperation holidays and summer camps); tax exemptions; access to university education and monthly allowances for disabilities linked to Chernobyl.

Unfortunately, the compensation policy has its own set of undesirable side-effects, one of the most striking being a pattern of behaviour described as 'Chernobyl Accident Victim Syndrome' (Petryna, 2002). Research carried out by the Institute of Sociology in Kiev indicates that 84% of resettlers expect special medical treatment and 71% claim unemployment allowance (UNDP, 2002). Poverty caused by resettlement, restrictions on land use and the effect of the collapse of the Soviet Union led to more and more people claiming Chernobyl-related benefits. The sparse information on radiation-related health effects (as radiation *per se* was treated as secret information in the fSU) resulted in large uncertainty, and an increased pressure to register as a victim for an ever-increasing number of people.

For many people, being a victim became the only means of access to an income and to vital aspects of health provision, including medicines. This led to a situation where resources were allocated, not on the basis of medical need, but rather on an individual's ability to register as a victim (UNDP, 2002). Many families returned to the contaminated areas in order to claim a higher level of Chernobyl-related benefits. The compensation system also created a situation that blocked economic initiatives and investments.

For example, tractor mechanics had turned down an opportunity to open their own workshop in fear of losing Chernobyl entitlements. Due to a system that promoted an exaggerated awareness of ill health and a sense of dependency, which has prevented people from taking part in economic and social life, resources available for mainstream provision have been further reduced, both in the affected areas and beyond (UNDP, 2002).

3.4. Economic costs

In 1986, the economic and social conditions for the environment and people living in Belarus, Russia and Ukraine were controlled by the Soviet Union. With the collapse of the Union in 1991, both Russia and Ukraine became independent nations and so-called 'economies in transfer'.[1] This represented a huge political alteration from a welfare system based on Soviet-plan economy to market-oriented solutions for employment, healthcare and infrastructure. The region still faces enormous economic challenges as a consequence of the political shift, and it is in this context that the social and economic consequences of the accident need to be interpreted (IAEA, 2006).

The exact numbers for economic losses due to the nuclear accident are unmeasurable, simply because of the scale and complexity of the political situation. The growing economic crisis and hyper-inflation exacerbated the effects of the Chernobyl accident. Nevertheless, even without these exceptional circumstances, the expenditures had a profound effect on the Belarus, Ukrainian and, to a lesser extent, Russian budgets. Authorities have published some broad estimates of the magnitude of costs (Table 2). All countries implemented so-called 'National programmes' in order to mitigate the social, economic and environmental interrelated consequences of the accident.

While Belarus and Ukraine implemented a Chernobyl tax on wages in all non-agricultural firms, Russia borrowed money to cover the extra expenditure. In 1994, the Chernobyl tax stood at a rate of 18%, but was heavily criticised for making Belarusian products uncompetitive. Hence, the tax was eventually reduced and in 1999 was set at 4% in both Belarus and Ukraine (Sahm, 1999). In Belarus, priority was given to improving conditions for people still living in the contaminated areas, whereas the Russian and especially Ukrainian governments put priority on resettlement (Petryna, 2002).

In recent years, money allocation has been moved from resettlement to compensation and social protection, which now dominate the budgets. This includes social and medical benefits such as monthly allowances, free/subsidised medical treatment, free meals for school children and students and respite holidays. In 2000, more than 80% of the expenditure in the

[1]Belarus is the last country in Europe to have a centrally planned economy. Executive power has rested with President Alexander Lukashenko since 1994 (www.chernobyl.info).

Table 2 Estimates of total expenditures, proportion of national budgets, numbers of newly built houses, schools and hospitals in the three most affected countries: Belarus, Russia and Ukraine (UNDP, 2002; EMERCOM, 2001; IEBNAS, 2001; Sahm, 1999).

	Ukraine	Belarus	Russia
Total expenditures estimates (billion USD)	148 (1986–2015)	235 (1986–)	3,8 (1992–1998)
Per cent of national budget in 1991	15	16.8	–
Per cent of national budget in 1996	6	10.9	–
Per cent of national budget in 2002	5	6	–

Ukraine went to 'social protection' which includes cash subsidies, family allowances, free medical care and education and pension benefits for sufferers and the disabled.

In addition to direct costs, a number of capital losses such as damage to buildings, agricultural land, mineral resources, labour, income, production and value of products have created major economic challenges.[2] In monetary terms, the largest economic losses were those associated with the removal of agricultural land and forests from use, and the closure of agricultural and industrial facilities. A total of 784,000 ha of agricultural land and 694,000 ha of forest were removed from use in the three countries (UNDP, 2002). Closure of agricultural and industrial facilities was another source of losses.

The agricultural sector has been worst hit by the effects of the accident. The rural areas are particularly vulnerable because of the great economic dependence on agriculture and forestry (UNDP, 2002). In Belarus alone, 282 rural settlements, including their economic, cultural and social structures, were closed down due to resettlement. The economic crisis in the 1990s made the financial situation of rural families extremely precarious. Many of the people lost not only their main livelihood from the possibility for forestry and agriculture, but also their secondary source of income, namely hunting, fishing and collection of wild berries and mushrooms.

The whole spectrum of additional 'side-effect' costs that might be associated with social and environmental costs such as loss of access to an amenity or disruption to livelihoods is discussed in more detail in the following section. These parameters are difficult to measure because this demands the transfer of social and environmental values into monetary values – controversial in itself.

[2]This may of course be regarded as two sides of the same case, as the costs in many cases constitute the compensation for the losses.

4. Ethical Aspects of Remediation

Acknowledgement of the multi-dimensional aspects of remediation has been an important part of the STRATEGY, Fifth Framework EU project – Sustainable Restoration and Long-Term Management of Contaminated Rural, Urban and Industrial Ecosystems (www.strategy-ec.org.uk), and the follow-up EURANOS project (www.euranos.fzk.de), both of which include a number of countermeasure evaluation criteria such as practicality and acceptability, socio-ethical aspects, environmental consequences and indirect side-effect costs (Howard et al., 2004; EURANOS, 2006). Stakeholder evaluation of remedial measures suggested that many options (especially in the United Kingdom) were as likely to be rejected on socio-ethical grounds as on technical and economic grounds (Nisbet, 2002).

Examples included a strong aversion to any measure that would bring about contamination of previously uncontaminated foods (e.g. mixing milk from different sources) or environments, and an awareness of the problems of contaminated foodstuffs appearing on the black market. Legal constraints also play an important role, particularly with respect to environmental legislation (e.g. habitat protection) and labour rights (Oughton et al., 2004). It follows that deciding on a remediation strategy is going to require a whole suite of trade-offs and value judgements, further complicated by the variety in which different remedial actions can impact different people, and a possible lack of agreement within society on what is practical or acceptable.

There are a number of general ethical issues that will be relevant for any risk assessment, including radiation protection (Shrader-Frechette, 1991; Oughton, 1996, 2000; Shrader-Frechette and Persson, 1997). These included questions such as whether (i) the distribution of cost and benefits is equitable; (ii) the risks are imposed or voluntary; (iii) the stakeholders[3] have been involved in the decision-making process and (iv) the action carries a risk of serious environmental damage. A short summary of these findings in relation to remedial measures is presented later, taken from previously published work carried out under the STRATEGY and EURANOS projects (Oughton et al., 2003, 2004). The list represents an overview of the types of questions and ethical criteria against which remedial measures might be evaluated. Obviously, the list is not exhaustive and can provide only an illustration of some of the issues that might be considered. The descriptions and examples are rather general since the actual issues will be site- and context-specific.

[3]Although increasingly referred to in risk management, the use of the term 'stakeholder' can raise problems since there are various definitions and interpretations of the word, including (in English) something akin to a legal claim. In this paper, we use the term very generally to indicate affected or interested parties.

4.1. Self-help

'Self-help considers the extent to which the affected persons themselves can implement the remediation and their degree of control or choice over the situation. Voluntary remedial actions that are carried out by the public or affected individuals themselves, or that increase personal understanding or control over the situation, are usually deemed positive as the action respects the fundamental ethical values of autonomy, liberty and dignity. Concrete examples include the provision of counting equipment, dietary advice and certain agricultural procedures that could be carried out by the farmer. On the contrary, imposed remedial actions that are highly disruptive, infringe upon liberty or restrict normal practices are usually judged to be negative. Examples would include relocation, bans on amenity use or a radical change in farming practice.

4.2. Free informed consent of workers (to risks of radiation exposure and/or chemical exposure) and consent of private owners to access to property

The issue of consent is strongly linked to the fundamental ethical value of autonomy. Employers have a duty to obtain the informed consent of any worker who may be exposed to chemical and/or radiation risk. This is particularly important if lower-paid workers (e.g. cleaners for industrial remediation) are employed to carry out the measure, as it has been suggested that the necessary conditions for free informed consent are often violated for these groups (Bullard, 1990; Shrader-Frechette, 2001). The increased risk may justify some form of compensation via higher wage premiums, but the opportunity for certain sub-groups of the population to make a profit at the expense of others can have negative social consequences (e.g. increased inequity – see Section 4.4). Furthermore, compensation itself can raise questions of whether or not this may coerce people into taking risks they would otherwise not have taken (Bullard, 1990; Rawles, 2002).

4.3. Informed consent regarding consumption of foodstuffs

In cases where foodstuffs are already contaminated due to the accident, consent of consumers can be an issue, but it is complicated by the question of who exactly has the obligation to obtain consent – authorities, farmers, producers, retailers – since they are not directly responsible for causing the contamination. Remedial measures that bring about a change in dose distribution and contaminate previously uncontaminated food can raise even more complex issues of consent and responsibility (e.g. mixing milk sources or feeding livestock with contaminated product). Those responsible for carrying out the remediation might be deemed to have a special obligation to obtain free informed consent from affected consumers/ producers. In both cases, informed consent may necessitate a specific need for labelling and other forms of information provision.

4.4. Distribution of dose, costs and benefits

The way in which the remediation may influence the distribution of costs, risks and benefits has significance due to the fundamental ethical values of equity, justice and fairness. Costs, benefits and risk may vary over both space and time, and between different members of a community. Dose distribution is obviously a main consideration for radiation protection, and many remedial measures that reduce collective dose (man-Sv) may also change the distribution of dose, for example from consumers/users/farmers to workers/consumers/populations around waste facilities. The question of who is paying the monetary and social costs of the remediation and who will receive the benefits must also be addressed. Some remedial actions, and sets of them, result in an equitable distribution of cost and dose reduction, such as investment by tax payers to reduce activity concentrations in a common food product; others are less equitable, for example when a reduction of dose to the majority is only possible at the expense of a higher dose, cost or welfare burden to a minority (e.g. banning all farm production in a small community). Also it may be possible that some sections of the population can make a profit from remedial actions (such as selling or hiring equipment), which can lead to further social inequity. As a parallel case, this effect was seen in the aftermath of the foot-and-mouth outbreak in the United Kingdom, when a minority of the affected communities made a large profit from the disaster (DEFRA, 2002). Other questions are: Whether the countermeasure has implications for vulnerable or already disadvantaged members of society (children, ethnic or cultural minorities)? Who is being affected? Who is paying?

4.5. Liability and/or compensation for unforeseen health or property effects

Employers usually hold legal and ethical responsibilities over their employees, and contractors or industries may be held legally or financially liable for any damage they may cause to public or private property. The matter of who bears liability is relevant both from the point of responsibility (moral and legal) and because of links to equity issues. Liability can become particularly important if outside contractors are paid to carry out the countermeasure, both for the contractors themselves – will I be sued if the countermeasure causes unforeseen damage? – and the workers/property-owners who may risk injury – will I be compensated if the countermeasure causes me damage?

4.6. Animal welfare issues

Animal welfare is concerned with the amount of suffering that remediation may inflict on non-human sentient animals. In the context of agricultural remedial measures, these will be most relevant to farm animals, but could also include zoo exhibits, pets or wild animals. For example, a ban on use of recreational areas may have implications for dogs. There are a number of philosophical debates around the question of why one should value non-human living beings, and whether or not they have moral standing (Regan and Singer, 1989; Oughton, 2003). Nevertheless, in many countries, animal welfare issues are covered by law and may result in both legal and ethical constraints on some remedial measures.

4.7. Change in public perception or use of an amenity

If a remedial measure has some effect on the public's use of a particular amenity (such as a park), then this will have an influence on the acceptability of that action. As was seen in the Chernobyl case, people place large value on places with strong community and personal ties, such as those having childhood memories. And the importance of such community values has been seen in many cultures (O'Neill, 2007). Such effects can have deeper relevance than whether or not people are able to use the amenity. Perceptions can include, for example that something has changed from being 'natural' to 'unnatural' or 'clean' to 'damaged'. An example might be the reaction to building a fast-food outlet at the base of Everest. Although the ethical and rational relevance of a distinction between 'natural' and 'unnatural' is a matter of contention between philosophers (Reiss and Straughan, 1996; Thompson, 1997), it is an issue with which the public has a strong tendency to attach moral relevance, and certainly impacts upon people's sense of their quality of life. In recent years, debates about the relevance of such perceptions have perhaps been most prominent within biotechnology, for example the contention that 'natural' selective breeding to obtain desired biological traits is acceptable but 'unnatural' genetic manipulation is not.

4.8. Uncertainty

Uncertainty in this context can be taken to refer to an evaluation of the risk (environmental, technical, social) associated with the countermeasure and relates to the question of what the actual consequences of the countermeasure implementation will be and the probability that that outcome will occur. Uncertainties can arise due to variability (e.g. environmental factors influencing ecosystem transfer of radionuclides, or differences in individual susceptibility to disease), statistical error or from data gaps and lack of knowledge (Oughton et al., 2008). It is usually impossible to predict with 100% certainty what the actual consequences of the countermeasure implementation will be and the probability that that outcome will occur. In some cases, uncertainties can be reduced by further research, and thus knowledge from previous experience of remediation will be important. However, here one must also consider the question of variability and the rationality of extrapolating from one set of conditions to another. In evaluating individual measures we need to consider: What are the main uncertainties associated with the remediation? What action might be taken to avoid or reduce these uncertainties, and are some inevitably indeterminate? What are the consequences of being wrong?

4.9. Ethical considerations for ecosystem changes

As discussed above, remedial actions that change or interfere with ecosystems (e.g. ploughing or fertilising unimproved pastures) may produce negative environmental consequences. In addition to the obvious questions of uncertainty, environmental risk raises a variety of ethical issues including consequences for future generations, sustainability, cross-boundary pollution and balancing harms to the environment/animals against benefits to humans. The ethical acceptability of the remediation will be highly dependent on the ecological status of the area and the degree to which the action diverges from usual practice. In most cases, environmental legislation must be considered.

4.10. Waste generation and treatment (chemical and radioactive)

Remedial actions that produce waste will raise both equity and environmental risk issues. Waste disposal can lead to a 'redistribution' of radiation exposures, and the environmental impact of disposal sites will need to address both ethical and legal implications, for example balancing human (present and future generations) and non-human (animal, environment) interests. Hence, any countermeasure involving the generation of waste and/or its treatment will have ethical relevance (and controversy) in itself. Treatment of waste *in situ* can be positive as it avoids problems arising from 'dilute and disperse' or the 'redistribution' of exposures to persons living close to the disposal site. Such issues were seen to be very important in some European countries after Chernobyl (Nisbet and Mondon, 2001; Nisbet and Woodman, 2000). But *in situ* treatment may also have negative side-effects due to complicating future waste removal or 'causing' contamination of soil. However, in many cases, the underlying soil will be unavoidably contaminated and the additional amount of activity incorporated would be relatively minor.

4.11. Doses, costs and side-effects

The averted dose and the calculated cost of countermeasure implementation have direct consequences for the welfare of society and/or individuals, and are thus also important ethically relevant aspects. A whole range of non-monetary consequences may influence the welfare, wellbeing or 'happiness' of people. In the STRATEGY project, a methodology was introduced that enabled an economic assessment of costs incurred due to some of the side-effects (Alvarez and Gil, 2003). Although it should be clear that not all side-effects are reducible to such calculations, some have already been noted in the evaluation of the Chernobyl accident, such as the rural breakdown and stigma of contaminated communities. Remediation may affect a community or cultural values in a number of ways (Hunt and Wynne, 2002). For example, it may specifically affect regional identity (with implications for people's sense of wellbeing). Additional benefits for the community can be gained from protecting such factors. Disruptions to existing social and cultural patterns – such as those requiring changes in employment or lifestyle – are generally taken as negative. Beneficial consequences might include the generation of employment opportunities. Implementing a strategy which benefits workers but disbenefits children (such as prioritising the clean-up of industrial plant rather than schools, streets, homes and parks) is likely to produce social divisions and antagonisms (Hunt and Wynne, 2002; Oughton et al., 2004). All these consequences should be addressed within an ethical evaluation of remediation strategies.

4.12. Non-dose-reducing remedial measures

Choice, control, familiarity, closeness and numerous other social and psychological factors play an important role in shaping perceptions towards hazards. It follows that both communication policy and remediation measures that are sensitive to these factors may stand a greater chance of success. For certain remedial actions, reduction in dose need not be the only benefit, or even the main benefit. Some measures will tend to increase personal control or choice regarding the risks (i.e. information on actions that can be taken to reduce exposures), while others (i.e. state-controlled interventions) might provoke feelings of helplessness (MacGregor, 1991). Also, actions need not be limited to those that reduce the exposure to radiation; for example, remedial measures might include better medical attention to reduce all illnesses (Morrey and Allen, 1996).

In the STRATEGY and EURANOS projects, a number of remedial measures were evaluated where dose reduction was not the primary aim. These included actions such as compensation, provision of medical check-up, setting up of a public information centre, instigation of education programmes, stimulation of the involvement of stakeholders in decision-making or provision of counting equipment (Howard et al., 2004; EURANOS, 2006). The problems associated with compensation schemes have been clearly discussed in Section 3.3. Nevertheless, ethically, most people would agree that affected persons have a right to some form of compensation for damages; either those brought about directly by the accident or as a result of remediation. The main problems associated with compensation are 'who is to pay?' (a particular challenge for cross-boundary accidents) and 'how to stop paying compensation when one has started?'.

4.13. Provision of counting equipment for self-help options

Provision of counting equipment is another remedial measure that was used in Chernobyl-affected communities, largely with great success (Hériard Dubreuil et al., 1999). The study in question was carried out in Belarusian villages under restrictions due to fallout from the Chernobyl accident. The basic idea was that experts should attempt to teach and advise populations on ways to deal with the situation, rather than adopt the paternalistic role of making decisions on behalf of the public. Thus, scientists instructed the villagers on how to use and interpret radiation-monitoring equipment, and outlined the possible 'self-help' ways of reducing radiation exposures. At least for this study, the conclusion was that this approach not only resulted in reducing exposures with minimal social and psychological side-effects, but also was more economically cost-effective than the standard bureaucratic 'top–down' management procedures (Hériard Dubreuil et al., 1999). Ethically, intervention procedures that involve the populations themselves help promote the principle of informed personal control over radiation risks. It should be noted that, in some cases, information policies that reduce anxiety over radiation might actually lead to an increase in doses, since people may eat food that they would have otherwise avoided.

5. Two-Way Communication and Information

Communication and information policies can be directed at both dose reduction and increased control and autonomy of populations. According to both IAEA (1991) and UNDP (2002), one of the most important factors influencing psycho-social effects on the population in Belarus, Ukraine and Russia was the quality of public information. Communication experts and psychologists also suggested that many of the conflicts and problems following the Chernobyl accident arose because of a lack of information and an inappropriate communication strategy

(Gould, 1990; Otway et al., 1988; Drottz and Sjöberg, 1990; Slovic, 1996). According to Sahm (1999), lack of knowledge seems to have been crucial for the information crisis to develop. She claims that 'Soviet policy in the first two years was clearly based on the assumption that the radioecological situation would largely return to normal during this period…'.

Criticism of the information policy has not only been restricted to the Soviet authorities. Many authorities in other parts of Europe were accused of gross incompetence, cover-up, providing false information or not providing full information regarding the risks from Chernobyl and ways of reducing them (Gould, 1990; Samuelsson, 1997). A number of psychologists and sociologists accused authorities of failing to keep the public properly informed after the Chernobyl accident (Wynne, 1989; Drottz and Sjöberg, 1990). On the other hand, scientists have suggested that those with a vested interest in radiation protection exaggerated the risks and stirred up media attention (Jaworowski, 1999). What most people would agree on was that the majority of authorities initially underestimated the effects of Chernobyl, and the impression given by all but a few government officials was that the radioactive releases would have neither serious nor lasting effects on human health or food production (Gould, 1990; Alloway and Ayers, 1997). The immediate response of many governments after Chernobyl was that there was no cause for alarm. In France, the public was informed that 'the cloud had turned at the Alps' (Gould, 1990).

The case in the United Kingdom serves as a classic example of the situation. When weather reports confirmed that the radioactive cloud would pass over the United Kingdom, the Minister for the Environment released a statement that 'the effects of the cloud have been assessed and none presents a risk to health in the UK' (Hansard, 1986). Less than seven weeks later, however, due to high levels of radiocaesium in lambs grazing mountain pastures in England, Scotland and Wales, the government imposed a 21-day restriction on the movement and slaughter of sheep from certain areas. Authorities were convinced that levels of radiocaesium in meat would drop quickly. Unfortunately that optimism proved unfounded. The number of animals under restriction rose quickly to nearly 5 million and, four years later, 600 farms and over half a million sheep were still under control (Rich, 1991).

The ways in which governments act in the immediate aftermath of a large accident have a major influence on trust between the public and the authorities, as well as compliance with subsequent measures in the following years. Public acceptance of a policy is often highlighted as one of the main benchmarks of a successful communication strategy; 'an informed and satisfied public is a compliant public' (Oughton et al., 2002). A relationship of trust between the public and the authorities depends crucially on the immediate responses of the authorities. The experience from Chernobyl shows us that it is essential that trust is generated, not dissipated, in the immediate aftermath (Hunt and Wynne, 2002).

6. Conclusions

Scientists from all areas of radiation research agree that there were lessons to be learnt from the Chernobyl accident, and that there was room for improvement both in how authorities acted to reduce risks and in radiation protection policy in general. Chernobyl helped to strengthen the growing awareness that effective management of radiation risks needs to be sensitive to the public's perception of those risks and that authorities are often faced with moral dilemmas when evaluating actions. The general problems in the aftermath of Chernobyl are attributed to a lack of preparedness, knowledge and clarity on the distribution of responsibility (Czada, 1990; Marples, 1991; Nohrstedt, 1991).

This chapter has tried to highlight some of the environmental, economical and social consequences for the Belarusian, Ukrainian and Russian societies in the aftermath of the Chernobyl accident. It appears that the system of Chernobyl-related benefits has created expectations of payments and advantages and has undermined the capacity of the individuals and communities concerned to cope with the economic and social problems and to search for solutions. This underlines the importance of assessing social and ethical aspects of all remediation measures, and not to simply focus on the question of economic cost per sievert reduction. A remediation strategy should include development and implementation of actions for environmental, economic and social problems, as well as for the associated health effects. Due to the complexity and timescale of such effects, experts in many kinds of disciplines should be involved in the design of such strategies.

The recommendations offered by UNDP emphasise the need for a holistic approach to remediation, integrating economic, ecological and health measures, rather than a blinkered preoccupation with 'dose-reduction' aspects of radiation protection (UNDP, 2002). This is supported by multidisciplinary research on long-term management of radioactive contamination carried out by the EU (Howard et al., 2002; Oughton et al., 2003, 2004) and fSU (Fesenko et al., 2007). All projects highlighted the importance of including the affected population with regard to self-help measures and consensus-driven decision-making processes. In addition to respecting people's fundamental rights to shape their own future and thereby increasing trust and compliance, such an approach may lead to proactive efforts made by the people themselves, which also improve factors like cost-effectiveness.

REFERENCES

Alexakhin, R. M., L. A. Buldakov, V. A. Gubanov, Y. G. Drozhko, L. A. Ilyin, I. I. Kryshev, I. I. Linge, G. N. Romanov, M. N. Savkin, M. M. Saurov, F. A. Tikhomirov, and Y. B. Kholina. (2004). In: *Large Radiation Accidents: Consequences and Protective Countermeasures* (Eds L. A. Ilyin and V. A. Gubanov). IzdAT Publisher, Moscow, 555pp.

Alexakhin, R. M., S. V. Fesenko, and N. I. Sanzharova. (1996). Serious radiation accidents and the radiological impact on agriculture. *Radiation Protection Dosimetry*, 64(1/2), 37–42.

Alloway, B. J., and D. C. Ayres. (1997). *Chemical Principles of Environmental Pollution*. Blackie Academic and Professional, London.

Alvarez, B. & J. M. Gil (2003). *Valuing Side-Effects Associated with Countermeasures for Radioactive Contamination*, STRATEGY Deliverable 7, SIA/ESAB, Barcelona.

Bay, I., and D. H. Oughton. (2005). Social and economic effects. In: *Chernobyl, Catastrophe and Consequences* (Eds J. Smith and N. A. Beresford). Springer-Verlag, Berlin, pp. 239–262, ISBN 3-540-23866-2.

Botsch, W. (2000). Untersuchungen zur Strahlenexposition von Einwohnern kontaminierter Ortschaften der nördlichen Ukraine, Universität Hannover, p. 16.

Bullard, R. D. (1990). *Dumping in Dixie: Race, Class and Environmental Quality*. Westview Press, Boulder, CT.

Czada, R. A. (1990). Politics and Administration during a 'nuclear-political' crisis: The Chernobyl disaster and radioactive fallout in Germany. *Contemporary Crises*, 14, 285–311.

DEFRA. (2002). *Surveys of the Economic Impact of Foot and Mouth Disease in Six Districts*. Department for Environment, Food and Rural Affairs (DEFRA): London. Available at http://www.defra.gov.uk/animalh/diseases/fmd/2001/index.htm

Drottz, B., and L. Sjöberg. (1990). Risk perception and worries after the Chernobyl accident. *Journal of Environmental Psychology*, 10, 135–149.

EMERCOM. (2001). *Chernobyl Accident, Results and Problems in Eliminating the Consequences in Russia 1986–2001*. Ministry of the Russian Federation for Civil Defence Affairs, Emergencies and Elimination of Consequences of Natural Disasters, Moscow, Russia, p. 3.

EURANOS. (2006). *Generic Handbook for Assisting in the Management of Contaminated Food Productions Systems in Europe Following a Radiological Emergency*. Deliverable From the EURANOS Project. EURANOS (CAT1)-TN(06)-06. Available at http://www.euranos.fzk.de/index.php?action = euranos&title = products

Fesenko, S. V., R. M. Alexakhin, M. I. Balonov, I. M. Bogdevich, B. J. Howard, V. A. Kashparov, N. I. Sanzharova, A. V. Panov, G. Voigt, and Y. M. Zhuchenko. (2006). Twenty years application of agricultural countermeasures following the Chernobyl accident: Lessons learned. *Journal of Radiological Protection*, 26, 351–359.

Fesenko, S. V., R. M. Alexakhin, M. I. Balonov, I. M. Bogdevich, B. J. Howard, V. A. Kashparov, N. I. Sanzharova, A. V. Panov, G. Voigt, and Y. M. Zhuchenko. (2007). An extended review of twenty years of countermeasures used in agriculture after the Chernobyl accident. *Science of the Total Environment*, 383, 1–24.

Flynn, J., P. Slovic, and H. Kunreuther. (2001). *Risk, Media and Stigma: Understanding Public Challenges to Science and Technology*. Earthscan, London.

Gould, P. (1990). *Fire in the Rain: The Democratic Consequences of Chernobyl*. Polity Press, Cambridge.

Hansard. (1986). 8th May 1986, House of Commons, UK.

Hériard Dubreuil, G. F., J. Lochard, P. Girard, J. F. Guyonnet, G. Le Cardinal, S. Lepicard, P. Livolsi, M. Monroy, H. Ollagon, A. Pena-Vega, V. Pupin, J. Rigby, I. Rolevitch, and T. Schneider. (1999). Chernobyl post-accident management: The ETHOS project. *Health Physics*, 77, 361–372.

Howard, B. J. (2000). The concept of radioecological sensitivity. *Radiation Protection Dosimetry*, 92, 29–34.

Howard, B. J., K. G. Andersson, N. A. Beresford, N. M. J. Crout, J. M. Gil, J. Hunt, A. Liland, A. Nisbet, D. H. Oughton, and G. Voigt. (2002). Sustainable restoration and long-term management of contaminated rural, urban and industrial ecosystems. *Radioprotection*, 37, 1067–1072.

Howard, B. J., A. Liland, N. A. Beresford, K. G. Andersson, G. Cox, J. M. Gil, J. Hunt, A. Nisbet, D. Oughton, and G. Voigt. (2004). A critical evaluation of the Strategy Project. *Radiation Protection Dosimetry*, 109, 63–67.

Hunt, J. and B. Wynne. (2002). *Social Assumptions in Remediation Strategies*. Deliverable 5, STRATEGY project (www.strategy-ec.org.uk), Lancaster University.

IAEA. (1991). *The International Chernobyl Project: Assessment of Radiological Consequences and Evaluation of Protective Measures*, IAEA, Vienna.

IAEA (1996). *One decade after Chernobyl: Summing up the Consequences of the Accident*. IAEA/WHO/EC International Conference, IAEA, Vienna.

IAEA. (2006). *Environmental Consequences of the Chernobyl Accident and their Remediation: Twenty Years of Experience*. Report of the Chernobyl Forum Expert Group Environment, Vienna, ISBN 92-0-114705.

ICRP. (1989). *Optimisation and Decision Making in Radiological Protection*. ICRP Publication 55. *Annals of the ICRP*. Oxford: Pergamon Press.

ICRP. (1991). *Principles for Intervention for Protection of the Public in a Radiological Emergency*. ICRP Publication 63. *Annals of the ICRP*, 22. Pergamon Press, Oxford.

Institute of Economics of the Belarusian National Academy of Sciences. (2001). Committee on the problems of the consequences of the catastrophe at the Chernobyl NPP 15 years after the Chernobyl disaster. Minsk, 538pp.

Jaworowski, Z. (1999). Radiation risk and ethics. *Physics Today (September)*, 24–29.

MacGregor, D. (1991). Worry over technological activities and life concerns. *Risk Analysis*, 11, 315–324.

Marples, D. R. (1991). Chernobyl: Five years later. *Soviet Geography*, 32, 291–313.

Morrey, M., and P. Allen. (1996). The role of social and psychological factors in radiation protection after accidents. *Radiation Protection Dosimetry*, 68, 267–271.

Nisbet, A. F. (2002). Management options for food production systems contaminated as a result of a nuclear accident. *Radioprotection*, 37(C1), 115–120.

Nisbet, A. F., and K. J. Mondon, (2001). *Development of Strategies for Responding to Environmental Contamination Incidents Involving Radioactivity*. The UK Agriculture and Food Countermeasures Working group 1997–2000. NRPB-R331.

Nisbet, A. F., and R. F. M. Woodman. (2000). Options for the Management of Chernobyl-restricted areas in England and Wales. *Journal of Environmental Radioactivity*, 51, 239–254.

Nohrstedt, S. A. (1991). The information crisis in Sweden after Chernobyl. *Media, Culture and Society*, 13, 477–497.

O'Neill, J. (2007). *Markets, Deliberation and Environment*. Routledge, London.

Otway, H., P. Haastrup, W. Connell, G. Gianitsopoulas, and M. Paruccini. (1988). Risk Communication in Europe after Chernobyl: A media analysis of seven countries. *Industrial Crisis Quarterly*, 2, 31–35.

Oughton, D. H. (1996). Ethical values in radiological protection. *Radiation Protection Dosimetry*, 68, 203–208.

Oughton, D. H. (2000). Ethical aspects of ICRP recommendations for radiation protection. In: *Ethical Issues in Radiation Protection* (Ed. L. Persson). SSI, Stockholm, pp. 30–32.

Oughton, D. H. (2003). Protection of the environment from ionising radiation: ethical issues. *Journal of Environmental Radioactivity*, 66, 3–18.

Oughton, D. H., A. Agüero, R. Avila, J. E. Brown, D. Copplestone, and M. Gilek. (2008). Addressing uncertainties in the ERICA integrated approach. *Journal of Environmental Radioactivity*, 99, 1384–1392.

Oughton, D. H., I. Bay, E.-M. Forsberg, J. Hunt, M. Kaiser, and D. Littlewood. (2002). Social and ethical aspects of countermeasure evaluation and selection – using an ethical matrix in participatory decision making, Deliverable 4 STRATEGY, www.strategy-ec.org.uk

Oughton, D. H., I. Bay, E.-M. Forsberg, J. Hunt, M. Kaiser, and D. Littlewood. (2003). Social and ethical aspects of countermeasure evaluation and selection – using an ethical matrix in participatory decision making. STRATEGY Deliverable 4. Agricultural University of Norway, Ås.

Oughton, D. H., I. Bay, E.-M. Forsberg, M. Kaiser, and B. Howard. (2004). An ethical dimension to sustainable restoration and long-term management of contaminated areas. *Journal of Environmental Radioactivity*, 74, 171–183.

Petryna, A. (2002). *Life Exposed.* Princeton University Press, Princeton.

Rawles, K. (2002). *Compensation in Radioactive Waste Management: Ethical Issues in the Treatment of Host Communities – A Report for Nirex.* Nirex, Harwell.

Regan, T., and P. Singer. (1989). *Animal Rights and Human Obligations.* Prentice-Hall, New Jersey.

Reiss, M. J., and R. Straughan. (1996). *Improving Nature? The Science and Ethics of Genetic Engineering.* Cambridge University Press, Cambridge.

Rich, V. (1991). An ill wind from Chernobyl. *New Scientist*, 130, 26–28.

Roed, J., and K. G. Andersson. (1996). Clean-up of urban areas in the CIS countries contaminated by Chernobyl fallout. *Journal of Environmental Radioactivity*, 33, 107–116.

Sahm, A. (1999). *Transformation im Shatten von Tschernobyl.* LIT, Münster, pp. 238–262.

Salt, C. A., H.S. Hansen, G. Kirchner, H. Lettner, S. Rekolainen, & M. Culligan Dunsmore (1999). *Impact Assessment Methodology for Side-effects of Countermeasures against Radionuclide Contamination in Food Products.* Research Report No. 1, Nord-Troendelag College, Steinkjer, Norway, ISBN 82-7456-119-8.

Salt, C. A., and B. Rafferty. (2001). Assessing potential secondary effects of countermeasures in agricultural systems: a review. *Journal of Environmental Radioactivity*, 56, 99–114.

Samuelsson, C. (1997). Radiation risk information to the public: Principles or common sense. In: *Proceedings of the 11th Meeting of Nordic Radiation Protection Society* (Eds T. Walderhaug and E. P. Gudlaugsson), Reykjavik 26th–29th August 1996. Geislavarnir Rikisins, Reykjavik, pp. 539–544.

Semioshkina, N., G. Voigt, and D. Tarsitano, (2004). *Evaluation and Network of EC-Decision Support Systems in the field of Terrestrial Radioecological Research.* Final Report of the EVANET Project. GSF, Neuherberg.

Shrader-Frechette, K. (1991). *Risk and Rationality.* University of California Press, California.

Shrader-Frechette, K. (2001). Risky business: Nuclear workers, ethics and the market efficiency argument. *Ethics and the Environment*, 7, 1–19.

Shrader-Frechette, K., and L. Persson. (1997). Ethical issues in radiation protection. *Health Physics*, 73, 378–382.

Slovic, P. (1996). Perception of risk from radiation. *Radiation Protection Dosimetry*, 68, 165–180.

Thompson, P. (1997). *Food Biotechnology in Ethical Perspectives.* Blackie, London.

United Nations Development Programme (UNDP) and UNICEF. (2002). *The Human Consequences of the Chernobyl Nuclear Accident – A Strategy for Recovery.* Available online: http://www.undp.org/dpa/publications/chernobyl.pdf

United Nations Office for the Coordination of Humanitarian Affairs (UN-OCHA). (1999). United Nations Appeal for International Cooperation on Chernobyl. Priority Projects Proposed by The Governments of Belarus, Russian Federation and Ukraine, April, http://www.reliefweb.int/ocha_ol/programs/response/cherno/cher.htm

United Nations Office for the Coordination of Humanitarian Affairs (UN-OCHA). (2000). *Chernobyl – A Continuing Catastrophe*. United Nations, New York and Geneva.

Voigt, G., K. Eged, B. J. Howard, Z. Kis, A. F. Nisbet, D. H. Oughton, B. Rafferty, C. A. Salt, J. T. Smith, and H. Vandenhove. (2000). A wider perspective on the selection of countermeasures. *Radiation Protection Dosimetry*, 92, 45–48.

Wynne, B. (1989). Sheep farming after Chernobyl: a case study in communicating scientific information. *Environment*, 31, 33–39.

Concluding Remarks

Contamination with radioactivity due to release of radioactive substances into the environment is a comprehensive problem and has been subject to investigations since the industrial use of nuclear energy began, but especially since the occurrence of radiation accidents and incidents. The ICRP, as one of the highest authorities for radiological protection, provides guidelines and recommendations for decision-makers and other stakeholders to prevent harm to the population and the environment and is consistently adapting those as new information becomes available for improving knowledge about the consequences and optimising strategies for not only prevention but also mitigation of events.

Remediation of contaminated environments is a very complex process. Initially the term 'remediation' was defined as 'any measures that may be carried out to reduce the radiation exposure from existing contamination of land areas through actions applied to the contamination itself (the source) or to the exposure pathways to humans', and such a definition is still in use in many international documents (IAEA, 2006a). However, it must be realised that remediation has many facets, and it thus has to be approached in a multi-disciplinary way. Already some years ago, efforts were undertaken to look into 'a wider perspective on the selection of countermeasures' (Voigt et al., 2000), and a special issue of the *Journal of Environmental Radioactivity* (edited by Voigt, 2001) addressed important features of remediation strategies which were not considered or only slightly considered in the past. The experience gained following the Chernobyl accident, summarised in the Chernobyl Forum report, has also confirmed a need to consider not only radiological and economic factors but also the perception of remediation by the general public and secondary ecological effects (IAEA, 2006b).

Much more information has evolved in the meantime, and new data and approaches have become available using new tools, but, more importantly, a change of mind-set has occurred with the emphasis on environmental protection, resulting in holistic and integrated approaches for remediation strategies. This comprises not only the cost-effectiveness and technically achievable reduction potential of a given countermeasure, but also long-term post-accident management considerations, secondary effects, socio-economic interactions and feasibility in a real situation. The attitudes and perceptions of a variety of different stakeholders that may include farmers, health officers, consumers, food distributors, and even tourists, city councils, NGOs and the public at large should be taken into consideration when implementing countermeasure actions (Nisbet et al., 2005). Here communication and public outreach for a better understanding of the problems and their solutions for all counterparts are essential. The nuclear scientific community has become aware of this responsibility and obligation to consider and respond to public concerns and provide transparent and understandable information. Specifically integrated approaches will now be applied from the planning and conceptualising phase during preparation to establish nuclear facilities and resource exploitations, until the consequent remediation after termination of activities and decommissioning and final release of the sites for other usage.

All these issues have been considered in this book, each chapter covering some specific topic from those mentioned above. Thus, a historical overview of contamination events and of radiological impacts due to accidents, nuclear weapons testing, nuclear legacy and natural radioactivity enhancement during mining and milling processes is given in the book. The most important issues for developing and deciding on optimised remediation strategies are presented, such as site characterisation, the factors influencing the behaviour of radionuclides in different ecosystems and pathways, and the feasibility and cost-effectiveness of countermeasures. These are illustrated with examples and case studies within the individual chapters. The identification of the most important and driving parameters finally determining the dose to man and biota and the environmental and health impacts is one of the most important tools for remediation focus and will involve multi-criteria analysis and sensitivity analysis.

One special chapter is devoted to the ethical and environmental side-effects of countermeasures which are often difficult to value in monetary terms and often cannot be directly related to radiation effects, but might, however, have serious unexpected impacts on social, economic and ecological sustainability and development and might lead to distress and health and psychological problems within the affected population.

Advancement of human development and security is a huge challenge to humankind and governments in the 21st century. The United Nations predicts that the world population will grow from its current 6.5 billions to some 8 billions in 2030. Taking this factor into account as well as the influence of climate change, the shortage of natural resources and the present food crisis will trigger the use of a variety of nuclear applications in the fields of human health, food and agriculture and environmental management, including energy production which is already noted to be on the rise. Some even speak about the *Nuclear Renaissance*, and projections by the IAEA indicate that the nuclear component in energy production will increase from 15%–45% by 2020 and by 25%–95% by 2030 (IAEA, 2008). It is anticipated that much of this future growth will happen in the developing world; of 34 new reactors currently under construction, 16 are in developing countries, mainly in Asia. The prediction of such growth will enforce the need to prepare for intended or unintended, accidental or incidental, releases of radionuclides into the environment and to have environmental decision-support systems in place to rapidly and effectively respond. The world is changing consistently and rapidly, and a new threat has risen – nuclear terrorism. Remediation strategies will also be required to deal with malicious acts potentially leading to radioactive releases and radiation exposures; these are non-predictable but are becoming increasingly frequent, and they call for a preparedness to cope with such situations in order to minimise damage and return normality as fast as possible.

Although it has to be kept in mind that every release scenario will be different, that nuclides will behave differently and that their impact will greatly vary, lessons have been learned which can substantially contribute to mitigation of the consequences of any radioactive releases and environmental contamination in an optimised and efficient way.

REFERENCES

IAEA. (2006a). *IAEA Safety Glossary: Terminology Used in Nuclear, Radiation, Radioactive Waste and Transport Safety (Version 2.0).* IAEA Booklet. IAEA, Vienna.

IAEA. (2006b). *International Conference Chernobyl: Looking Back to Go Forward.* Vienna, 6–7 September 2005, Organized by IAEA on behalf of the Chernobyl Forum, http://www.iaea.org/NewsCenter/Focus/Chernobyl/pdfs/05-28601_Chernobyl.pdf

IAEA. (2008). *20/20 Visions for the Future. Background Report by the Director General for the Commission of Eminent Persons.* IAEA, Vienna, http://www.iaea.org/NewsCenter/News/PDF/20-20vision_220208.pdf

Nisbet, A., B. Howard, N. Beresford, and G. Voigt (Eds). (2005). Stakeholder involvement in post-accidental management. *Journal of Environmental Radioactivity*, 83, 259–435.

Voigt, G. (Ed.). (2001). Remediation strategies. special issue. *Journal of Environmental Radioactivity*, 56, 1–253.

Voigt, G., K. Eged, J. Hilton, B. J. Howard, Z. Kis, A. F. Nisbet, D. H. Oughton, B. Rafferty, C. A. Salt, J. T. Smith, and H. Vandenhove. (2000). A wider perspective on the selection of countermeasures. *Radiation Protection Dosimetry*, 92, 45–48.

G. Voigt
S. Fesenko
Editors

Author Index

Subject Index